COLLOQUIA MATHEMATICA
SOCIETATIS JÁNOS BOLYAI, 9.

PROGRESS IN STATISTICS

European Meeting of Statisticians
Budapest (Hungary) 1972.

Vol. II.

Edited by

J. GANI

K. SARKADI

I. VINCZE

NORTH-HOLLAND PUBLISHING COMPANY
AMSTERDAM-LONDON

Budapest, Hungary, 1974

ISBN North-Holland: 0 7204 2808 4

Joint edition published by

JÁNOS BOLYAI MATHEMATICAL SOCIETY

and

NORTH-HOLLAND PUBLISHING COMPANY

Amsterdam-London

Technical editor:

P. BÁRTFAI

Printed in Hungary

ÁFÉSZ, VÁC

Sokszorosító üzeme

CONTENTS

G.R. Laha – E. Lukacs, On some problems connected with the characterization of distributions by constant regression 461

D.V. Lindley, A Bayesian solution for two-way analysis of variance 475

J. Loris-Teghem, An algebraic approach to the waiting time process in GI/M/S 497

P. Mandl, On the asymptotic normality of the reward in a controlled Markov chain 499

M. Metivier, Stochastic integral and vector valued measures 507

V. Miké, Robust estimation of location: A Monte Carlo study ... 523

M.A. Mirzahmedov – S.A. Hasimov, On some properties of density estimation 535

S.G. Mohanty – S. Vellore, On some distributions of a generalized restricted random walk 547

E. Mohn – I. Holme – R. Volden, On stepwise regression in orthogonal models 557

V.V. Nalimov, Systematization and codification of the experimental designs –
The survey of the works of soviet statisticians 565

D. Oakes, A *k*-fold quasi-Poisson process 583

A. Obretenov, An estimation for the renewal function of an IFR-distribution 587

F. Österreicher, An information-type measure of difference of probability distributions based on testing statistical hypotheses 593

W.R. Pabst, Statistical studies of the costs of six-man versus twelve-man juries 601

V. Paulauskas, On sums of a random number of random multidimensional vectors 615

A. Perez, Generalization of Chernoff's result on the asymptotic discernibility of two random processes 619

J. Pergel, On the statistical examination of Poisson processes with randomly changing intensity 633
V.V. Poznjakov, Approximate formulas for probability of non-crossing of level by stationary process 637
J. Radcliffe, The convergence of an age-dependent branching process allowing immigration 641
T. Rolski, Some inequalities in queuing theory 653
M. Ruda – L. Szeidl – G. Tusnády, Time series analysis on CDC 3300 661
L.E. Schwartz, A globally convergent algorithm for nonlinear least squares 667
R. Sibson, D_A-optimality and duality 677
Z. Šidák, A chain of inequalities for some types of multivariate distributions 693
R.S. Silverman – A. Nádas, Optimal tests for the non-existence of immortals 701
A. Simonovits, On discretionary queuing discipline 725
A.N. Širjaev, Statistics of diffusion processes 737
J. Stene, Hierarchical procedures 753
J. Štěpán, Probability limit identification function 765
M. Stone, Large deviation connections 769
J.M. Stoyanov, Optimal estimation of some stochastic processes by incomplete data 773
D.O.H. Szász, On the rate of convergence in Levy's metric for random indiced sums 781
G.J. Székely, Statistical theory of topological groups 789
W.Y. Tan – I. Guttman, A Bayesian approach to the comparison of sensitivities of two experiments 793
J. Tankó, A simple algorithm to determine the ergodic classes of a Markov chain 803
E. Thorp – R. Whitley, Concave utilities are distinguished by their optimal strategies 813
P. Thyregod, Bayesian single sampling acceptance plans for life-testing 831

J. Tomkó, On the rarefaction of multivariate processes 843
V. Tzonev, Contribution to Girault's lecture 861
K. Urbanik, On the concept of information 863
I. Vincze, On the maximum probability principle in statistical physics 869
K.W. Wachter, Exchangeability and asymptotic random matrix spectra 895
T. Williams, The diffusion approximation to a branching process . . . 909

COLLOQUIA MATHEMATICA SOCIETATIS JÁNOS BOLYAI
9. EUROPEAN MEETING OF STATISTICIANS, BUDAPEST (HUNGARY), 1972.

ON SOME PROBLEMS CONNECTED WITH THE CHARACTERIZATION OF DISTRIBUTIONS BY CONSTANT REGRESSION

R.G. LAHA – E. LUKACS*

Dedicated to the memory of Ju. V. Linnik

1. INTRODUCTION

A random variable Y, which has finite expectation $\mathsf{E}(Y)$, is said to have constant regression on a random variable X if the relation

$$\mathsf{E}(Y \mid X) = \mathsf{E}(Y) \tag{1.1}$$

holds almost everywhere. Here $\mathsf{E}(Y \mid X)$ denotes the conditional expectation of Y, given the value of X. It is known [1] that Y has constant regression on X if, and only if, the relation

$$\mathsf{E}(Ye^{itX}) = \mathsf{E}(Y)\mathsf{E}(e^{itX}) \tag{1.2}$$

holds for all real t.

Let $x_1, x_2, \ldots, x_n$ be a sample of size n (i.e. $x_1, \ldots, x_n$ are identically and independently distributed random variables) taken from a certain population. We consider a polynomial statistic

*Research supported by the National Science Foundation through grant GP-35724-X.

(1.3) $$P(x_1, \ldots, x_n) = \sum A_{j_1,\ldots,j_n} x_1^{j_1} \ldots x_n^{j_n}$$

where the summation is extended over all $j_1, \ldots, j_n$ which satisfy the conditions

(1.3a) $$j_1 + j_2 + \ldots + j_n \leqslant k; \qquad 0 \leqslant j_s \leqslant m \qquad (s = 1, 2, \ldots, k) .$$

Then $P(x_1, \ldots, x_n)$ is a polynomial of degree k such that the highest power in which each variable occurs does not exceed m $(m \leqslant k)$.

Suppose that the moments of the population exist up to order m and that the statistic $P(x_1, \ldots, x_n)$ has constant regression on the sample mean. The characteristic function $f(t)$ of the population satisfies then the differential equation

(1.4) $$\sum A_{j_1,\ldots,j_n} i^{-(j_1+\ldots+j_n)} f^{(j_1)}(t) \ldots f^{(j_n)}(t) = C[f(t)]^n$$

[sum extended over all subscripts satisfying (1.3a)]

where $$f^{(s)}(t) = \frac{d^s f(t)}{dt^s} \qquad (s = 1, 2, \ldots, m) .$$

A.A. Zinger and Ju.V. Linnik ([3], [4]) considered the differential equation (1.4) and called it positive-definite if the polynomial

$$A(x_1, \ldots, x_n) = \sum_{(j_1,\ldots,j_n)} P(x_{j_1}, \ldots, x_{j_n})$$

is non-negative. The summation is here extended over all permutations $(j_1, \ldots, j_n)$ of the first n integers. These authors show that every positive definite solution $f(t)$ of a positive definite equation of the form (1.4) is necessarily an entire function, provided that $m \geqslant n - 1$.

This is a very interesting result concerning the analytic properties of the solutions of certain differential equations. If one tries to use this theorem in connection with characterization problems, then one is greatly handicapped by the severe restrictions contained in its assumptions. This is illustrated by the following facts. The positive definiteness of the equation excludes the k-statistics of order exceeding two. If $P(x_1, \ldots, x_n)$ is the

k-statistic of order two then the condition $m \geqslant n-1$ restricts the applicability of the theorem to sample of size $m \leqslant 3$, even though it is known that the normal population is characterized by the property that the k-statistic of any order $p \geqslant 2$ has constant regression on the sample mean whatever be the sample size. Moreover, the theorem can not be used in the case of some known characterizations of the Poisson and of the Gamma populations [1].

In the present paper we derive similar results for a wider class of polynomial statistics.

2. THEOREM ON THE EXISTENCE OF MOMENTS

We consider a polynomial statistic $P(x_1, \ldots, x_n)$ of the form (1.3) and assume that the highest power in which each variable occurs in $P(x_1, \ldots, x_n)$ is the same, we denote it by m. We can then write

$$P(x_1, \ldots, x_n) = x_j^m \Phi_{0j}(x_1, \ldots, x_n) + \Phi_j(x_1, \ldots, x_n) + + \Theta(x_1, \ldots, x_n) . \tag{2.1}$$

The subscript j is here any of the integers $1, 2, \ldots, n$. The function Φ_{0j} does not contain the variable x_j and is a polynomial of degree not exceeding m in the other variables. The polynomial Φ_j contains terms of the form $A_{s_1,\ldots,s_n} x_1^{s_1} \ldots x_n^{s_n}$ where $s_j \leqslant m-1$ while at least one s_k with $k \neq j$ is equal to m. The polynomial Θ consists of all terms $A_{s_1,\ldots,s_n} x_1^{s_1} \ldots x_n^{s_n}$ where all $s_k \leqslant m-1$.

We say that $P(x_1, \ldots, x_n)$ is an *ordinary polynomial of order* m if it has the form (2.1) and if the functions $\Phi_{0j}(x_1, \ldots, x_n)$ and $\Phi_j(x_1, \ldots, x_n)$ $(j = 1, \ldots, n)$ are non-negative for non-negative values of the arguments.

We prove the following theorem.

Theorem 1. *Let* $x_1, x_2, \ldots, x_n$ *be* n *independently (but not necessarily identically) distributed non-negative random variables. Suppose* *that each* x_j *has a finite moment of order* m. *Assume further that an ordinary polynomial statistic* $P = P(x_1, \ldots, x_n)$ *of order* m *has constant regression on the sum* $\Lambda = x_1 + x_2 + \ldots + x_n$. *Then each random variable* x_j *has finite moments of all orders.*

Proof. We see from the assumptions of the theorem and (1.2) that

$$\mathsf{E}(Pe^{it\Lambda}) = \mathsf{E}(P)\mathsf{E}(e^{it\Lambda}) . \tag{2.2}$$

Let $G_j(x)$ be the distribution function of x_j $(j = 1, 2, \ldots, n)$ and write $C = \mathsf{E}(P)$. Then (2.2) can be rewritten as

$$\int_0^\infty \ldots \int_0^\infty P(x_1, \ldots, x_n) e^{it\Lambda} dG_1(x_1) \ldots dG_n(x_n) =$$

$$= C \int_0^\infty \ldots \int_0^\infty e^{it\Lambda} dG_1(x_1) \ldots dG_n(x_n) .$$

It follows then from (2.1) that

$$\int_0^\infty \ldots \int_0^\infty [x_j^m \Phi_{0j} + \Phi_j] e^{it\Lambda} dG = \int_0^\infty \ldots \int_0^\infty (C - \Theta) e^{it\Lambda} dG \tag{2.3}$$

$(j = 1, \ldots, n)$. For the sake of simplicity we omitted the argument in $\Phi_{0j}(x_1, \ldots, x_n)$ and in $\Phi_j(x_1, \ldots, x_n)$ and $\Theta(x_1, \ldots, x_n)$ and wrote dG instead of $dG_1(x_1) \ldots dG_n(x_n)$. We denote by $f_j(t)$ the characteristic function of x_j $(j = 1, \ldots, n)$ and note that this function can be differentiated at least m times so that

$$f_j^{(k)}(t) = \frac{d^k f_j(t)}{dt^k} = i^k \int_0^\infty x^k e^{itx} dG_j(x) \qquad (k \leqslant m, \; j = 1, \ldots, n) . \tag{2.4}$$

We give an indirect proof of Theorem 1 and assume therefore that at least one of the random variables $x_1, \ldots, x_n$ has only moments up to a certain finite order. It is no restriction to assume that x_1 has moments up to order $m + p$ $(p \geqslant 0)$ (but not to any higher order) while all other random variables have moments at least up to order $m + p$. Then $f_1(t)$ can be differentiated $m + p$ times, while the functions $f_2(t), \ldots, f_n(t)$

can be differentiated at least $m+p$ times. We put now $j=1$ in equation (2.3). Since the polynomial $C-\Theta$ has in each variable a degree not exceeding $m-1$ we see that the integral on the right side of (2.3) can be differentiated $p+1$ times at the origin. The polynomials Φ_{01} and Φ_1 are non-negative in the range of integration; we can therefore apply Fatou's lemma and show that the left hand side of (2.3) can be differentiated $p+1$ times under the integral sign at $t=0$. Thus we obtain

$$(2.5) \qquad \int_0^\infty \dots \int_0^\infty [x_1^m \Phi_{01} + \Phi_1]\Lambda^{p+1} dG = \int_0^\infty \dots \int_0^\infty (C-\Theta)\Lambda^{p+1} dG < \infty .$$

Since $x_1 \leqslant \Lambda$ we see from (2.5) that the integral

$$\int_0^\infty \dots \int_0^\infty [x_1^m \Phi_{01} + \Phi_1] x_1^{p+1} dG$$

exists and is finite. It follows from the definition of Φ_1 that the integral $\int_0^\infty \dots \int_0^\infty \Phi_1 x_1^{p+1} dG$ is also finite so that the same is true for

$$I = \int_0^\infty \dots \int_0^\infty x_1^{m+p+1} \Phi_{01} dG .$$

Since Φ_{01} does not contain x_1 we can write I in the form

$$(2.6) \qquad I = \int_0^\infty x_1^{m+p+1} dG_1 \int_0^\infty \dots \int_0^\infty \Phi_{01}(x_2, \dots, x_n) dG_2 \dots dG_n .$$

It is possible to choose a closed bounded subset Ω of the positive octant $x_2 \geqslant 0, \dots, x_n \geqslant 0$ of the $(n-1)$-dimensional space of the variables $x_2, \dots, x_n$ so that

$$\begin{cases} \int_\Omega dG_2 \dots dG_n = \alpha > 0 \\ \min_\Omega \Phi_{01}(x_2, \dots, x_n) \geqslant c > 0 . \end{cases}$$

Therefore

$$\int_0^\infty \ldots \int_0^\infty \Phi_{01} dG_2 \ldots dG_n > \int_\Omega \Phi_{01} dG_2 \ldots dG_n > 0 \tag{2.7}$$

so that we can conclude from (2.6) that the integral

$$\int_0^\infty x_1^{m+p+1} dG_1(x_1)$$

exists and is finite. But this means that the moment of order $m + p + 1$ of x_1 exist which is a contradiction. This completes the proof of the Theorem.

We consider next a particular case which is of some interest.

We suppose that the polynomial $P(x_1, \ldots, x_n)$ has the form

$$P(x_1, \ldots, x_n) = \sum_{s=1}^{n} x_s^m \psi_s(x_1, \ldots, x_n) + \Theta(x_1, \ldots, x_n) \tag{2.8}$$

where

(i) $\Theta(x_1, \ldots, x_n)$ consists of all monomials of $P(x_1, \ldots, x_n)$ which do not contain the m-th power of any variable

(ii) the polynomial ψ_s does not contain the variable x_s and is of degree less than m in the other variables

(iii) $\psi_s(x_1, \ldots, x_n) \geqslant 0$ for positive values of the argument $(s = = 1, \ldots, n)$.

Then $\Phi_{0j} = \psi_j$ and $\Phi_j = \sum_{s \neq j} x_s^m \psi_s$ and all the conditions of Theorem 1 are satisfied.

A simplification is obtained if one assumes that $k = m$ in (1.3) and (1.3a). Then

$$P(x_1, \ldots, x_n) = \sum_{s=1}^{n} A_s x_s^m + \sum_{\substack{j_1 + \ldots + j_n = m \\ 0 \leqslant j_s < m}} A_{j_1, \ldots, j_n} x_1^{j_1} \ldots x_n^{j_n} . \tag{2.9}$$

Then $\Phi_{0j} = A_j$, $\Phi_j = \sum_{\substack{s=1 \\ s \neq j}}^{n} A_s x_s^m + \sum_{\substack{j_1+\ldots+j_n=m \\ 0 \leqslant j_s < m}} A_{j_1,\ldots,j_n} x_1^{j_1} \ldots x_n^{j_n}$ and the conditions of Theorem 1 are satisfied if $A_s > 0$ $(s = 1, \ldots, n)$ and $A_{j_1,\ldots,j_n} > 0$ $(j_1 + \ldots + j_n = m;\ 0 \leqslant j_s < m)$. Here $\Theta(x_1, \ldots, x_n) \equiv 0$.

We discuss next another application of Theorem 1. Let $P = P(x_1, \ldots, x_n)$ be a homogeneous polynomial of degree m and $\Lambda = x_1 + \ldots + x_n$. We consider a rational function of the form

$$S = S(x_1, \ldots, x_n) = \frac{P}{\Lambda^m} .$$

We note first that S is bounded for positive values of the argument so that the expectation

$$\mathrm{E}(S) = \mathrm{E}\left(\frac{P}{\Lambda^m}\right) = K$$

exists, provided that the random variables $x_1, x_2, \ldots, x_n$ are positive.

If S has constant regression on Λ then according to (1.2) the relation

$$\mathrm{E}\left(\frac{P}{\Lambda^m} e^{it\Lambda}\right) = K\mathrm{E}(e^{it\Lambda})$$

holds. Suppose that each random variable $x_1, \ldots, x_n$ has moments up to the order m, then this relation can be differentiated m times and we obtain

$$\mathrm{E}[(P - K\Lambda^m) e^{it\Lambda}] = 0 . \tag{2.10}$$

Equation (2.10) means that the polynomial $P - K\Lambda^m$ has constant regression (with $C = 0$) on Λ so that the following Corollary is an immediate consequence of Theorem 1.

Corollary. *Let $x_1, \ldots, x_n$ be n independently (but not necessarily identically) distributed positive random variables and suppose that each of these has a finite moment of order m. Let P be a homogeneous polynomial of degree m and $\Lambda = x_1 + \ldots + x_n$ and write* $K = \mathrm{E}\left(\frac{P}{\Lambda^m}\right)$.

Suppose that $S = \dfrac{P}{\Lambda^m}$ *has constant regression on* Λ *and that* $P - K\Lambda^m$ *is an ordinary polynomial. Then for each of the random variables* x_j *all the moments exist.*

3. ANALYTICITY OF THE CHARACTERISTIC FUNCTION

In this section, we impose an additional restriction on the polynomial statistic studied in Theorem 1. We then show that each random variable has an analytic characteristic function.

We consider a polynomial of the form

$$(3.1) \qquad P = P(x_1, \ldots, x_n) = \sum_{j=1}^{n} a_j x_j^m + Q(x_1, \ldots, x_n)$$

where the following conditions are satisfied

(3.1a) $\quad a_j > 0 \quad$ for $\quad j = 1, \ldots, n;$

(3.1b) $Q \geqslant 0$ and the degree of $Q(x_1, \ldots, x_n)$ does not exceed $m - 1$ in each variable. Then $k = m$ in (2.9) while the representation (2.1) holds with

$$\Phi_{0j} = a_j, \quad \Phi_j = \sum_{\substack{s=1 \\ s \neq j}}^{n} a_s x_s^m, \qquad \Theta \equiv Q .$$

We formulate the result of this section.

Theorem 2. *Let* $x_1, \ldots, x_n$ *be* n *independently (but not necessarily identically) distributed non-negative random variables. Suppose that a polynomial statistic* P *is of the form* (3.1) *and satisfies* (3.1a) *and* (3.1b). *Assume that* P *has constant regression on* $\Lambda = x_1 + \ldots + x_n$. *Then the random variables* $x_1, \ldots, x_n$ *have analytic characteristic functions.*

Proof. We remark first that the conditions of Theorem 1 are satisfied so that $x_1, \ldots, x_n$ have moments of all orders. Hence all moments of Λ exist. Under the assumptions of the theorem relation (2.2) holds and can be differentiated with respect to t any number of times. We differentiate (2.2) N times and put $t = 0$ and obtain.

$$\mathsf{E}(P\Lambda^N) = \mathsf{E}(P)\mathsf{E}(\Lambda^N) .$$

Writing $C = \mathsf{E}(P)$ and using (3.1) this becomes

$$\mathsf{E}\left[\Lambda^N \sum_{j=1}^{n} a_j x_j^m\right] = \mathsf{E}[\Lambda^N (C - Q)] . \tag{3.2}$$

Our aim is to obtain an upper bound for the integral $\int \ldots \int \Lambda^{m+N} dG$. It follows from Hölder's inequality that

$$\Lambda^m = (x_1 + \ldots + x_n)^m \leqslant n^{m-1}(x_1^m + \ldots + x_n^m) .$$

It is no restriction to assume that a_1 is the smallest of the coefficients $a_1, a_2, \ldots, a_n$ so that

$$\int_0^\infty \ldots \int_0^\infty \Lambda^{m+N} dG \leqslant \frac{n^{m-1}}{a_1} \int_0^\infty \ldots \int_0^\infty (a_1 x_1^m + \ldots + a_n x_n^m)\Lambda^N dG . \tag{3.3}$$

We derive next an upper bound for $\mathsf{E}[\Lambda^N (C - Q)]$. For this purpose we write the polynomial $C - Q$ as

$$C - Q = \sum_{j_1 + \ldots + j_n \leqslant m-1} A_{j_1,\ldots,j_n} x_1^{j_1} \ldots x_n^{j_n}$$

so that for non-negative values of the $x_1, \ldots, x_n$

$$|C - Q| \leqslant \sum_{p=0}^{m-1} \sum_{j_1 + \ldots + j_n = p} |A_{j_1,\ldots,j_n}| x_1^{j_1} \ldots x_n^{j_n} .$$

Therefore

$$|\mathsf{E}[\Lambda^N (C - Q)]| \leqslant \int_0^\infty \ldots \int_0^\infty \sum_{p=0}^{m-1} \sum_{j_1 + \ldots + j_n = p} |A_{j_1,\ldots,j_n}| \times$$
$$\times\, x_1^{j_1} \ldots x_m^{j_n} (x_1 + \ldots + x_n)^N dG$$

so that

$$|\mathsf{E}[\Lambda^N(C-Q)]| \leqslant \tag{3.4}$$
$$\leqslant \sum_{p=0}^{m-1} \int_0^\infty \cdots \int_0^\infty \sum_{j_1+\ldots+j_n=p} |A_{j_1,\ldots,j_n}| (x_1+\ldots+x_n)^{N+p} dG .$$

For the proof of Theorem 2 we need the following lemma

Lemma 1. *Let*

$$P = P(x_1, \ldots, x_n) = \sum_{j_1+\ldots+j_n=p} A_{j_1,\ldots,j_n} x_1^{j_1} \ldots x_n^{j_n}$$

be a homogeneous polynomial in n *variables* $x_1, \ldots, x_n$ *of degree* p. *Then the number of terms in* P *can not exceed* $\binom{n+p-1}{n-1}$.

Proof. Clearly, P has at most as many terms as the polynomial $\sum_{j_1+\ldots+j_n=p} x_1^{j_1} \ldots x_n^{j_n}$. This number is the number of ways in which p can be partitioned into n non-negative integers. This is the coefficient of x^p in the polynomial

$$(1+x+\ldots+x^p)^n = \frac{(1-x^{p+1})^n}{(1-x)^n} =$$
$$= (1-x)^{-n} \sum_{j=0}^{n} (-1)^j \binom{n}{j} x^{(p+1)j} .$$

The coefficient of x^p in this expression is equal to the coefficient of x^p in the expansion of $(1-x)^{-n}$ and hence equals $\binom{n+p-1}{p} = \binom{n+p-1}{n-1}$ as stated in the Lemma.

We denote by

$$A = \max_{j_1+\ldots+j_n \leqslant m-1} |A_{j_1,\ldots,j_n}| .$$

Using Lemma 1 and (3.4) we obtain the inequality

$$|\mathsf{E}[\Lambda^N(C-Q)]| \leqslant A \sum_{p=0}^{m-1} \binom{p+n-1}{n-1} \int_0^\infty \cdots \int_0^\infty \Lambda^{N+p} dG . \tag{3.5}$$

From (3.2), (3.3) and (3.5) we obtain

$$(3.6)\qquad \int_0^\infty \dots \int_0^\infty \Lambda^{m+N}\, dG \leqslant \frac{An^{m-1}}{a_1} \sum_{p=0}^{m-1} \binom{p+n-1}{n-1} \int_0^\infty \dots \int_0^\infty \Lambda^{N+p}\, dG\,.$$

Since all moments of Λ exist it is possible to choose a finite positive number M_1 such that the relation $\mathsf{E}(\Lambda^s) \leqslant s!M_1^s$ holds for $s = 1, 2, \dots$ $\dots, m$. We prove the following Lemma.

Lemma 2. *Let* $M_0 = \max\left[1, Aa_1^{-1}\binom{m+n-1}{n} n^{m-1}\right]$. *Then the inequality*

$$(3.7)\qquad \mathsf{E}(\Lambda^s) \leqslant s!M^s$$

holds for all s *provided that* $M = \max\,[M_0, M_1]$.

Proof. The relation (3.7) obviously holds for all $s \leqslant m + N - 1$ when $N = 1$. We prove the Lemma by induction on N. We suppose that (3.7) is valid for $s = 1, 2, \dots, m + N - 1$ and show that this implies the validity for $s = m + N$ also. We have

$$(3.8)\qquad \begin{gathered} \int_0^\infty \dots \int_0^\infty \Lambda^{p+N}\, dG \leqslant (p+N)!M^{p+N} \\ \text{for} \quad p = 0, 1, \dots, (m-1)\,. \end{gathered}$$

Combining (3.6) and (3.8) we get

$$\int_0^\infty \dots \int_0^\infty \Lambda^{m+N}\, dG \leqslant \frac{An^{m-1}}{a_1} \sum_{p=0}^{m-1} \binom{p+n-1}{n-1} (p+N)!M^{p+N} \leqslant$$

$$\leqslant \frac{An^{m-1}}{a_1} (m+N)!M^{m+N-1} \sum_{p=0}^{m-1} \binom{p+n-1}{n-1}\,.$$

It is not difficult to show by induction on m that $\sum_{p=0}^{m-1} \binom{p+n+1}{n-1} =$ $= \binom{m+n-1}{n}$;

hence $\int_0^\infty \dots \int_0^\infty \Lambda^{m+N}\, dG \leqslant (m+N)!M^{m+N-1}M_0$

or

$$(3.9) \qquad \int_0^\infty \ldots \int_0^\infty \Lambda^{m+N} dG \leqslant (m+N)! M^{m+N} .$$

This completes the proof of Lemma 2.

It follows from Lemma 2 that the distribution of Λ has a characteristic function $f_\Lambda(t)$ which can be expanded into a power series about the origin. The radius of convergence of this series is at least equal to $1/M$. Thus $f_\Lambda(t)$ is an analytic characteristic function. Hence according to a theorem of D.A. Raikov [2] the distribution of each x_j has also an analytic characteristic function.

Corollary. *Under the conditions of Theorem 2, the characteristic functions of the x_j are regular in a half plane which contains the upper halfplane in its interior.*

The Corollary follows immediately from the assumption that the random variables $x_1, \ldots, x_n$ are non-negative.

In conclusion we remark that the questions discussed in this paper can also be regarded as problems in the theory of ordinary differential equations. We study here primarily some analytic properties of the positive definite solutions of differential equations of the form (1.4). We had to impose certain suitable restrictions which were motivated partly by applications in statistics, partly by analytical difficulties.

REFERENCES

[1] E. Lukacs, Characterization of populations by properties of suitable statistics, *Proceedings of the Third Berkeley Symposium*, Univ. of Calif. Press, Berkeley, 2 (1956), 195-214.

[2] D.A. Raikov, On the decomposition of Gauss and Poisson laws, (in Russian; English summary), *Izv. Akad. Nauk SSSR. Ser. Mat.*, 2 (1938), 91-124.

[3] A.A. Zinger – Yu.V. Linnik, On a theorem of the theory of differential equations and the invariance of statistics from the mean, (in Russian), *Dokl. Akad. Nauk SSSR.*, 108 (1956), 577-579.

[4] A.A. Zinger – Yu.V. Linnik, On a class of differential equations and its applications to some questions of regression theory, (in Russian), *Vestnik Leningrad. Univ.*, 7 (1957), 121-130.

COLLOQUIA MATHEMATICA SOCIETATIS JÁNOS BOLYAI
9. EUROPEAN MEETING OF STATISTICIANS, BUDAPEST (HUNGARY), 1972.

A BAYESIAN SOLUTION FOR TWO-WAY ANALYSIS OF VARIANCE

D.V. LINDLEY

INTRODUCTION

Perhaps the most widely used of all statistical techniques is that of least squares particularly in the context of a linear model. It has a history going back as far as Gauss, and since his day has been extensively used in disciplines as diverse as numerical analysis and biology: we cite the example of fitting a straight line to bivariate data. As part of the flowering of modern statistics in this century, least squares has been significantly extended by the distributional results of Fisher, leading to analysis of variance concepts. The F-test and the minimum variance unbiased estimate are two cornerstones of statistical theory that have found many applications.

But the time has come when we should seriously consider whether these twin ideas are sound in theory or useful in practice.

It is now 17 years since Stein [7] first announced the result that the sample mean is inadmissible as an estimator of the normal population mean, variance known, in three or more dimensions, using squared-error loss. Brown [1] has significantly extended the scope of results of this type. Since the linear model can always be reduced, by appropriate transformations, to data $\{x_i\}$ with $E(x_i) = \theta_i$, for $i \leqslant p$ and otherwise $E(x_i) = 0$, and since the sample mean $(x_1, x_2, \ldots, x_p)$ is the resulting least-square estimate of $(\theta_1, \theta_2, \ldots, \theta_p)$, it follows that except when one or two parameters only are involved the Gaussian estimates are inadmissible. It is worth noting that the estimates can be seriously inefficient, improved ones having mean-square error which is sometimes only a fraction $\frac{2}{p}$ of that of the least-squares value. The possible improvement over the classical technique by using other estimates is therefore important.

The F-test, the other aspect of modern least-squares theory, can be criticized as part of the general comment on significance tests that their use violates the likelihood principle. In a formal development of statistical theory, Lindley [3], there seems to be no place for arguments that, after the data are to hand, inquire what the date might have been but weren't. Furthermore the general emphasis in the literature seems to be away from significance tests and towards confidence interval types of statement.

Many practitioners of the analysis of variance have argued that the major point in performing such an analysis is to obtain a valid estimate of error. If so, it should be noted that even that estimate is inadmissible, Brown [2]; but ignoring this, the error is going to be attached to a least squares estimate which itself is seriously inefficient, and will consequently exaggerate the uncertainty in the data.

It therefore seems appropriate to have a fresh look at the analysis of variance, least-squares complex of statistical tools. In the present paper a special linear model, the two-way layout with replication, is discussed, so that interactions can be considered. The emphasis is on obtaining usable results rather than on formal theory.

1. TWO-WAY MODEL

In this paper, we consider the analysis of data (x_{ijk}) having the following probability structure. For given parameter values (θ_{ij}) and (σ^2_{ij}), the random variables x_{ijk} are independent and normally distributed with $\mathrm{E}(x_{ijk}) = \theta_{ij}$ and $\mathrm{var}(x_{ijk}) = \sigma^2_{ij}$: here $i = 1, 2, \ldots, m$; $j = 1, 2, \ldots, n$; and $k = 1, 2, \ldots, r_{ij}$.

An example where this model for data might be appropriate is where x_{ijk} is the performance of a subject in an educational test, the subject having been to School i and College j, there being r_{ij} such subjects and the suffix k serving to enumerate them. There, θ_{ij} would correspond to the true score of subjects from School i and College j on the test, and σ^2_{ij} would measure their variability. Any analysis of the data would investigate what effects the school and college attended had on performance. We shall confine attention to the case where the variabilities σ^2_{ij} are all the same, equal to σ^2; and there are the same numbers of subjects in each group, so that $r_{ij} = r$, say. This is usually referred to as the orthogonal case, and its analysis is rather simpler than that for the general situation. Rather than refer to schools and colleges, we shall use the neutral terms "rows" and "columns"; x_{ijk} is then the kth observation in Row i and Column j, the data being conveniently laid out on the page in such a row and column formation.

Let us first recall how such data are traditionally analyzed. Any good textbook on statistics that deals with the two-way analysis of variance, with interaction, will provide details beyond the summary which follows: for example, Snedecor [6], Chapter 11. We use the familar "dot" notation for averages. Thus, $x_{i\,..} = \frac{1}{nr}\sum_{j,k} x_{ijk}$, the mean of the data in Row i, the dots replacing the suffixes j and k over which summation has taken place. The usual analysis breaks up the total sum of squares about the overall mean, $\sum_{i,j,k}(x_{ijk} - x_{...})^2$, into at least four components. Firstly, there is the main effect of rows

$$(1) \qquad nr \sum_i (x_{i\,..} - x_{...})^2 ,$$

and secondly, that of columns

(2) $$mr \sum_j (x_{.j.} - x_{...})^2 .$$

The third is the interaction between rows and columns

(3) $$r \sum_{i,j} (x_{ij.} - x_{i..} - x_{.j.} + x_{...})^2 ,$$

and the last is the residual, or within groups, sum of squares

(4) $$\sum_{i,j,k} (x_{ijk} - x_{ij.})^2 .$$

On division by their appropriate degrees of freedom, each of the first three may be tested against the last using the familiar F-test. If, for example, only the first test is significant, then the column and interaction effects are supposed zero and θ_{ij}, for all j, is estimated by $x_{i..}$. Comparisons between these means are effected by multiple-comparison procedures of which Scheffé's is, perhaps, the most popular.

This analysis, apart from being open to the usual criticisms that can be levelled against significance tests, is unsatisfactory in that it forces one into the position of having to be dogmatic about whether a particular effect exists, or not. Thus, several estimates of θ_{ij} are available depending on the results of the tests. Two are $x_{i..}$ (mentioned above) and $x_{i..} + x_{.j.} - x_{...}$ (if row and column, but no interaction, effects exist). A better procedure would be to estimate the size of each of these effects and estimate θ_{ij} accordingly. The methods developed below do just this and for example, weight the row in which θ_{ij} appears heavily if the row effect appears to be large. Significance tests are, thereby, avoided.

For the one-way classification, where $\mathsf{E}(x_{ik}) = \theta_i$, such an analysis has been given by Lindley [3] and extended to other situations in the context of a general theory by Lindley and Smith [4]. In this paper, we apply the results of the latter reference to obtain an estimate of θ_{ij} that uses $x_{ij.}, x_{i..}, x_{.j.}$, and $x_{...}$ in a balance that depends on the relative sizes of the main effects and interaction. In order to utilize this theory, it is necessary to describe the prior probability distribution of the (θ_{ij}) (and also σ^2, but in the first analysis this will be supposed known).

In the one-way case, it was suggested that the joint distribution might reasonably have the property of exchangeability; that is, be invariant under any permutation of the suffixes. This property is clearly inappropriate in the two-way case as is seen by considering the joint distribution of a pair, θ_{ij} and θ_{rs}. Under exchangeability, this distribution is the same for any pair of (different) θ's, whereas it would be reasonable for the relation between θ_{ij} and θ_{is} $(j \neq s)$ in the same row to be different from that between θ_{ij} and θ_{rs} $(i \neq r)$ in different rows (and columns). In our example, knowledge of the performance of subjects at School i and College j might affect knowledge of subjects from the *same* school at another college, whereas it might say little about those from a *different* school at the college. We, therefore, have to express the prior ideas other than through exchangeability. We use, instead, a modified form of it.

Our prior opinions might lead us to think that the value of θ_{ij} is influenced both by the row and the column that it is in. If these effects are assumed additive, we might suppose

$$\theta_{ij} = \mu + \alpha_i + \beta_j + \gamma_{ij}$$

where μ is an overall mean, (α_i) and (β_j) respectively describe row and column effects, and (γ_{ij}) represent independent error terms, say, normal with zero mean and variance σ_c^2. Alternatively expressed, we could say; given μ, (α_i), (β_j), and σ_c^2, the θ's are independent and normally distributed with

$$\text{(5)} \qquad \mathrm{E}(\theta_{ij}) = \mu + \alpha_i + \beta_j$$

and variance σ_c^2. The rows and columns might reasonably be exchangeable; and hence, given $\mu_a, \mu_b, \sigma_a^2, \sigma_b^2$, we might suppose the α's and β's independent and normally distributed with $\mathrm{E}(\alpha_i) = \mu_a$, $\mathrm{E}(\beta_j) = \mu_b$, $\mathrm{var}(\alpha_i) = \sigma_a^2$, and $\mathrm{var}(\beta_j) = \sigma_b^2$.

This model fits conveniently into the framework developed by Lindley and Smith. In their terminology, it is a four-stage model; the first stage decribes the dependence of the x's on the θ's; the second, that of the θ's on the α's and β's; the third describes the structure of the α's and β's; and a fourth stage is necessary to describe the prior distributions of μ_a and μ_b. As in earlier examples, this distribution can be supposed diffuse and the variances for μ_a and μ_b allowed to tend to infinity. It is possible to proceed with the analysis of the four-stage form, but it is convenient to reduce it first to a three-stage version with a diffuse prior at the third and final stage: the two analyses are equivalent, except for one point to be discussed later in considering the variance estimation.

To derive the three-stage model, consider the distribution of the θ's, given μ, but not the α's and β's. From (5), it is clear that the covariances are given by

(6a) $\quad \operatorname{cov}(\theta_{ij}, \theta_{rs}) = 0, \qquad i \neq r,\ j \neq s\,,$

(6b) $\quad \operatorname{cov}(\theta_{ij}, \theta_{is}) = \sigma_a^2, \qquad j \neq s\,,$

(6c) $\quad \operatorname{cov}(\theta_{ij}, \theta_{rj}) = \sigma_b^2, \qquad i \neq r\,,$

and

(6d) $\quad \operatorname{cov}(\theta_{ij}, \theta_{ij}) = \sigma_a^2 + \sigma_b^2 + \sigma_c^2\,.$

(The last is just the variance of θ_{ij}.) For example, the difference between (6a) and (6b) is just the distinction we were discussing above concerning subjects from the same School (row) i. Consequently, a second stage, which replaces the second and third stages of the first model, supposes (θ_{ij}) has a multivariate normal distribution with covariance structure given by equation (6) and constant mean μ (now incorporating μ_a and μ_b). The third (and final) stage says the knowledge of μ is diffuse.

This is the model we suggest might be appropriate for some two-way analyses. We must emphasize that there may well exist two-way situations in which the above prior specification (in the second and third stages) is quite unsuitable. Before performing an analysis of the type suggested below, it must first be checked that the model is reasonably suitable. Our second- and third-stage forms are assumptions that may not always be realistic. For example, suppose the rows (schools) were of two types, say urban and rural, then the α's (in the four-stage form) would not be exchangeable for all i – perhaps, only within-urban and within-rural schools.

With this caution, let us summarize the model:

First stage. Given (θ_{ij}), σ^2; the (x_{ijk}) are normal and independent with $\mathrm{E}(x_{ijk}) = \theta_{ij}$ and variance σ^2.

Second stage. Given $\mu, \sigma_a^2, \sigma_b^2, \sigma_c^2$; the (θ_{ij}) have a multivariate normal distribution with dispersion matrix, given by equations (6), and $\mathrm{E}(\theta_{ij}) = \mu$.

Third stage. The prior knowledge of μ is diffuse.

2. ESTIMATES FOR THE MEANS

Our first object is, for given $\sigma^2, \sigma_a^2, \sigma_b^2$ and σ_c^2, to find the posterior distribution of the (θ_{ij}). It is easy to see that it will be multivariate normal; the means will then provide estimates of the (θ_{ij}), and the dispersion matrix will enable standard errors to be attached to these estimates. We later relax the conditions on the knowledge of the four variances and show how they too may be estimated, thereby providing revised estimates and standard errrors for the θ's.

The algebraic derivation of the estimates θ_{ij}^* of θ_{ij} is given in Appendix 1. It is there shown that

$$\theta_{ij}^* = \frac{\frac{r}{\sigma^2}}{\frac{r}{\sigma^2} + \frac{1}{\sigma_c^2}} (x_{ij.} - x_{i..} - x_{.j.} + x_{...}) +$$

$$(7) \qquad + \frac{\frac{r}{\sigma^2}}{\frac{r}{\sigma^2} + \frac{1}{\sigma_c^2 + n\sigma_a^2}} (x_{i..} - x_{...}) +$$

$$+ \frac{\frac{r}{\sigma^2}}{\frac{r}{\sigma^2} + \frac{1}{\sigma_c^2 + 'm\sigma_b^2}} (x_{.j.} - x_{...}) + x_{...} \, .$$

This is the main result of this paper. The form of this estimate is interesting. It depends on four aspects of the data: $x_{ij.}$, the mean of the observations in cell (i, j); $x_{i..}$, and $x_{.j.}$, the corresponding row and column means; and $x_{...}$, the overall mean. It is a weighted combination of this last, $x_{i..} - x_{...}$, the effect of the row, $x_{.j.} - x_{...}$, the effect of the column, and $x_{ij.} - x_{i..} - x_{.j.} + x_{...}$, the interaction effect. The weights depend on the variances σ_a^2, σ_b^2, and σ_c^2 in addition to the residual variance (from the data) σ^2. Some special cases are interesting. Suppose $\sigma_c^2 = 0$ so that, equation (5), θ_{ij} is a linear combination of the row and column effects and no interaction exists. Then, the first term in (7) vanishes, there is no contribution from the data-interaction effect, and θ_{ij}^* uses only $x_{i..}, x_{.j.}$ and $x_{...}$. This is an extreme case corresponding to the assumed lack of an interaction as indicated in the usual approach by a non-significant F-test for the interaction. If, in addition to $\sigma_c^2 = 0$, $\sigma_a^2 = 0$, the second term in (7) also vanishes and only the column effect appears from the data. If $\sigma_a^2 = 0$ without σ_c^2 vanishing, the first and second terms in (7) combine to give a multiple of $(x_{ij.} - x_{.j.})$. These results generalize in a natural way those of Lindley [3] for the one-way case in which similar weighted combinations occurred. Later, we shall see how to estimate the four variances and, hence, the weights.

To obtain the posterior variances and covariance of these estimates, write the weights in (7) as

$$w_c = \frac{\frac{r}{\sigma^2}}{\frac{r}{\sigma^2} + \frac{1}{\sigma_c^2}},$$

(8) $$w_a = \frac{\frac{r}{\sigma^2}}{\frac{r}{\sigma^2} + \frac{1}{\sigma_c^2 + n\sigma_a^2}},$$

$$w_b = \frac{\frac{r}{\sigma^2}}{\frac{r}{\sigma^2} + \frac{1}{\sigma_c^2 + m\sigma_b^2}}.$$

Then, (7) becomes

$$\theta^*_{ij} = w_c(x_{ij.} - x_{i..} - x_{.j.} + x_{...}) + w_a(x_{i..} - x_{...}) +$$
$$+ w_b(x_{.j.} - x_{...}) + x_{...} \ .$$

If we further put

(9) $$nW_a = w_a - w_c, \quad mW_b = w_b - w_c, \quad mnW = w_c - w_a - w_b + 1$$

and put $w_c = W_c$, for symmetry, (7) can be written

(10) $$\theta^*_{ij} = W_c x_{ij.} + nW_a x_{i..} + mW_b x_{.j.} + mnWx_{...} \ .$$

For reasons given in Appendix 1, the dispersion matrix for θ_{ij} is given by [compare equations (6)]

(11a) $$\operatorname{cov}(\theta_{ij}, \theta_{rs}) = W \frac{\sigma^2}{r}, \qquad i \neq r, \ j \neq s,$$

(11b) $$\operatorname{cov}(\theta_{ij}, \theta_{is}) = (W_a + W) \frac{\sigma^2}{r}, \qquad j \neq s,$$

(11c) $$\operatorname{cov}(\theta_{ij}, \theta_{rj}) = (W_b + W) \frac{\sigma^2}{r}, \qquad i \neq r.$$

(11d) $$\operatorname{cov}(\theta_{ij}, \theta_{ij}) = (W_a + W_b + W_c + W) \frac{\sigma^2}{r} .$$

These expressions are somewhat cumbersome since the W's are fairly complicated, but some results are a little easier. For example, consider the posterior variance of $\theta_{ij} - \theta_{is}$ $(j \neq s)$, that is the difference between Columns j and s in the same Row i. It is $2 \operatorname{var}(\theta_{ij}) - 2 \operatorname{cov}(\theta_{ij}, \theta_{is})$, which, from (11b) and (11d) is $(W_b + W_c) \frac{2\sigma^2}{r}$. For the means of rows (or columns), the results are easier still. For example, the variance of $\theta_{i.} - \theta_{r.}$, the difference between two rows (schools) averaged over columns (colleges) is $(i \neq r, \ j \neq s)$

$$\operatorname{var}(\theta_{i.} - \theta_{r.}) = \frac{1}{n^2} \operatorname{var}\left[\sum_j \theta_{ij} - \sum_s \theta_{rs}\right] =$$

$$= \frac{1}{n^2} [2n \operatorname{var}(\theta_{ij}) - 2n \operatorname{cov}(\theta_{ij}, \theta_{rj})] +$$

$$+ \frac{1}{n^2} [2n(n-1) \operatorname{cov}(\theta_{ij}, \theta_{is}) - 2n(n-1) \operatorname{cov}(\theta_{ij}, \theta_{rs})] =$$

$$= \frac{2\sigma^2}{rn} [W_a + W_b + W_c + W - (W_b + W) +$$

$$+ (n-1)(W_a + W) - (n-1)W]$$

from (11), and using (9), this is finally equal to

$$\frac{2\sigma^2}{rn} w_a = \frac{2}{n} \left(\frac{r}{\sigma^2} + \frac{1}{\sigma_c^2 + n\sigma_a^2} \right)^{-1} . \tag{12}$$

Since θ_{ij}^* [equation (7)] is the posterior mean, the mean of $\theta_{i.}$ is $\theta_{i.}^*$, which, from (7), is easily seen to be

$$w_a(x_{i..} - x_{...}) + x_{...} = w_a x_{i..} + (1 - w_a) x_{...} ,$$

a weighted average of $x_{i..}$ and $x_{...}$. Had $x_{i..}$ been used as an estimate, as standard theory would suggest, then the variance for $\theta_{i.} - \theta_{r.}$ quoted would be $\frac{2\sigma^2}{rn}$ rather than this times w_a, given by (12). Hence, our estimate is pulled toward the overall mean and has smaller variances when compared with other values. It follows that the usual multiple comparison procedures, such as Scheffé's, are unnecessary in our approach. The shift toward the mean and the reduced standard errors perform exactly the function that these orthodox procedures are designed to provide.

3. ESTIMATES FOR THE VARIANCES

These estimates (and standard errors) depend upon knowledge of the four variances $\sigma^2, \sigma_a^2, \sigma_b^2$, and σ_c^2. In any application, these are typically unknown but can be estimated from the data. This is obvious for σ^2 but is also true for the others since there is replication of rows and columns. We proceed to discuss their estimation.

Lindley and Smith, in discussing the general theory, show that if we are content with posterior modes for estimates (rather than posterior means), we can continue to estimate θ_{ij} by equations (7) provided we insert, for the four variances, modal estimates of them. It will, therefore, suffice to find the posterior modes for the variances. It is inconvenient to do this within the context of the three-stage model because the compression of two stages into one results in σ_c^2 (from the original second stage) being combined with σ_a^2 and σ_b^2 (from the third stage) in expressions like $\sigma_c^2 + n\sigma_a^2$, and we have the difficulties familiar in components of variance problems (or what is sometimes called Type II analysis of variance) of having to estimate σ_c^2 and $\sigma_c^2 + n\sigma_a^2$ separately, and hence σ_a^2 by subtraction, so leading to the possibility of negative estimates for σ_a^2, or even within the Bayesian framework, to difficult calculations. This can be avoided by using the four-stage model, when the procedure is essentially to estimate (α_i) and (β_j) and, hence, σ_a^2 by a multiple of $\sum_i (\alpha_i^* - \alpha_.^*)^2$; similarly, σ_b^2. Also, σ_c^2 can be found from the sums of squares of $\theta_{ij}^* - \alpha_i^* - \beta_j^*$ [see equation (5)]. Finally, σ^2 can be found, although the usual within sum of squares is not enough since θ_{ij} is, within the present theory, not estimated by $x_{ij.}$ as is usual. Hence, the within-sum underestimates the total variation that contributes to σ^2. All these ideas are straightforward generalizations of ideas contained in the papers to which reference has already been made.

The details of the calculation of the posterior modes are given in Appendix 2. Equations (2.3) and (2.4) provide estimates of (α_i) and (β_j), respectively. Notice that only the deviation from the mean is estimated, which is all that is necessary. Distinction should be made between the estimate of, for example, α_i by [equation (2.3)]

$$(\alpha_i - \alpha_.)^* = \frac{rn\sigma_a^2}{rn\sigma_a^2 + r\sigma_c^2 + \sigma^2}(x_{i..} - x_{...})$$

and that of $\theta_{i.}$ by

$$(\theta_{i.} - \theta_{..})^* = \frac{rn\sigma_a^2 + r\sigma_c^2}{rn\sigma_a^2 + r\sigma_c^2 + \sigma^2}(x_{i..} - x_{...})$$

[from (1.18), or (7) on summing over j, and a little simplification.] The difference is that $\theta_{i.}$ is the average for Row i over the columns used in the experiment, whereas α_i is a similar average not confined to the columns of the experiment. In particular, α_i^* is shrunk more towards the overall mean than is $\theta_{i.}^*$, since the coefficient of the deviation $(x_{i..} - - x_{...})$ is smaller in the former.

Equations (2.5) provide the estimates of the variances, using the estimates for (α_i) and (β_j) just obtained as well as those for (θ_{ij}) already calculated. Those estimates, in turn, depend on the variances, and so some iterative procedure has to be used. We suggest the following: Obtain initial estimates of the four variances from the usual analysis of variance expressions, equations (1) to (4), divided by their respective degrees of freedom. These will be unsatisfactory estimates but will serve to provide weights to be used to estimate the θ's [equation (7)] and the α's and β's. With these estimated, new values for the variances can be found from equations (2.5) and the cycle repeated until convergence.

Notice that the estimates (2.5) involve quantities derived from the prior distributions of the variances. There is no objection to putting ν, corresponding to σ^2, equal to zero; but the remaining values ν_a, ν_b, ν_c cannot be ignored. The difficulty is that if σ_a^2, σ_b^2, or σ_c^2 are small in comparison with σ^2 (or more correctly $\frac{\sigma^2}{r}$), there is little information in the data from which to estimate them since the variation in the $(x_{ij.})$ is mostly due to σ^2. In this case, the prior knowledge is clearly important and so naturally arises in any estimation procedure. If σ_a^2, for example, is large; its estimation is easier, and in (2.5b), the sum of squares for α_i^* will dominate $\nu_a \lambda_a$ unless the latter is large: that term and ν_a in the denominator may be ignored.

Whilst the estimates for θ_{ij}, given the variances, are almost certainly satisfactory; it may be possible to improve the estimation of the variances in comparison with the methods given in this paper; and we hope to study the problem in more detail later. In the meantime, it might be reasonable to guess that the term mn in the denominator of (2.5d) might be replaced by the degrees of freedom $(m-1)(n-1)$. In deriving modes, rather than means, the usual integrations that remove degrees of freedom do not take place, and hence, the divisor always involves the total number, here mn, of parameters. Another way of looking at it is to appreciate that the modes of marginal distributions are not the components of the modes of the whole distribution.

4. NUMERICAL EXAMPLE

In this example, we describe the results of analyzing a simple case using the methods developed in the paper. Richmers and Todd [5] give the following data, in their Table (8.21), taken from an experiment on the breaking strength of three fabrics at four temperatures with two replicates at each of the twelve combinations. We therefore have the case of constant numbers of replicates, and we assume that

Fabric	Temperature 210	215	220	225
A	1.8	2.0	4.6	7.5
	2.1	2.1	5.0	7.9
B	2.2	4.2	5.4	9.8
	2.4	4.0	5.6	9.2
C	2.8	4.4	8.7	13.2
	3.2	4.8	8.4	13.0

σ_{ij}^2 is also fixed but unknown at σ^2. The prior distribution suggested seems appropriate except that exchangeability of the column values (temperatures) ignores the fact that they are in sequence. But such information on ordering is neglected in the usual analysis of variance technique, so we do the same for comparison purposes. In the standard method, the 3 degrees of freedom associated with temperature would be broken up into linear and perhaps, quadratic terms: a parallel Bayesian analysis could easily be developed. We took $\nu = 0$, $\nu_t = 1$ $(t = a, b, c)$ in equations (2.5). These correspond to weak prior knowledge without causing convergence problems. (Values $\nu_t = 3$ were also tried with only a small effect on the results).

The next table gives for each of the 12 cells the estimate θ^*_{ij} of the cell mean obtained from equation (7) with estimates from (2.5) of the variance components replacing the σ's. Also, included in brackets is the mean of the two original readings for that cell for comparison purposes. For each row and column there are similarly given the estimates α^*_i and β^*_j from (3.2) and (3.3) together with the data means in brackets for comparison.

Fabric	Temperature 210	215	220	225	
A	1.39	2.41	5.11	8.80	4.31
	(1.95)	(2.05)	(4.80)	(7.70)	(4.13)
B	2.24	3.49	6.02	9.85	5.38
	(2.30)	(4.10)	(5.50)	(9.60)	(5.38)
C	3.63	4.85	7.72	11.62	7.10
	(3.00)	(4.60)	(8.55)	(13.10)	(7.31)
	2.53	3.66	6.26	9.95	
	(2.42)	(3.58)	(6.28)	(10.13)	

The estimates of the variances are $s^2 = 0.495$, $s_a^2 = 0.991$, $s_b^2 = 5.591$, $s_c^2 = 0.098$. These show a large effect of temperature, a smaller effect of fabric, and a small interaction term. The estimates θ^*_{ij} are, therefore, dominated by the additive effect of the two factors. These, displayed in the borders of the table, show the usual shift towards the overall mean. For example, the value of β^*_1, the mean breaking strength at 210 is 2.53, greater than the observed mean of 2.42. The shift with the cell means is greater because of the almost complete removal of the interaction component. Thus, fabric A at 225 is estimated at 8.80 against an observed value of 7.70 which is a shift *away* from the mean. Notice that as a result of these shifts, the estimate of residual variance is at 0.495, much larger than the conventional value of 0.056 obtained from the 12 within-cell differences.

I am most grateful to D. Christ and G. Isaacs who wrote the computer program and ran the above example. Their enthusiasm and expertise was most helpful and provided an illuminating insight into the merits of interactive computing.

This work has been carried out as part of the research activities of the American College Testing Program under the guidance of M. R. Novick, whose stimulus at all stages of the work is gratefully acknowledged. The research was performed pursuant to Grant No. OEG-0-72-0711 with the Office of Education, U.S. Department of Health, Education and Welfare. The invitation of the Bolyai János Mathematical Society to Budapest, and the financial support offered are much appreciated, even though illness prevented my attending the meetings in person.

APPENDIX 1: POSTERIOR DISTRIBUTION OF THE CELL MEANS ASSUMING THE VARIANCES KNOWN

When writing out vectors of elements depending on two or more suffixes, we shall use a lexicographical order: thus,

$$\theta_1^T = (\theta_{11}, \theta_{12}, \ldots, \theta_{1n}, \theta_{21}, \theta_{22}, \ldots, \theta_{mn}).$$

The three-stage model is exactly in the linear framework developed by Lindley and Smith, and their Corollary 2 [equations (16) and (17)] shows that the posterior distribution of (θ_{ij}) is normal with first and second moments there stated. Their notation is

First stage. $\mathsf{E}(x) = A_1\theta_1$, dispersion matrix C_1.

Second stage. $\mathsf{E}(\theta_1) = A_2\mu$, dispersion matrix C_2.

Then, the posterior distribution of θ_1 is $N(Dd, D)$ with

$$D^{-1} = A_1^T C_1^{-1} A_1 + C_2^{-1} - C_2^{-1} A_2 (A_2^T C_2^{-1} A_2)^{-1} A_2^T C_2^{-1} \tag{1.1}$$

and

$$(1.2) \qquad d = A_1^T C_1^{-1} x\,.$$

We proceed to evaluate (1.1) and (1.2). The matrix C_2 is given in equations (6): thus, the element in the row corresponding to θ_{ij} and column corresponding to θ_{is} $(j \neq s)$ is σ_a^2, and others similarly. The inversion required for (1.1) is most easily accomplished by solving the equations in z, $C_2 z = a$. Written out in full, these are

$$(1.3) \qquad \sigma_c^2 z_{ij} + n\sigma_a^2 z_{i\,.} + m\sigma_b^2 z_{.\,j} = a_{ij}\,,$$

using the "dot" notation. Summing over i and j, we have

$$(\sigma_c^2 + n\sigma_a^2 + m\sigma_b^2) z_{..} = a_{..}$$

or

$$(1.4) \qquad z_{..} = \frac{a_{..}}{v_{mn}}$$

where

$$(1.5) \qquad v_{mn} = \sigma_c^2 + n\sigma_a^2 + m\sigma_b^2\,.$$

Summing (1.3) over j, we similarly obtain

$$(\sigma_c^2 + n\sigma_a^2) z_{i\,.} + m\sigma_b^2 z_{..} = a_{i\,.}$$

which, on using (1.4), can be written

$$(1.6) \qquad z_{i\,.} = \frac{a_{i\,.} - m\sigma_b^2 \dfrac{a_{..}}{v_{mn}}}{v_n}$$

where

$$(1.7) \qquad v_n = \sigma_c^2 + n\sigma_a^2\,.$$

Similarly,

$$(1.8) \qquad z_{.\,j} = \frac{a_{.\,j} - n\sigma_a^2 \dfrac{a_{..}}{v_{mn}}}{v_m}$$

where

$$(1.9) \qquad v_m = \sigma_c^2 + m\sigma_b^2 .$$

Substitution of (1.6) and (1.7) into (1.3) gives

$$(1.10) \qquad z_{ij} = \sigma_c^{-2}\left[a_{ij} - \frac{n\sigma_a^2}{v_n}\left(a_{i.} - m\sigma_b^2 \frac{a_{..}}{v_{mn}}\right) - \frac{m\sigma_b^2}{v_m}\left(a_{.j} - n\sigma_a^2 \frac{a_{..}}{v_{mn}}\right)\right].$$

Since $z = C_2^{-1}a$, identification of terms on the right-hand shows that C_2^{-1} has the same structure as C_2 itself [equations (6)]. For example, all the terms in rows (i, j) and columns (r, s) with $i \neq r$, $j \neq s$ are the same. From (1.10), the terms are

(1.11a) $\quad i \neq r,\ j \neq s: \qquad h,$

(1.11b) $\quad i = r,\ j \neq s: \qquad f + h,$

(1.11c) $\quad i \neq r,\ j = s: \qquad g + h,$

(1.11d) $\quad i = r,\ j = s: \qquad e + f + g + h,$

where h is the coefficient of $mna_{..}$ in (1.10). That is,

$$(1.12a) \qquad h = \frac{\sigma_a^2\sigma_b^2}{\sigma_c^2 v_{mn}}\left(\frac{1}{v_n} + \frac{1}{v_m}\right),$$

f is the coefficient of $na_{i.}$ in (1.10); namely,

$$(1.12b) \qquad f = -\frac{\sigma_a^2}{\sigma_c^2 v_n}.$$

Similarly, g is the coefficient of $ma_{.j}$, so

$$(1.12c) \qquad g = -\frac{\sigma_b^2}{\sigma_c^2 v_m},$$

and e corresponds to a_{ij}; namely,

$$(1.12d) \qquad e = \frac{1}{\sigma_c^2}.$$

We note for future reference that summation of (1.10) over i and j gives

$z_{..} = a_{..}(e + nf + mg + mnh)$ so that, on comparison with (1.4),

$$(1.13) \qquad e + nf + mg + mnh = \frac{1}{v_{mn}}.$$

Having evaluated C_2^{-1}, we now return to (1.1). A_2 is easily seen to be a vector, all of whose elements are unity. Hence, $A_2^T C_2^{-1}$ is a (row) vector, all of whose elements are $e + nf + mg + mnh = \frac{1}{v_{mn}}$ [from (1.13)]. Hence, $A_2^T C_2^{-1} A_2 = \frac{mn}{v_{mn}}$. Simple calculation shows that $C_2^{-1} A_2 \times \times (A_2^T C_2^{-1} A_2)^{-1} A_2^T C_2^{-1}$ is a matrix, every element of which is $\frac{1}{mnv_{mn}}$.

Reference to the first stage of the model shows easily that $A_1^T C_1^{-1} A_1$ is a diagonal matrix with every diagonal element equal to $\frac{r}{\sigma^2}$. Consequently, D^{-1} [equation (1.1)] is a matrix of the same form as C_2^{-1} [equations (1.11)]; but with e replaced by $e + \frac{r}{\sigma^2} = e'$, say, and h by $h - \frac{1}{mnv_{mn}} = h'$, say. The values of f and g are unaltered. Further consideration of the first stage of the model shows that d is a vector whose (i, j) element is $x_{ij\cdot}\frac{r}{\sigma^2}$.

If θ_{ij}^* denotes the estimate of θ_{ij}, that is, the posterior mean of their joint distribution; the corollary quoted above shows that $\theta^* = Dd$, or $D^{-1}\theta^* = d$. Inserting the values of D^{-1} and d just obtained and writing these equations out in full, we have

$$(1.14) \qquad e'\theta_{ij}^* + nf\theta_{i.}^* + mg\theta_{.j}^* + mnh'\theta_{..}^* = x_{ij\cdot}\frac{r}{\sigma^2}$$

[compare equations (1.3)]. These equations are most easily solved by writing

$$(1.15) \qquad \begin{cases} \varphi_{ij}^* = \theta_{ij}^* - \theta_{i.}^* - \theta_{.j}^* + \theta_{..}^*, \\ \varphi_{i.}^* = \theta_{i.}^* - \theta_{..}^*, \\ \varphi_{.j}^* = \theta_{.j}^* - \theta_{..}^*, \end{cases}$$

and

$$
(1.16)\quad \begin{cases} y_{ij} = x_{ij\,.} - x_{i\,..} - x_{.j\,.} + x_{...}\,, \\ y_{i\,.} = x_{i\,..} - x_{...}\,, \\ y_{.j} = x_{.j\,.} - x_{...}\,. \end{cases}
$$

We can then rewrite (1.14) as

$$
(1.17)\quad \begin{aligned} &e'\varphi^*_{ij} + (e' + nf)\varphi^*_{i\,.} + (e' + mg)\varphi^*_{.j} + (e' + nf + mg + mnh')\theta^*_{..} = \\ &= (y_{ij} + y_{i\,.} + y_{.j} + x_{...})\frac{r}{\sigma^2}\,. \end{aligned}
$$

We note, from (1.13), and the fact that $e' = e + \dfrac{r}{\sigma^2}$, $h' = h - \dfrac{1}{mnv_{mn}}$; that $e' + nf + mg + mnh' = \dfrac{r}{\sigma^2}$.

Summation of (1.17) over i and j gives $\theta^*_{..} = x_{...}$, over j alone gives $(e' + nf)\varphi^*_{i\,.} = y_{i\,.}\dfrac{r}{\sigma^2}$ or

$$
(1.18)\quad \varphi^*_{i\,.} = \frac{\dfrac{r}{\sigma^2}}{\dfrac{r}{\sigma^2} + \dfrac{1}{v_n}}\, y_{i\,.}
$$

on inserting the values for $e' = e + \dfrac{r}{\sigma^2}$, e, from (1.12d) and f, from (1.12b). Similarly,

$$
(1.19)\quad \varphi^*_{.j} = \frac{\dfrac{r}{\sigma^2}}{\dfrac{r}{\sigma^2} + \dfrac{1}{v_m}}\, y_{.j}\,,
$$

and inserting these values into (1.17),

$$
(1.20)\quad \varphi^*_{ij} = \frac{\dfrac{r}{\sigma^2}}{\dfrac{r}{\sigma^2} + \dfrac{1}{\sigma^2_c}}\, y_{ij}\,.
$$

Returning to the original form in terms of θ^*_{ij} and $x_{ij\,.}$, we easily obtain the expressions given in (7).

The dispersion matrix for these estimates (that is, the dispersion matrix of the posterior normal distribution) is, by the corollary, $\boldsymbol{D}$. The equations just solved are $\theta^* = \boldsymbol{D}d$, so $\boldsymbol{D}$ may be found by taking the coefficients of the elements, $x_{ij\,.}\frac{r}{\sigma^2}$, of d in the solutions. For example, to obtain the covariance of θ^*_{ij} and θ^*_{rs} with $i \neq r$, $j \neq s$, it is only necessary to take the coefficient of $x_{rs\,.}\frac{r}{\sigma^2}$ in the expression for θ^*_{ij}. In the notation given by (8) and (9), this is easily to be seen from (10), since $x_{rs\,.}$ only occurs in $x_{...}$, there with coefficient W. All the expressions given in equations (11) can be obtained in the same way.

APPENDIX 2: ESTIMATION OF THE VARIANCE COMPONENTS

In the four-stage model, described by (5) and the following sentence, the joint probability distribution of all the random quantities (x_{ijk}), (θ_{ij}), (α_i), and (β_j), after integration with respect to the diffuse priors of μ, μ_a, and μ_b, is easily seen to be proportional to

$$\frac{1}{\sigma^{mnr}\sigma_c^{mn-1}\sigma_a^{m-1}\sigma_b^{n-1}} \exp\left\{-\frac{1}{2\sigma^2}\sum_{i,j,k}(x_{ijk}-x_{ij\,.})^2 - \right.$$

$$(2.1) \qquad -\frac{r}{2\sigma^2}\sum_{i,j}(x_{ij\,.}-\theta_{ij})^2 - \frac{1}{2\sigma_c^2}\sum_{i,j}(\theta_{ij}-\theta_{..}-\alpha_i +$$

$$\left. + \alpha_{.} - \beta_j + \beta_{.})^2 - \frac{1}{2\sigma_a^2}\sum_i(\alpha_i - \alpha_{.})^2 - \frac{1}{2\sigma_b^2}\sum_j(\beta_j-\beta_{.})^2\right\}.$$

There, the total sum of squares for the data has been broken into the two components within- and between-cells. Differentiation with respect to the θ's, α's, and β's, and equating the results to zero gives modal estimates for these parameters. It is not difficult to verify that for θ_{ij} is exactly θ^*_{ij} given by the three-stage model in equation (7). We proceed to find the corresponding modes (α^*_i) and (β^*_j). The result of differentiating (2.1) with respect to α_i is easily seen to be

$$(2.2) \qquad \frac{n}{\sigma_c^2}\varphi_{i\,.} - (\alpha_i - \alpha_{.})\left(\frac{n}{\sigma_c^2}+\frac{1}{\sigma_a^2}\right),$$

where $\varphi_{i.} = \theta_{i.} - \theta_{..}$ [cf (1.15)]. Equating this to zero and using the estimate of $\varphi^*_{i.}$ [equation (1.18)], we easily obtain

$$(2.3)\qquad (\alpha_i - \alpha_.)^* = \frac{rn\sigma_a^2}{rn\sigma_a^2 + r\sigma_c^2 + \sigma^2}(x_{i..} - x_{...}).$$

Similarly,

$$(2.4)\qquad (\beta_j - \beta_.)^* = \frac{rm\sigma_b^2}{rm\sigma_b^2 + r\sigma_c^2 + \sigma^2}(x_{.j.} - x_{...}).$$

With these estimates, it is an easy matter to obtain equations for the modal estimates of the four variance components σ^2, σ_a^2, σ_b^2, and σ_c^2. Suppose these have independent prior distributions which are all inverse $-\chi^2$. Specifically, let

$$\frac{\nu\lambda}{\sigma^2} \sim \chi^2_\nu, \quad \frac{\nu_t\lambda_t}{\sigma_t^2} \sim \chi^2_{\nu_t} \qquad (t = a, b, c).$$

Multiplication of the distribution (2.1), by this prior, gives the posterior distribution of all the parameters, including the variances, apart from constant factors. The modal equations for the variances are straightforward since the expression factors into four parts, each depending on one of the variances. The results are (we use s^2 for an estimate of σ^2 rather than the asterisk notation used with the other parameters)

$$(2.5a)\qquad s^2 = \frac{\nu\lambda + S_w + r\sum_{i,j}(x_{ij.} - \theta^*_{ij})^2}{mnr + \nu + 2}$$

$$(2.5b)\qquad s_a^2 = \frac{\nu_a\lambda_a + \sum_i(\alpha^*_i - \alpha^*_.)^2}{m + \nu_a + 1}$$

$$(2.5c)\qquad s_b^2 = \frac{\nu_b\lambda_b + \sum_j(\beta^*_j - \beta^*_.)^2}{n + \nu_b + 1}$$

$$(2.5d)\qquad s_c^2 = \frac{= \nu_c\lambda_c + \sum_{i,j}(\theta^*_{ij} - \theta^*_{..} - \alpha^*_i + \alpha^*_. - \beta^*_j + \beta^*_.)^2}{mn + \nu_c + 1}$$

where $S_w = \sum_{i,j,k} (x_{ijk} - x_{ij.})^2$, the usual within-cells sum of squares. For reasons given in the main text, mn in the denominator of (2.5d) can probably be replaced by $(m-1)(n-1)$.

REFERENCES

[1] L. Brown, On the admissibility of invariant estimators of one or more location parameters, *Ann. Math. Statist.*, 37 (1966), 1087-1136.

[2] L. Brown, Inadmissibility of the usual estimators of scale parameters in problems with unknown location and scale parameters, *Ann. Math. Statist.*, 39 (1968), 29-48.

[3] D.V. Lindley, The estimation of many parameters, *Foundations of Statistical Inference,* Toronto, Holt, Rinehart and Winston, 435-455.

[4] D.V. Lindley – A.F.M. Smith, Bayes estimates for the linear model, *J. Roy. Statist. Soc. Ser. B,* 34 (1972), 1-41.

[5] A.D. Richmers – H.N. Todd, *Statistics: An Introduction,* McGraw-Hill, 1967.

[6] G.W. Snedecor, *Statistical Methods,* Iowa, State College Press, 1956.

[7] Ch. Stein, Inadmissibility of the usual estimator for the mean of a multivariate normal distribution, *Proc. Third Berkeley Symp. Math. Statist. Prob.,* 1 (1956), 197-206.

COLLOQUIA MATHEMATICA SOCIETATIS JÁNOS BOLYAI
9. EUROPEAN MEETING OF STATISTICIANS, BUDAPEST (HUNGARY), 1972.

AN ALGEBRAIC APPROACH TO THE WAITING TIME PROCESS IN GI/M/S

J. LORIS-TEGHEM*

We consider a queuing system GI/M/S in which the customers, denoted by $\mathscr{C}_n$, are served in the order of their arrival. The waiting time process $\{w_n\}$ does not seem to be well-suited for a direct algebraic treatment, but it can be studied by means of the two-dimensional process $\{(\nu_n, l_n)\}$, where ν_n is the time of full occupation of the system immediately after the arrival instant T_n of $\mathscr{C}_n$, and where $l_n = [\lambda_n - 1]^+$, with λ_n denoting the number of idle servers at $(T_n + 0)$. Note that this process takes values only on the coordinate axes and that for $S = 2$, it is but an one-dimensional process.

For the generating function $h_{(z)} = \sum_{n=0}^{\infty} z^n h_n$ of the distribution functions h_n of the variables (ν_n, l_n)-considered, for $-1 < z < +1$, as an element of the Banach algebra κ_2 of linear combinations of two-dimensional distribution functions – it is possible to obtain an equation of

*Chargé de Recherches at the Fonds National Belge de la Recherche Scientifique.

the form

$$h_{(z)} = h_0 - z \sum_{i=0}^{S-2} {}_i\xi_{(z)} \, {}_i\tilde{\beta} + z \epsilon_{S\mu} \Pi[u^* h_{(z)}] ,$$

where Π is the linear operator induced on κ_2 by the measurable function $(x, x') \to (x^+, 0)$, and where $\epsilon_{S\mu}$, u^* and the ${}_i\tilde{\beta}$ are known elements of κ_2, while the ${}_i\xi_{(z)}$ are $(S-1)$ unknown real numbers which can be expressed in terms of the unknown function $h_{(z)}$. Using the fact that Π is a Wendel projection, it is possible to determine $h_{(z)}$ for $|z|$ sufficiently small.

The generating function $g_{(z)}$ of the distribution functions g_n of the waiting times w_n can be derived by the aid of the relation

$$w_n = [v_{n-1} - u_{n-1}]^+$$

where u_{n-1} denotes the inter-arrival time $(T_n - T_{n-1})$.

REFERENCES

J. Loris-Teghem, Un traitement algébrique du modèle d'attente GI/M/2, *Cah. Centre Et. Rech. Op.*, 13 (1971), 57-62.

J. Loris-Teghem, An algebraic approach to the waiting time process in GI/M/S, *J. Appl. Prob.*, 10 (1973), 181-191.

COLLOQUIA MATHEMATICA SOCIETATIS JÁNOS BOLYAI
9. EUROPEAN MEETING OF STATISTICIANS, BUDAPEST (HUNGARY), 1972.

ON THE ASYMPTOTIC NORMALITY OF THE REWARD IN A CONTROLLED MARKOV CHAIN

P. MANDL

In the paper, we consider two sequences of random variables: the state variables $\{X_n, n = 0, 1, \dots\}$ ranging in a finite set I and the control variables $\{Z_n = z_n(X_0, \dots, X_n), n = 0, 1, \dots\}$, $Z_n \in \mathscr{Z}(X_n)$, where $\mathscr{Z}(j)$, $(j \in I)$ are closed bounded sets in R^s. The sequence of functions $\omega = [z_n(j_0, \dots, j_n), n = 0, 1, \dots\}$ represents the controller's policy, briefly the control. ω is called stationary if $z_n(j_0, \dots, j_n) = z(j_n)$ $(n = = 0, 1, \dots)$. The hypothesis that the random sequences form a controlled Markov chain reads

$$\mathsf{P}(X_{n+1} = k \mid X_n, Z_n, \dots, X_0, Z_0) = p(X_n, k;\ Z_n)$$
$$(k \in I;\ n = 0, 1, \dots)$$

$p(j, k;\ z)$ $(z \in \mathscr{Z}(j);\ j, k \in I)$ are the transition probabilities from state j to state k if the control variable equals z. Further, we introduce the reward from the chain up to time n

$$C_n = \sum_{m=0}^{n-1} c(X_m, X_{m+1}; Z_m) \qquad (n = 1, 2, \ldots).$$

The functions

$$p(j, k; z), \ c(j, k; z), \qquad (z \in \mathscr{Z}(j);\ j, k \in I)$$

are assumed to be continuous. We make also

Assumption. For arbitrary $z(j) \in \mathscr{Z}(j)$ $(j \in I)$, the states which are recurrent for the Markov chain with transition probabilities $p(j, k; z(j))$ $(j, k \in I)$, form only one irreducible set.

From the central limit theorem for Markov chains it follows that under a stationary control ω the asymptotic distribution of $\dfrac{C_n - n\theta(\omega)}{\sqrt{n}}$ for $n \to \infty$ is normal $N(0, \zeta(\omega))$ for appropriate constant $\theta(\omega)$, $\zeta(\omega)$. (We admit the degenerate distribution $N(0, 0)$). The maximum of the mean

$$\hat{\theta} = \max \{\theta(\omega):\ \omega \text{ stationary}\}$$

is a widely used target for the choice of a control. The following characterization of $\hat{\theta}$ comes from R. Bellman [1]: *$\hat{\theta}$ is the unique constant to which there exist numbers w_{1j} $(j \in I)$ so that*

$$(1) \qquad w_{1j} + \hat{\theta} = \max_{z \in \mathscr{Z}(j)} \left\{ \sum_k p(j, k; z)\,[c(j, k; z) + w_{1k}] \right\} \qquad (j \in I).$$

(1) can be solved by using Howard's iteration procedure.

Next we define

$$(2) \qquad \hat{\zeta} = \min\{\zeta(\omega):\ \omega \text{ stationary},\ \theta(\omega) = \hat{\theta}\}.$$

This definition is motivated by the fact that among the controls, for which the maximal mean is attained, those with minimal variance are preferable. To obtain a characterization of $\hat{\zeta}$ similar to (1) we introduce

$$\varphi_1(j, z) = \sum_k p(j, k; z)\,[c(j, k; z) + w_{1k}] - w_{1j} - \hat{\theta}$$

$$(j \in I;\ z \in \mathscr{Z}(j)),$$

$$\mathscr{Z}'(j) = \{z \in \mathscr{Z}(j): \varphi_1(j, z) = 0\} \qquad (j \in I).$$

The counterpart of (1) for $\hat{\xi}$ reads

$$(3) \qquad \begin{aligned} w_{2j} + \hat{\xi} = \min_{z \in \mathscr{Z}'(j)} \sum_k p(j, k;\ z)[(c(j, k;\ z) - \hat{\theta})^2 + \\ + 2(c(j, k;\ z) - \hat{\theta})w_{1k} + w_{2k}] \qquad (j \in I). \end{aligned}$$

The proof that $\hat{\xi}$ obtained from (3) coincides with that defined by (2) can be based on the methods developed in [2]. (3) does not loose its sense if each $\mathscr{Z}'(j)$ contains one point only. We define

$$\begin{aligned} \varphi_2(j, z) = \sum_k p(j, k;\ z)\,[(c(j, k;\ z) - \hat{\theta})^2 + \\ + 2(c(j, k;\ z) - \hat{\theta})w_{1k} + w_{2k}] - w_{2j} - \hat{\xi} \qquad (j \in I;\ z \in \mathscr{Z}(j)). \end{aligned}$$

The limitation to stationary controls is restrictive e.g. in control problems with unknown parameters because the estimates of such parameters are usually made from the entire past trajectory. A sufficient condition for $\frac{C_n - n\hat{\theta}}{\sqrt{n}}$ to be asymptotically $N(0, \hat{\xi})$ under ω arbitrary is contained in the following theorem:

Theorem. *Let the control* ω *be such that*

$$(4) \qquad \begin{aligned} \lim_{n \to \infty} \frac{1}{\sqrt{n}} \sum_{m=0}^{n-1} \mathsf{E}\varphi_1(X_m, Z_m) = 0 = \\ = \lim_{n \to \infty} \frac{1}{n} \sum_{m=0}^{n-1} \mathsf{E}\,|\varphi_2(X_m, Z_m)|. \end{aligned}$$

Then $\frac{C_n - n\hat{\theta}}{\sqrt{n}}$ *has asymptotically normal distribution* $N(0, \hat{\xi})$ *as* $n \to \infty$.

Proof. Denote

$$\psi_l(j, z) = \sum_k p(j, k;\ z)w_{lk} - w_{lj} \qquad (j \in I;\ z \in \mathscr{Z}(j);\ l = 1, 2),$$

$$\chi_n(u) = \exp\{iu(C_n - n\hat{\theta})\} \qquad (n = 0, 1, \ldots;\ u \in (-\infty, \infty)),$$

$$e_1(x) = e^{ix} - ix - 1\,,$$

$$e_2(x) = e^{ix} + \frac{x^2}{2} - ix - 1\,.$$

Hence,

$$|e_1(x)| \leqslant \frac{x^2}{2}\,,$$

$$|e_2(x)| \leqslant \frac{|x|^3}{6}\,.$$

To demonstrate the theorem we shall verify

$$\lim_{n\to\infty} \mathrm{E}\chi_n\left(\frac{u}{\sqrt{n}}\right) = e^{-\frac{1}{2}\hat{\xi}u^2}\,. \tag{5}$$

Noting that $\psi_l(X_m, Z_m)$ is the conditional expectation of $w_{lX_{m+1}} - w_{lX_m}$ given $X_0, X_1, \ldots, X_m$, $Z_0, Z_1, \ldots, Z_m$ we obtain

$$0 = iu\mathrm{E}\left[\sum_{m=0}^{n-1} \chi_m \psi_1(X_m, Z_m) - \sum_{m=0}^{n-1} \chi_m (w_{1X_{m+1}} - w_{1X_m})\right], \tag{6}$$

$$0 = \frac{u^2}{2}\,\mathrm{E}\left[\sum_{m=0}^{n-1} \chi_m \psi_2(X_m, Z_m) - \sum_{m=0}^{n-1} \chi_m (w_{2X_{m+1}} - w_{2X_m})\right]. \tag{7}$$

Furthermore,

$$\mathrm{E}\chi_n - 1 = \mathrm{E}\sum_{m=0}^{n-1} (\chi_{m+1} - \chi_m) =$$

$$= \mathrm{E}\sum_{m=0}^{n-1} (e^{iu(c(X_m, X_{m+1};\, Z_m) - \hat{\theta})} - 1)\chi_m = \tag{8}$$

$$= \mathrm{E}\sum_{m=0}^{n-1} [iu(c - \hat{\theta}) - \tfrac{1}{2}u^2(c - \hat{\theta})^2 + e_2(u(c - \hat{\theta}))]\chi_m\,,$$

$$- iu\mathrm{E}\sum_{m=0}^{n-1} \chi_m (w_{1X_{m+1}} - w_{1X_m}) = \tag{9}$$

$$= iu\mathrm{E}\left[w_{1X_0} - \chi_n w_{1X_n} + \sum_{m=0}^{n-1} w_{1X_{m+1}}(\chi_{m+1} - \chi_m)\right] =$$

$$= iu\mathsf{E}\Big[w_{1X_0} - \chi_n w_{1X_n} + \sum_{m=0}^{n-1} w_{1X_{m+1}}(iu(c-\hat{\theta}) +$$

$$+ e_1(u(c-\hat{\theta})))\chi_m\Big] ,$$

$$-\frac{u^2}{2}\,\mathsf{E}\sum_{m=0}^{n-1}\chi_m(w_{2X_{m+1}} - w_{2X_m}) =$$

(10)

$$= \frac{u^2}{2}\,\mathsf{E}\Big[w_{2X_0} - \chi_n w_{2X_n} + \sum_{m=0}^{n-1} w_{2X_{m+1}}(e^{iu(c-\hat{\theta})} - 1)\chi_m\Big] .$$

We recall that

$$\mathsf{E}[c(X_m, X_{m+1}; Z_m) - \hat{\theta} + \psi_1(X_m, Z_m)]\chi_m =$$

$$= \mathsf{E}\varphi_1(X_m, Z_m)\chi_m ;$$

$$\mathsf{E}[(c(X_m, X_{m+1}; Z_m) - \hat{\theta})^2 +$$

$$+ 2(c(X_m, X_{m+1}; Z_m) - \hat{\theta})w_{1X_{m+1}} +$$

$$+ \psi_2(X_m, X_m)]\chi_m = \mathsf{E}[\varphi_2(X_m) + \hat{\xi}]\chi_m .$$

Consequently, adding (6)-(10) we get

$$\mathsf{E}\chi_n = 1 - \frac{1}{2}\hat{\xi}u^2 \sum_{m=0}^{n-1}\mathsf{E}\chi_m + \kappa'(n,u) , \tag{11}$$

where

$$\kappa'(n,u) = iu\mathsf{E}\Big[\sum_{m=0}^{n-1}\chi_m\varphi_1(X_m, Z_m) + w_{1X_0} -$$

$$- \chi_n w_{1X_n} + \sum_{m=0}^{n-1} w_{1X_{m+1}} e_1(u(c-\hat{\theta}))\chi_m\Big] -$$

$$- \frac{u^2}{2}\,\mathsf{E}\Big[\sum_{m=0}^{n-1}\chi_m\varphi_2(X_m, Z_m) + w_{2X_0} - \chi_n w_{2X_n} +$$

$$+ \sum_{m=0}^{n-1} w_{2X_{m+1}}(e^{iu(c-\hat{\theta})} - 1)\chi_m\Big] + \mathsf{E}\sum_{m=0}^{n-1} e_2(u(c-\hat{\theta}))\chi_m .$$

More suitable form of (11) is

$$(12) \qquad \mathsf{E}\chi_n = 1 + (e^{-\frac{1}{2}\hat{\xi}u^2} - 1) \sum_{m=0}^{n-1} \mathsf{E}\chi_m + \kappa(n, u),$$

with

$$\kappa(n, u) = \kappa'(n, u) + \left[1 - \frac{1}{2}\hat{\xi}u^2 - e^{-\frac{1}{2}\hat{\xi}u^2}\right] \sum_{m=0}^{n-1} \mathsf{E}\chi_m .$$

The modulus of the expression in the square bracket is majorized by $\frac{\hat{\xi}^2 u^4}{8}$.

From (12) one calculates that

$$\sum_{m=0}^{n-1} \mathsf{E}\chi_m = \sum_{m=0}^{n-1} e^{-\frac{1}{2}\hat{\xi}u^2(n-1-m)} (\kappa(m, u) + 1).$$

Hence, reinserting the result into (12),

$$(13) \qquad \begin{aligned} \mathsf{E}\chi_n &= e^{-\frac{1}{2}\hat{\xi}nu^2} + \\ &+ \left(e^{-\frac{1}{2}\hat{\xi}u^2} - 1\right) \sum_{m=0}^{n-1} e^{-\frac{1}{2}\hat{\xi}u^2(n-1-m)} \kappa(m, u) + \kappa(n, u). \end{aligned}$$

Relation (5) and consequently the assertion of the Theorem follows from (13), if we show that (4) implies

$$\max_{1 \leqslant m \leqslant n} \left|\kappa\left(m, \frac{u}{\sqrt{n}}\right)\right| \to 0 \qquad \text{as} \qquad n \to \infty .$$

This relation is obtained without difficulties from an inspection of the different terms of $\kappa\left(m, \frac{u}{\sqrt{n}}\right)$.

Remark. If

$$(14) \qquad \varphi_2(j, z) = 0 \qquad (z \in \mathscr{X}'(j);\ j \in I)$$

then the first of relations (4) implies the second one. (14) holds whenever each $\mathscr{X}'(j)$ consists of one point only.

Paper [3] includes the continuous time analogue of the Theorem. The Markov chain case treated here requires a more elaborate estimation technique.

REFERENCES

[1] R. Bellman, A Markovian decision process, *J. of Math. and Mech.*, 6 (1957), 679-684.

[2] P. Mandl, On the variance in controlled Markov chains, *Kybernetika,* 7 (1971), 1-12.

[3] P. Mandl, On the adaptive control of finite state Markov processes, *Z. Wahrscheinlichkeitstheorie verw. Geb.*, 27 (1973), 263-276.

Paper [3] includes the continuous time analogue of the Theorem. The Markov chain case presented here requires a more elaborate estimation technique.

REFERENCES

[1] R. Bellman, A Markovian decision process, J. Math. and Mech. 6 (1957) 679–684.

[2] P. Mandl, On the variance in controlled Markov chains, Kybernetika 7 (1971) [illegible].

[3] P. Mandl, On the adaptive control of finite state Markov processes, Z. Wahrscheinlichkeitstheorie verw. Geb. 27 (1973) 263–276.

COLLOQUIA MATHEMATICA SOCIETATIS JÁNOS BOLYAI
9. EUROPEAN MEETING OF STATISTICIANS, BUDAPEST (HUNGARY), 1972.

STOCHASTIC INTEGRAL AND VECTOR VALUED MEASURES

M. METIVIER

INTRODUCTION

The purpose of this paper is to sketch a treatment of stochastic integration, which integrates it fully as a part of vector valued measure theory.

Let us recall the main features of stochastic integration. Let $(\Omega, \mathscr{F}, (\mathscr{F}_t)_{t \in R^+}, \mathsf{P})$ be a probability space, with an increasing family of sub σ-algebras $\mathscr{F}_t$ of $\mathscr{F}$. Let $\mathscr{S}$ be the class of real valued stochastic processes $Z = V + M$ where V is a process, adapted to the family $(\mathscr{F}_t)$, the trajectories of which are with bounded variation and continuous to the right, and M is a martingale (or more generally a local martingale) with respect to $(\mathscr{F}_t)$, equally continuous to the right. Let us suppose for example, that, for each t: $V_t \in L^1(\Omega, \mathscr{F}, \mathsf{P})$ and $M_t \in L^2(\Omega, \mathscr{F}, \mathsf{P})$. $\mathscr{C}$ being the σ-algebra of subsets of $R^+ \times \Omega$, called previsible, the stochastic integration with respect to X defines a linear mapping $X \rightsquigarrow \int X dZ$ of a suitable subspace of $\mathscr{C}$-measurable processes (i.e. the mapping $(t, \omega) \rightsquigarrow X(t, \omega)$ is measurable on $(R^+ \times \Omega, \mathscr{C})$) in $\mathscr{S}$. This linear mapping is usually defined in a direct way, without reference to any measure, despite is possesses the properties of an integral.

We think that, revealing the very nature of stochastic integration illuminates the theory, and gives new tools (one of which is the dominated convergence theorem, available in the Bartle theory of integration [1]). It is to be expected that those tools are useful to deal with the case of integration with respect to vector valued processes (this case will not be treated here).

Another feature of this new presentation is that it requires no prerequisite on the Doob – Mayer decomposition theory of super martingales. The Doob – Meyer decomposition theorem is got here as part of the theory.

Many results here presented (and refinements) are due to M. Pellaumail and will be developped in his thesis work which is beeing completed [10]. They have been announced in a Note ([9]). A rather systematic treatment is given in an author's self content seminar (preprints soon available: [6]).

1. NOTATIONS – DEFINITIONS

1.1. Algebras of subsets of $(R^+ \times \Omega)$. $(\Omega, \mathscr{F}, \mathrm{P})$ is a probability space. $(\mathscr{F}_t)_{t \in R^+}$ is an increasing family of sub-algebras of $\mathscr{F}$.

$\mathscr{R}$ is the set of subsets of $R^+ \times \Omega$ of the form $F \times \,]s, t]$ where $F \in \mathscr{F}_s$, and $s < t$.

$\mathscr{R}$ is a semi-ring and we denote by $\mathscr{A}$ the ring generated by $\mathscr{R}$, by $\mathfrak{A}$ the δ ring generated by it and $\mathscr{C}$ the σ-algebra generated by $\mathscr{A}$. The σ-algebra $\mathscr{C}$ is called usually the σ-algebra of previsible sets. This σ-algebra is generated as well by the "stochastic intervalls" $]S, T] = \{(t, \omega) \colon S(\omega) < t \leqslant (\omega)\}$, where $S \leqslant T$ are any two stopping time of the family $(\mathscr{F}_t)$. This σ-algebra is generated too by the processes with trajectories continuous to the left. (We abreviate by saying continuous to the left or c.t.l.).

We call $\mathscr{W}$ the σ-algebra generated by the "stochastic intervalls" $[S, T[= \{(t, \omega): S(\omega) \leqslant t < T(\omega)\}$ where $S \leqslant T$ are two stopping time of the family $(\mathscr{F}_t)$. This σ-algebra is called usually the σ-algebra of well measurable sets.

In what follows, in order to deal with non bounded vector measures, we define, as we did in [5], vector measures on δ-rings. The considered δ-ring will be $\mathfrak{A}$, as long as we treat the integration with respect to a square integrable martingale. We will indicate at the end, which is the suitable δ-ring to consider (in an evident way), when the martingale is replaced by a local martingale.

1.2. Stochastic processes. A stochastic process will be a mapping of $(R^+ \times \Omega)$ in R.

It will be called *continuous to the right* if P-almost all trajectories (i.e.: mapping $t \curvearrowright X(t, \omega)$)· are continuous to the right (c.t.r.).

It will be called with bounded variation (resp. increasing) if its trajectories are P-almost surely with bounded variation (resp. increasing). When considering processes with bounded variation (b.v.), *we will always suppose them null for* $t = 0$, and adapted to $(\mathscr{F}_t)$.

We need almost only to know that a martingale is: a process M, such that for every $t \in R^+$, $M_t \in L^2$ $(\Omega, \mathscr{F}_t, \mathsf{P})$ and $E(M_t \mid \mathscr{F}_s) = M_s$ for every $s < t$, and to know that such a martingale admits a modification (i.e. a M' such that for every t $M_t = M'_t$ a.s.) with trajectories continuous to the right (resp. to the left).

A process M is a local martingale with respect to $(\mathscr{F}_t)$ if there exists an increasing sequence $\sigma = (\sigma_n)$ of stopping times such that, $\lim_n \sigma_n = +\infty$, a.s., and $(M_{t \wedge \sigma_n})_{t \in R^+}$ is a martingale with respect to $(\mathscr{F}_t)$ for every n.

We recall that two processes X and X' are said *indistinguishable* if, with probability one, the mappings $t \curvearrowright X(t, \omega)$ and $t \curvearrowright X'(t, \omega)$) are identical.

2. STOCHASTIC MEASURES

2.1. Definition 1. We call stochastic measure on $(R^+ \times \Omega)$, adapted to the $(\mathscr{F}_t)_{t \in R^+}$, with values in L^p, any measure m, on the δ-ring $\mathfrak{A}$, with values in $L^p(\Omega, \mathscr{F}, \mathrm{P})$, such that

$$\forall]s, t] \times F \in \mathscr{R} \qquad m(]s, t] \times F) = 1_F \cdot m(]s, t] \times \Omega) \in \in L^p(\Omega, \mathscr{F}_t, \mathrm{P}) . \tag{3}$$

2.2. Examples. 1°. A rather trivial (but important) example is the following: Z is an increasing L^1 process and for every $A \in \mathfrak{A}$, such that

$A \subset [0, T] \times \Omega$ we define

$$m(A)(\omega) = \int_0^T 1_A(s, \omega) dZ(s, \omega)$$

where the integral is taken with respect to the random Stieltjes measure defined for each ω by the increasing function $s \rightsquigarrow Z(s, \omega)$.

2°. If (T_n) is a sequence of stopping times and (Z_n) is a sequence of random variables such that for each n Z_n is $\mathscr{F}_{T_n}$ measurable and the series $\sum_n Z_n \cdot 1_{[T_n \leqslant t]}$ converges in $L^p(\Omega, \mathscr{F}, \mathrm{P})$. Considering the measures m_n, associated to the increasing processes $Z_n \cdot 1_{[T_n \leqslant t]}$, and denoting their semi-variation by $\| m_n \|$, it is easy to see that for every $A \subset [0, t] \times \Omega$

$$\| m_n \|(A) \leqslant 2 \sup_{B \subset A} \sqrt[p]{\int | 1_B(T_n(\omega), \omega) Z_n(\omega)|^p \mathrm{P}(d\omega)} \leqslant$$

$$\leqslant 2 \| Z_n(\omega) 1_{[T_n \leqslant t]} \|_p .$$

The space of vector measures beeing complete for the semi-norms of semi-variation, the series $\sum_n m_n$ is normally convergent for those semi-norms. From this, we get the existence of a stochasic measure m, such that for every $s < t$, $F \in \mathscr{F}_s$

$$m(]s, t] \times F) = 1_F \cdot \sum_n Z_n 1_{]s < T_n \leqslant t]}$$

(convergence in L^p, not necessarily a.s.).

3. STOCHASTIC MEASURES GENERATED BY A STOCHASTIC PROCESS

The purpose of this § is to put into evidence a one to one correspondence between stochastic measures and a class of stochastic process.

3.1. Theorem 1. *To every stochastic measure m on $(R^+ \times \Omega, \mathfrak{A})$, with values in L^p, is associated a stochastic process Z_m, continuous to the right, null in zero, unique up to indistinguability, such that for every $s < t$, $F \in \mathscr{F}_s$:*

(3.1.1) $\quad m(]s, t] \times F) = 1_F \cdot (Z_m(t) - Z_m(s)) \qquad a.s.$

The process has a.s. limits to the left.

Proof. The unicity is trivial for necessarily $Z_m(t) = m(]0, t] \times \Omega)$ a.s. We then consider such a process Z_m and prove its "regularity". The proof goes along the same line as the regularity theorem for martingales. Using the fact that every vector measure on $\mathfrak{A}$ is bounded when restricted to subsets of $]0, t] \times \Omega$, and introducing for any set $S = \{t_1 < \ldots \ldots < t_{2n}\}$ and any rationals $a < b$ the stopping times

$$\sigma_1^S = t_1; \quad \sigma_{2k+1}^S = \inf\{t: t \in S, \ t > \sigma_{2k}^S, \ Z_m(t) \geqslant b\}$$

$$\sigma_{2k}^S = \inf\{t: t \in S, \ t \geqslant \sigma_{2k-1}^S, \ Z_m(t) \leqslant a\}$$

it is easily seen that for some constant K

$$K \geqslant \int \sum_{k=1}^{n-1} |Z_m(\sigma_{2k+1}^S) - Z_m(\sigma_{2k}^S)| \, dP .$$

From that, if $F_{S,j}^{(a,b)}$ denotes the set of trajectories having at least j "upcrossings" of $[a, b]$ on S, we get

$$P(F_{S,j}^{(a,b)}) \leqslant \frac{K}{j(b-a)^p} .$$

The set of trajectories having an infinity of upcrossings of $[a, b]$ on the rationals is of probability zero. From there on, using standard arguments, we get the existence of a modification of Z_m, which is continuous to the right.

3.2. Definition 2. A process Z_m, continuous to the right, such that (3.1.1) is true is said to be *associated* with the stochastic measure m. We say also that m is *generated* by Z_m.

3.3. Remark 1. It is rather easy to see that if Z_m is associated with m, for any two stopping times σ, τ, $\sigma \leqslant \tau$ one has

$$m(]\sigma, \tau]) = Z_m(\tau) - Z_m(\sigma) .$$

We try now to make more precise the class of those processes associated with a stochastic measure. We need first a theorem, which is of the Doob – Meyer's decomposition theorem type.

3.4. Theorem 2 (cf. [10]). *Let* α *be a positive finite measure on* $\mathfrak{A}$, *such that* $(A \in \mathfrak{A}, \forall u \ 1_A(u, \cdot) = 0$ a.s.) *implies* $\alpha(A) = 0$.

Then, there exists an increasing process (c.t.l.), unique up to indistinguability, such that $\forall s < t \ \forall F \in \mathscr{F}_t$

$$(3.4.1) \qquad E[1_F \cdot (V_t - V_s)] = \int_{]s,t] \times \Omega} E(1_F \mid \mathscr{F}_{u-}) d\alpha$$

denoting by $E(1_F \mid \mathscr{F}_{u-})$ *a c.t.l. (then previsible) version of the martingale* $(E(1_F \mid \mathscr{F}_u))_{u \leqslant t}$. *The stochastic measure* m *in* L^1, *defined by* $\forall A \in \mathfrak{A}$

$$m(A) = \int 1_A(s, \cdot) dV_s(\cdot)$$

is with bounded variation α.

Moreover, for every previsible positive (resp. bounded) process Y

$$(3.4.2) \qquad E\left(\int Y dm\right) = \int Y d\alpha .$$

Proof. The unicity, up to indistinguability, is quite trivial, V_t being necessarily such that

$$(3.4.3) \qquad \forall F \in \mathscr{F}_t \qquad \mathrm{E}(1_F \cdot V_t) = \int_{]0,1] \times \Omega} E(1_F \mid \mathscr{F}_{u^-}) d\alpha .$$

We consider the following function on $\mathscr{F}_t$

$$\alpha_t : F \to \int_{]0,t] \times \Omega} \mathrm{E}(1_F \mid \mathscr{F}_{u^-}) d\alpha .$$

Using the martingale inequality and the Borel – Cantelli lemma, we prove in a standard way that from any decreasing sequence g_n of $\mathscr{F}_t$ measurable functions, such that

$$\lim_n g_n = 0 \qquad \text{a.s.}$$

we can extract a subsequence (g_{n_k}) such that, if

$$Y_k(u) = \mathrm{E}(g_{n_k} \mid \mathscr{F}_{u^-}) ,$$

we have

$$\lim_{k \to \infty} \sup_{0 \leqslant u \leqslant t} | Y_k(u, \omega) | = 0 \qquad \text{a.s.}$$

The σ-additivity of α_t follows from this, and, denoting by $\widetilde{A}_t$ an expression of the Radon – Nikodym derivative $\left(\frac{d\alpha_t}{d\mathrm{P}} \right) \mathscr{F}_t$ one gets easily the following:

$$\forall f \in \mathscr{L}^1(\Omega, \mathscr{F}_t, \mathrm{P}), \qquad \mathrm{E}(f \cdot \widetilde{A}_t) = \int_{]0,t]} \mathrm{E}(f \mid \mathscr{F}_{u^-}) d\alpha ,$$

$$\forall f \in \mathscr{L}^1(\Omega, \mathscr{F}_t, \mathrm{P}), \qquad \forall s < t,$$

$$\mathrm{E}(f \cdot (\widetilde{A}_t - \widetilde{A}_s)) = \mathrm{E}(f \cdot \widetilde{A}_t) - \mathrm{E}(\mathrm{E}(f \mid \mathscr{F}_s) \cdot \widetilde{A}_s) =$$

$$= \int_{]s,t] \times \Omega} \mathrm{E}(f \mid \mathscr{F}_{u^-}) d\alpha .$$

One gets then easily a modification V of $\widetilde{A}$ having all the required properties.

Finally, if $A = \sum_k]s_k, u_k] \times F_k \in \mathscr{A}$, one has

$$\| m(A) \|_1 = \mathsf{E} | m(A) | = \mathsf{E}(m(A)) = \sum_k \int_{]s,t] \times \Omega} 1_{F_k} d\alpha = \alpha(A)$$

from where it is seen that $\operatorname{var}(m) = \alpha$. Moreover, the relation (3.4.2) beeing trivially true for $Y = 1_{]s,t] \times F}$ one goes over to previsible processes by using a classical measurability extension argument.

3.5. Definition 3. The increasing process ν associated with the positive measure α by Theorem 2 will be called the natural increasing process of α.

In case α is real (then with bounded variation), the natural process of α will be the process with bounded variation $V^+ - V^-$ where V^+ is the natural process of α^+ and V^- is the natural process of α^-.

3.6. Remark 1. Using the relation

$$\mathsf{E}(Y_t \cdot V_t) = \mathsf{E} \int_0^t Y_s dV_s$$

for a positive martingale Y and in increasing process V (cf. [8] Ch. VIII), one gets immediately

$$\int_{]0,t] \times \Omega} Y_{s^-} d\alpha = \int_{]0,t] \times \Omega} \mathsf{E}(Y_t \mid \mathscr{F}_{s^-}) d\alpha =$$

$$= \mathsf{E}(Y_t V_t) = \mathsf{E} \int_0^t Y_s dV_s$$

which proves the "naturality" of the process V in the sense of P.A. Meyer (cf. [8] chap. VIII).

3.7. Theorem 3 *(Decomposition Theorem). If m is a stochastic measure, with values in L^p and if Z is the associated process, there exists a martingale M in L^1, and a process V with bounded variation V such that*

(3.7.1) $\quad Z = M + V$.

One can take for V *the natural process of the real measure*

$$\nu: A \rightsquigarrow \mathsf{E}(m(A)).$$

Proof. Once V has been defined as the natural process of ν we get immediately for the process $M = Z - V$:

$$\mathsf{E}(1_H \cdot (M_t - M_s)) = \mathsf{E}(1_H \cdot (Z_t - Z_s)) - \int_{]s,t]\times\Omega} 1_H \, d\alpha = 0$$

$$\forall s < t \qquad \forall M \in \mathcal{F}_s$$

which expresses the martingale property for M.

3.8. Remark 3. Reasoning as in Remark 2, we see that if we impose to the process V to be natural in the sense of Meyer, the decomposition (3.7.1) is unique, up to indistinguability.

The following theorem is of some use:

3.9. Theorem 4. *Let* (m_n) *be a sequence of stochastic measures converging to* m *for the semi-norms of semi-variation.*

Then m *is a stochastic measure and there exists a subsequence* (m_{n_k}) *such that the associated processes* (Z_{n_k}) *have the following property: a.s. the trajectories of* $(Z_{n_k})_{k\in N}$ *converge uniformly an compact intervalls towards the trajectories of* Z_m.

Proof. We use an extracted subsequence (m_{n_k}) such that

$$\|(m - m_{n_k})(B)\| \leqslant \frac{1}{k^{2p}}$$

and a standard argument based on Borel – Cantelli lemma.

4. EXISTENCE OF THE GENERATED STOCHASTIC MEASURE

In this paragraph, we try to prove the converse of Theorem 2: considering a process $Z = V + M$, where V is with bounded variation, and M is a martingale, does there exist a stochastic measure m generated by Z (cf. Definition 2)?

As the case of V is trivial, (cf. Example 2.2), we have only to deal with the case of a martingale M.

We start with a fundamental result due to Pellaumail (cf. [9] and [6]).

4.1. Theorem 5. *Let* m *be an additive mapping of* $\mathscr{A}$ *in* $L^p(\Omega, \mathscr{F}, \mathrm{P})$ *verifying the condition* (3) *of Definition* 1.

Let us suppose that there exists a positive finite simply additive function α *on* $\mathscr{A}$ *such that*

(j) $\forall F \in \mathscr{F}_s \quad s < t$

$$\lim_{t' \downarrow t} \alpha(]t, t'] \times F) = 0 ,$$

(jj) $\forall F_n \in \mathscr{F}_\infty$, if $F_n \downarrow \phi$ *then*

$$\lim_{n \to \infty} \sup \{\alpha(A)\colon A \in \mathscr{A},\ A \subset]0, t] \times F_n\} = 0 ,$$

(jjj) $\forall A \in \mathscr{A},\ \forall \epsilon > 0 \quad \exists \eta > 0$ *such that*

$$\forall H \in \mathscr{A},\ H \subset A,\ \alpha(H) \leqslant \eta \Rightarrow \|m(H)\|_p \leqslant \epsilon .$$

Then m *can be extended to a stochastic measure.*

Proof. The proof rests in a straightforward way on the following lemma.

Lemma. *If* α *is a simply additive positive function on* $\mathscr{A}$, *for which* (j) *and* (jj) *are true, then* α *can be extended to a* σ*-additive measure on* $\mathfrak{A}$.

Proof. $\mathscr{R}$ beeing a semi-ring it is enough to prove the σ-additivity of α on $\mathscr{R}$. Considering then a disjoint family (H_n) in $\mathscr{R}$ with

$$H = \bigcup_n H_n \in \mathscr{R} ,$$

$$H_n =]s_n, t_n] \times F_n ,$$

$$H =]s, t] \times F ,$$

we define, along the same line as in the classical real Stieltjes case,

$$H'_n =]s_n, t'_n] \times F_n \supset H_n ,$$

$$H' =]s, t'] \times F_n \subset H ,$$

in such a way that (using (j))

$$\alpha(H'_n) \leqslant \alpha(H_n) + \frac{\epsilon}{2^n}$$

$$\alpha(H') \geqslant \alpha(H) - \epsilon .$$

Using then the compactity of $[s', t]$ it is easy to prove that

$$H - \bigcup_{n=1}^{k} H'_n \subset]s', t] \times \Omega_k$$

where (Ω_k) is a decreasing sequence with $\bigcap_k \Omega_k = \phi$. The lemma follows from condition (jj).

To simplify we deal only with the case of a martinagle in L^2.

4.2. Theorem 6. *Let* M *be a martingale in* $L^2(\Omega, \mathscr{F}, \mathsf{P})$, *continuous to the right. Let us define*

$$\forall F \in \mathscr{F}_s,\ s < t , \qquad m(]0, t] \times F) = 1_F \cdot (M(t) - M(s)) ,$$

$$\alpha(]s, t] \times F) = \int_F (M(t) - M(s))^2 d\mathsf{P} .$$

1°. *Then* m *can be extended in a stochastic measure in* L^2, *and* α *in a positive measure on* $\mathscr{C}$.

2°. *Every process in* $L^2(\Omega, \mathscr{C}, \alpha)$ *is integrable with respect to* m *(in the sense of Bartle) and the mapping* $X \rightsquigarrow \int X dm$ *is an isometry of* $L^2(R^+ \times \Omega, \mathscr{C}, \alpha)$ *into* $L^2(\Omega, \mathscr{F}, \mathsf{P})$.

3°. *The natural increasing process* V *of* α *has the property that* $M^2 - V$ *is a martingale.*

Proof. From the martingale property of M, it is seen that

$$\alpha(]s, t] \times \mathsf{P}) = \mathsf{E}(1_F \cdot (M_t^2 - M_s^2)) .$$

The additivity of α follows easily. The continuity to the right of M implies property (j) for α.

One shows then for any

$$A_n = \sum_{k \leq p}]s_k^n, u_k^n] \times F'_{n,k} \subset]0, t] \times F_n$$

with $F'_{n,k} \in \mathscr{F}_{s_k^n}$, that

$$\alpha(A_n) \leq \int_{F_n} M_t^2 d\mathsf{P} .$$

The property (jj) follows.

From the martingale property again we see that if $A \in \mathscr{A}$ with

$$A = \sum_{k \leq p}]s_k, u_k] \times F_k$$

(4.2.1) $$\| m(A) \|_2^2 = \sum_{k \leq p} \alpha(]s_k, u_k] \times F_k) = \alpha(A) .$$

The property (jjj) follows.

We get then 1° of Theorem 6 by applying Theorem 5. The possibility of extending $A \rightsquigarrow m(A)$ to a linear isometry of $L^2(R^+ \times \Omega, \mathscr{C}, \alpha)$ into $L^2(\Omega, \mathscr{F}, \mathsf{P})$ follows easily from (4.2.1). We get 2° of Theorem 6 as an immediate consequence.

Finally, for every $]s, t] \times F \in \mathfrak{A}$, we get from the definition of V

$$0 = \alpha(]s, t] \times F) - \mathsf{E}(1_F \cdot (V_t - V_s)) =$$
$$= \mathsf{E}(1_F \cdot [(M_t^2 - V_t) - (M_s^2 - V_s)]) .$$

From then we get 3° of Theorem 6.

4.3. Remark. The process V is usually denoted by $\langle M, M \rangle$ or $\langle M \rangle$. We call α the quadratic variation of M.

4.4. Theorem 6b. *If the martingale M is continuous, the stochastic measure m can be extended to a stochastic measure on the δ-ring of well-measurable sets included in some $A \in \mathfrak{A}$.*

Proof. It is immediately seen that α does not charge any graph of stopping time, and then defines a σ-additive positive measure on the algebra generated by the stochastic intervals $[S, T[$.

5. LOCAL STOCHASTIC MEASURE GENERATED BY A LOCAL MARTINGALE

Let $\sigma = (\sigma_n)$ be an increasing sequence of stopping times such that $\lim_n \sigma_n = +\infty$ a.s. We will denote by $\mathfrak{A}_\sigma$ the δ-ring of previsible sets included in one of the stochastic intervals $[0, \sigma_n]$. We give the following

Definition 4. A local stochastic measure, with value in L^p, is any measure on a δ-ring $\mathfrak{A}_\sigma$, with values in $L^p(\Omega, \mathscr{F}, \mathrm{P})$, such that

$$(\delta') \qquad \forall \sigma_n \in \sigma, \qquad \forall]s, t] \times F \in \mathscr{R}$$

$$m(]s, t] \times F] \cap [0, \sigma_n]) = 1_F \cdot m(]s, t] \times \Omega] \cap [0, \sigma_n]) \in$$

$$\in L^p(\Omega, \mathscr{F}, \mathrm{P}) .$$

If M is a local martingale (cf. 1), with the associated $\sigma = (\sigma_n)$ family of stopping times, it is immediate to see from the Theorem 6 that we can state an analogous theorem, replacing the generated stochastic measure on $\mathfrak{A}$ by a generated local stochastic measure.

6. DEFINITION OF THE STOCHASTIC INTEGRAL

6.1. Definition 5. Let m be a stochastic measure, with values in L^p. Let X be a previsible process. One says that X is integrable with respect to m if the mapping $(t, \omega) \rightsquigarrow X(t, \omega)$ is strongly integrable on each $A \in \mathfrak{A}$ in the sense of Bartle (cf. [1]).

The process, with trajectories continuous to the right, unique up to indistinguability, associated with the stochastic measure.

$$A \rightsquigarrow \int_A X dm$$

will be called the integral process of X and denoted $\left(\int_{]0,t] \times \Omega} X dm \right)_{t \in R^+}$ (or $\left(\int_0^t X dZ \right)_{t \in R^+}$ if m is the measure generated by the process $\langle Z \rangle$, or more briefly $\left(\int X dm \right)$.

6.2. Theorem 7. *If the process associated with m has continuous paths the process $\left(\int X dm \right)$ is equally with continuous paths.*

In this case, a well-measurable process is said to be integrable if X is strongly integrable on each $A \in \mathfrak{A}$ in the sense of Bartle with respect to the extension (cf. Theorem 6b) of m to well-measurable sets. The integral process $\left(\int X dm \right)$ is then continuous.

Proof. The first part of theorem is immediate when $X = 1_A$, $A \in \in \mathscr{R}$. One can then apply standard measure theoretic arguments, since there is a dominated convergence theorem for the Bartle integral, and we can then use classical measurability extension lemmas.

From there on we derive immediately all the usual properties of stochastic integrals.

We mention the special case when M is a martingale in $L^2(\Omega, \mathscr{C}, \mathrm{P})$ and α its quadratic variation (cf. Remark 4.3). From the dominated convergence theorem for Bartle integrable functions, it is immediately seen that every $X \in L^2(R^+ \times \Omega, \mathscr{F}, \alpha)$ is strongly integrable with respect to the measure generated by M, and that $X \rightsquigarrow \int X dm$ is an isometry of $L^2(R^+ \times \Omega, \mathscr{C}, \alpha)$ into $L^2(\Omega, \mathscr{F}, \mathrm{P})$. One has easily:

6.3. Theorem 8. *If M is a martingale in $L^2(\Omega, \mathscr{F}, \mathrm{P})$, and if $X \in L^2(R^+ \times \Omega, \mathscr{C}, \alpha)$, then $Y = \int X dM$ is a martingale in $L^2(\Omega, \mathscr{F}, \mathrm{P})$. The quadratic variation α_Y of the stochastic measure generated by Y is given by*

We stop here sketching this theory of stochastic integral, referring for further details to [6] or [10].

$$\alpha_Y(A) = \int_A |X|^2 d\alpha$$

and

$$\langle Y \rangle_t = \int_0^t |X_s|^2 d\langle M \rangle_s \, .$$

We stop here sketching this theory of stochastic integral, referring for further details to [6] or [10].

REFERENCES

[1] R.G. Bartle, A general bilinear vector integral, *Studia Math.*, 15 (1956), 337-352.

[2] C. Doleans-Sade – P.A. Meyer, *Intégrales stochastiques par rapport aux martingales locales,* Séminaire de Probabilités IV, Lecture Notes in Math., Vol. 124, Springer Verlag, 1970.

[3] K. Ito, *Lecture notes on Markov processes,* Tata Institute, Bombay.

[4] H.P. MacKean, *Stochastic integral,* Academic Press, 1968.

[5] M. Mctivier, Limites projectives de mesures, *Annali di mat. pura ad aplicata,* 63 (1963), 225-351.

[6] M. Metivier, *Mesures vectorielles et intégrale stochastique,* Séminaire Rennes, Rennes, 1972.

[7] P.A. Meyer, *Intégrales stochastiques* I *et* II, Séminaire de Probabilités I, Lecture Notes in Math., Vol. 39., Springer Verlag, 1967.

[8] P.A. Melyer, *Probabilités et potentiel,* Hermann.

[9] M.J. Pellaumail, Un exemple d'intégrale d'une fonction réelle par rapport à une mesure vectorielle: l'intégrale stochastique, *C.R.A.S. Paris,* 274 (1972), 1369-1372.

[10] M.J. Pellaumail, Thèse, Rennes, (to appear).

COLLOQUIA MATHEMATICA SOCIETATIS JÁNOS BOLYAI
9. EUROPEAN MEETING OF STATISTICIANS, BUDAPEST (HUNGARY), 1972.

ROBUST ESTIMATION OF LOCATION: A MONTE CARLO STUDY

V. MIKÉ

INTRODUCTION

We assume given n independent, identically distributed random variables $X_1, X_2, \ldots, X_n$ with continuous, symmetric density

$$\frac{1}{\sigma_\lambda} f\left(\frac{x-\theta}{\sigma_\lambda} \,\Big|\, \lambda\right) = f(u \mid \lambda) ,$$

with unknown parameters $\theta \in \Theta = \{\theta: -\infty < \theta < \infty\}$, $0 < \sigma_\lambda < \infty$, and $\lambda \in \Lambda$, a specified finite set of shapes. We consider the problem of finding estimators of the location parameter θ that are in some sense optimum uniformly over Λ.

Birnbaum [2] introduced the mixture model

$$f(u \mid G) = \int_\Lambda f(u \mid \lambda) dG(\lambda) = \sum_{\lambda=1}^{m} g_\lambda f(u \mid \lambda) ,$$

where $G = \{g_1, \ldots, g_m\}$ is a prior distribution on Λ. This model may be used to construct families of efficiency-robust estimators of θ. The properties of one such estimator will be discussed below.

THE GENERALIZED PITMAN ESTIMATOR

Pitman [11] defined the estimator of location

$$\theta^*(x) = \frac{\int_{-\infty}^{\infty} \theta f(x, \theta) d\theta}{\int_{-\infty}^{\infty} f(x, \theta) d\theta},$$

where $x = (x_1, \ldots, x_n)$ and $f(x, \theta) = \prod_{i=1}^{n} f(x_i, \theta)$, and proved that it has minimum variance in the class of translation invariant estimators. Its admissibility was established, e.g., by Stein [12].

Application of Birnbaum's method to Pitman's estimator yields the "generalized Pitman estimator" (σ_λ known)

$$\theta_G^*(x) = \frac{\int_{-\infty}^{\infty} \theta \prod_{i=1}^{n} f(x_i - \theta \mid G) d\theta}{\int_{-\infty}^{\infty} \prod_{i=1}^{n} f(x_i - \theta \mid G) d\theta} =$$

$$= \frac{\sum_{\lambda=1}^{m} g_\lambda \int_{-\infty}^{\infty} \theta \prod_{i=1}^{n} f(x_i - \theta \mid \lambda) d\theta}{\sum_{\lambda=1}^{m} g_\lambda \int_{-\infty}^{\infty} \prod_{i=1}^{n} f(x_i - \theta \mid \lambda) d\theta} = \sum_{\lambda=1}^{m} g_\lambda^*(x) \theta_\lambda^*(x),$$

where $\theta_\lambda^*(x)$ is Pitman's estimator for shape λ. $\theta_G^*(x)$ was discussed by Birnbaum and Laska [3] and shown to be unbiased and admissible over $\Theta \times \Lambda$ with variance as risk function. It is an efficiency-robust estimator of θ for arbitrary n.

It was proved by Miké [10] that under suitable regularity conditions θ_G^* is asymptotically fully efficient uniformly over Λ; that is, it is asymptotically equivalent, under each shape, to the maximum likelihood estimator for that shape.

MONTE CARLO STUDY

A Monte Carlo study has been carried out to estimate the efficiencies of $\theta_G^* = \theta^*(\Lambda, G, n)$, when Λ consists of the normal and double exponential distributions, that is, when $\Lambda = N + DE$, with

$$f(u \mid N) = \frac{1}{\sqrt{2\pi}}\, e^{-\frac{u^2}{2}}$$

and

$$f(u \mid DE) = \frac{1}{\sqrt{2}}\, e^{-\sqrt{2}|u|}.$$

The performance of θ_G^* was assessed also under the logistic and contaminated normal shapes, that is, under

$$f(u \mid L) = \frac{\pi \exp\left(-\frac{\pi}{\sqrt{3}} u\right)}{\sqrt{3}\left(1 + \exp\left(-\frac{\pi}{\sqrt{3}} u\right)\right)^2}$$

and

$$f(u \mid CN(\gamma)) = (1-\gamma) f(u \mid N) + \frac{\gamma}{3} f\left(\frac{u}{3} \middle| N\right),$$

for $\gamma = 0.01$, 0.05, and 0.10.

θ_G^* was studied for $n = 5$, 10, 15, 20, 25, 30, 40, 50, 100, 200, and $G = (g, 1-g)$, with $g = 0, 0.1, 0.2, \ldots, 1$. The procedure involved estimating the variances under N and DE of the 11 estimators determined by G, for each value of n. Efficiencies were then computed for each shape and the maximin efficiency over G and Λ determined graphically. The efficiencies of these estimators were also examined under the other four shapes.

The criteria used in computing the efficiencies of the estimators were as follows: for N, Var (mean); for DE, Var (Pitman) as obtained in this study; for L, Var (BLUE) for $n \leqslant 25$, Cramér – Rao bound for $n \geqslant 30$, and for $CN(\gamma)$, Var (BLUE) for $n \leqslant 20$, Cramér – Rao bound for $n \geqslant 25$. The details of the Monte Carlo procedures are discussed fully in Miké [10]. Approximate 95% confidence limits on the maximin efficiencies ranged between 0.3% and 1.3%.

The estimated variance of Pitman's estimator for DE, with $\sigma^2 = 1$, is shown in Table 1 for the sample sizes examined. Not previously computed,

Table 1.

Estimated variance of Pitman's estimator for the double exponential distribution, $\sigma^2 = 1$

n	nV
5	0.786
10	0.698
15	0.659
20	0.636
25	0.620
30	0.609
40	0.594
50	0.583
100	0.558
200	0.541
Cramér – Rao bound	0.500

this is a useful baseline criterion for robustness studies. Table 2 summarizes the maximin efficiencies of θ_G^* obtained over $\Lambda = N + DE$; the value of g yielding the maximin estimator is also given for each n. The maximin efficiency approaches 1 with increasing n, while there is a decrease in the corresponding value of g specifying the maximin estimator; g here indicates the prior weight on DE. ($g = 0$ and $g = 1$ define the Pitman estimator for DE and N, respectively.)

Table 2.

Maximin efficiencies of the generalized
Pitman estimator for $\Lambda = N + DE$

n	g corresponding to maximin estimator	Maximin efficiency over Λ
5	0.46	0.960
10	0.44	0.945
15	0.43	0.942
20	0.40	0.942
25	0.40	0.945
30	0.38	0.948
40	0.37	0.956
50	0.35	0.962
100	0.28	0.985
200	0.27	0.998

Figures 1-3 show, for selected sample sizes, the efficiency curves of θ^*_G under the various shapes studied. They illustrate 1) the case where the maximin estimator for Λ is also maximin for the other shapes of interest, that is, where there is complete "bridging" ($n = 10$), 2) the case where there is partial bridging ($n = 25$), and finally 3) where there is no bridging. The maximin efficiency for a finite family Λ serves as an upper bound, not sharp in general, on the maximin efficiency attainable over a more inclusive family of shapes. In some cases these maximin values will coincide, as we see, for example, in Fig. 1. The problem becomes more difficult with increasing n, requiring the inclusion of more shapes in Λ.

EFFICIENCIES OF $\theta^*(g, \Lambda, n)$, $\Lambda = N + DE$, $n = 10$, UNDER SELECTED SHAPES

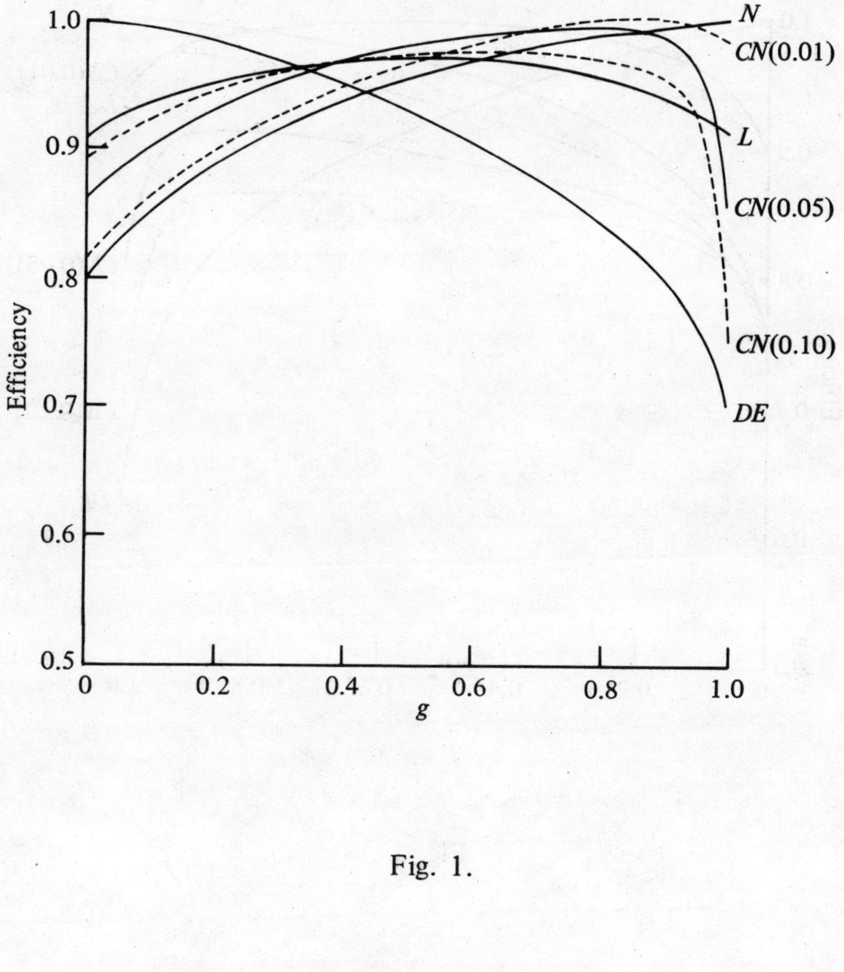

Fig. 1.

EFFICIENCIES OF $\theta^*(g, \Lambda, n)$, $\Lambda = N + DE$, $n = 25$,
UNDER SELECTED SHAPES

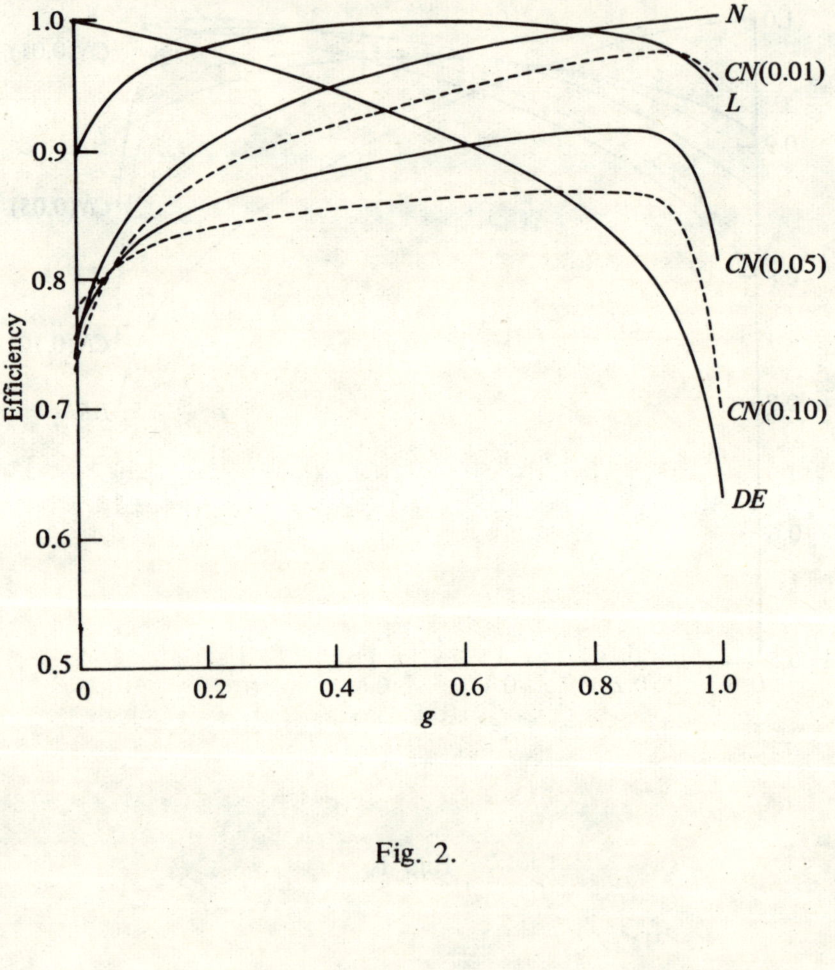

Fig. 2.

EFFICIENCIES OF $\theta^*(g, \Lambda, n)$, $\Lambda = N + DE$, $n = 100$,
UNDER SELECTED SHAPES

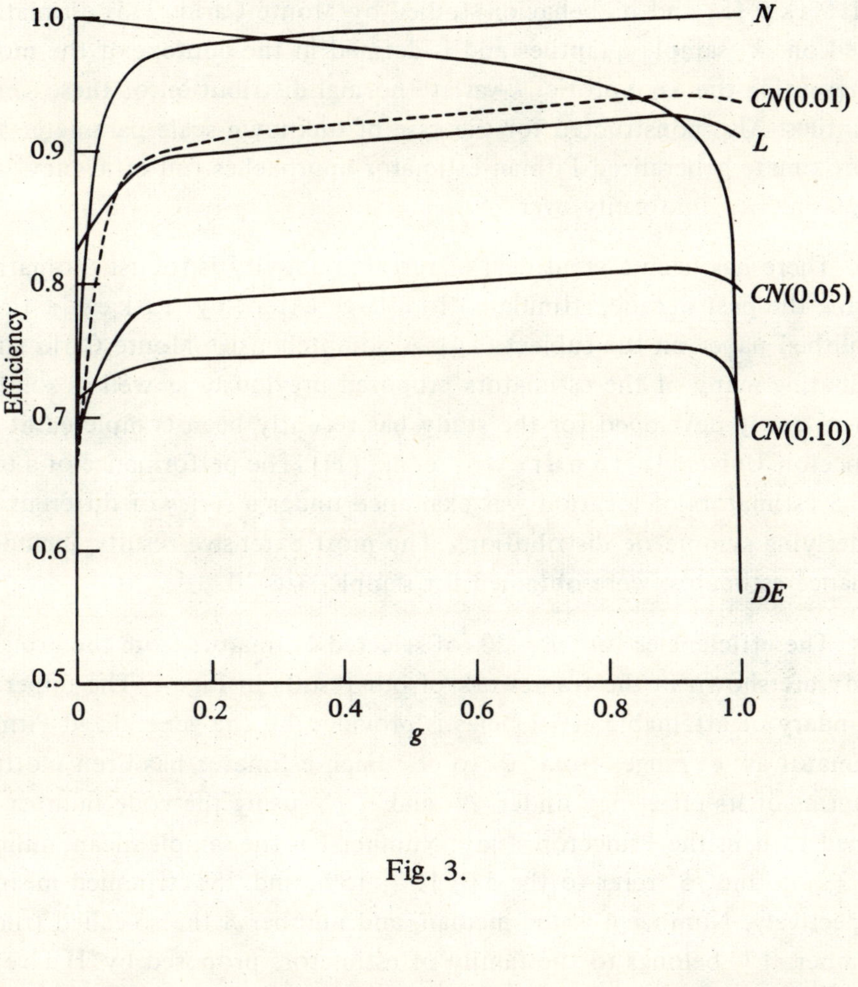

Fig. 3.

DISCUSSION

Birnbaum's mixture model has been applied to different types of linear estimators of location; the results have been reported in Birnbaum and Laska [3], Birnbaum, et al., [4] and Miké [9]. In addition, a computationally tractable large sample approximation of θ_G^*, the "approximate generalized Pitman estimator", has been developed by Birnbaum and Miké [5] and its behavior studied by Monte Carlo. This estimator is based on k sample quantiles and is defined in the context of the model representing the asymptotic k-variate normal distribution of these sample quantiles. Also constructed for the case of unknown scale parameter the approximate generalized Pitman estimator approaches full efficiency, with increasing k, uniformly over Λ.

There has been a great deal of research activity is robust estimation during the past decade, stimulated to a large extent by Tukey's first published paper on the subject [14]. A comprehensive Monte Carlo study evaluating many of the estimators proposed previously as well as some others newly developed for the study has recently been completed at Princeton University (Andrews, et al., [1]). The performance of a total of 65 estimators of location was examined under a series of different underlying symmetric distributions. The most extensive results, including variance estimates, were obtained for sample size 20.

The efficiencies for $n = 20$ of selected estimators from the Princeton study are shown in the framework of our results in Fig. 4. The upper boundary of attainable efficiencies is provided by the generalized Pitman estimator as g ranges from 0 to 1. Each estimator has been plotted in terms of its efficiency under N and DE, using the code number assigned to it in the Princeton study. Number 1 is the sample mean; numbers 2, 3, 4, and 5 refer to the 5%, 10%, 15%, and 25% trimmed means, respectively. Number 6 is the median, and number 8 the so-called trimean. Number 17 belongs to the family of estimators proposed by Huber [7]; numbers 49 and 51 were constructed by Takeuchi [13] and Johns [8], respectively. Number 57 is the Hodges – Lehmann [6] estimator. The estimator attaining maximin efficiency among the 65

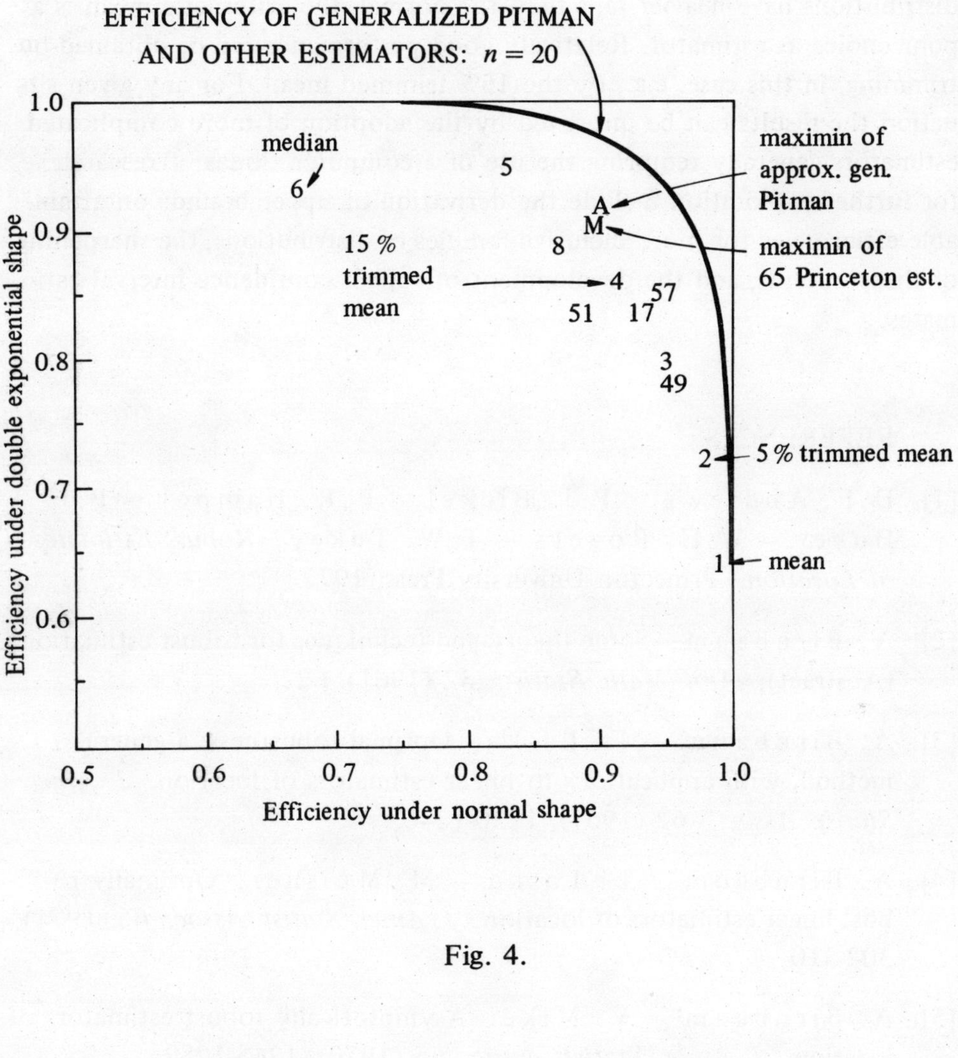
EFFICIENCY OF GENERALIZED PITMAN
AND OTHER ESTIMATORS: $n = 20$
1.0
0.9
0.8
0.7
0.6
0.5
median
6
5
15 %
trimmed
mean
8
4
57
51
17
A
M
3
49
2
1
maximin of
approx. gen.
Pitman
maximin of
65 Princeton est.
5 % trimmed mean
mean
Efficiency under double exponential shape
Efficiency under normal shape

Fig. 4.

included in the Princeton study has been plotted as "M". "A" refers to the maximin approximate generalized Pitman estimator for $k = 10$.

What is evident from this graph illustrates basic results of research in robust estimation of location. Assuming that other likely underlying distributions have heavier tails than the normal, the arithmetic mean is a poor choice as estimator. Relatively good performance can be obtained by trimming, in this case, e.g., by the 15% trimmed mean. For any given situation the results can be improved by the adoption of more complicated estimators generally requiring the use of a computer. Areas of research for further exploration include the derivation of upper bounds on attainable efficiencies for more inclusive families of distributions, the sharpening of these bounds, and the development of robust confidence interval estimates.

REFERENCES

[1] D.F. Andrews – P.J. Bickel – F.R. Hampel – P.J. Huber – W.H. Rogers – T.W. Tukey, *Robust Estimates of Location,* Princeton University Press, 1972.

[2] A. Birnbaum, Some theory and techniques for robust estimation, (Abstract), *Ann. Math. Statist.*, 32 (1961), 622.

[3] A. Birnbaum – E. Laska, Optimal robustness: a general method, with applications to linear estimators of location, *J. Amer. Statist. Assoc.*, 62 (1967), 1230-1240.

[4] A. Birnbaum – E. Laska – M. Meisner, Optimally robust linear estimators of location, *J. Amer. Statist. Assoc.*, 66 (1971), 302-310.

[5] A. Birnbaum – V. Miké, Asymptotically robust estimators of location, *J. Amer. Statist. Assoc.*, 65 (1970), 1265-1282.

[6] J.L. Hodges – E.L. Lehmann, Estimates of location based on rank tests, *Ann. Math. Statist.*, 34 (1963), 598-611.

[7] P.J. Huber, Robust estimation of a location parameter, *Ann. Math. Statist.*, 35 (1964), 73-101.

[8] M.V. Johns, Nonparametric estimation of location, Technical Report No. 41, Department of Statistics, Stanford University, 1971.

[9] V. Miké, Efficiency-robust systematic linear estimators of location, *J. Amer. Statist. Assoc.*, 66 (1971), 594-601.

[10] V. Miké, Robust Pitman-type estimators of location, *Ann. Inst. Math. Statist.*, 25 (1973), 65-86.

[11] E.J.G. Pitman, The estimation of location and scale parameters of a continuous population of any form, *Biometrika,* 30 (1939), 391-421.

[12] C. Stein, The admissibility of Pitman's estimator of a single location parameter, *Ann. Math. Statist.*, 30 (1959), 970-979.

[13] K. Takeuchi, A uniformly asymptotically efficient estimator of a location parameter. *J. Amer. Statist. Assoc.*, 66 (1971), 292-301.

[14] J. Tukey, A survey of sampling from contaminated distributions, *Contributions to Probability and Statistics,* (ed. by I. Olkin, et al.), Stanford University Press, 1960.

COLLOQUIA MATHEMATICA SOCIETATIS JÁNOS BOLYAI
9. EUROPEAN MEETING OF STATISTICIANS, BUDAPEST (HUNGARY), 1972.

ON SOME PROPERTIES OF DENSITY ESTIMATION

M.A. MIRZAHMEDOV – S.A. HAŠIMOV

INTRODUCTION

Let $x_1, x_2, \ldots, x_n$ be a sample of independent observations of the random variable ξ having distribution function $F(x)$ and density function (d.f.) $f(x)$. There are two distinct basic methods to estimate an unknown d.f. $f(x)$:

a) the method of weight function,

b) the method of "expanding of $f(x)$".

By the first method the estimation for the d.f. $f(x)$ is constructed using the so-called "weight function" which satisfies some given conditions. A wide class of such estimations for d.f. $f(x)$ are considered in Parzen [5].

Further properties of Parzen's type estimations were discussed by any authors [2], [3], [6], [7], [8]. By the second method the estimation of the unknown d.f. is constructed by using orthonormal functions [4].

S c h w a r t z [9] has considered the following estimation for the unknown d.f. $f(x)$:

$$f_n(x) = \sum_{j=0}^{q(n)} a_{jn} \varphi_j(x) \tag{1}$$

where $q(n) \to \infty$,

$$a_{jn} = \frac{1}{n} \sum_{i=1}^{n} \varphi_j(x_i)$$

and $\varphi_j(x)$ are the orthonormal Hermite functions. He deals the mean-square

$$\sigma_n^2(x) = \mathrm{E}[f_n(x) - f(x)]^2 \to 0$$

and the integrated mean-square convergence of $f_n(x)$ to $f(x)$

$$I_n = \int_{-\infty}^{\infty} \sigma_n^2(x)dx \to 0 .$$

A general work on nonparametric denisty estimation is in [11]. In the paper [10] a sufficient condition is given for the convergence $f_n^{(l)}(x)$ to $f^{(l)}(x)$ with probability 1, where l denotes the order of the derivates and $f_n(x)$ is defined as in [5].

RESULTS

1. Let $x_1, x_2, \ldots, x_n$ be a sample of independent observations of the random variables ξ having density function (d.f.) $f(x)$. Let

$$f(x) = \sum_{j=0}^{\infty} a_j \varphi_j(x) \tag{2}$$

and suppose that the above expansion is convergent uniformly to $f(x)$, where $\varphi_j(x)$ is an orthonormal system of functions, and

$$a_j = \int_{-\infty}^{\infty} \varphi_j(x)f(x)dx .$$

Suppose that

(3) $$|\varphi_j(x)| \leqslant C$$

(C does not depend on x and j),

(4) $$\sum_{n=1}^{\infty} q(n) e^{-\frac{n\beta}{q^2(n)}} < \infty \qquad \text{for each} \qquad \beta > 0 .$$

Consider the following estimation of $f(x)$

(5) $$f_n(x) = \sum_{j=0}^{q(n)} a_{jn} \varphi_j(x)$$

where $a_{jn} = \frac{1}{n} \sum_{i=1}^{n} \varphi_j(x_i)$.

Theorem 1. *Suppose that conditions* (3) *and* (4) *hold. Then a necessary and sufficient condition for*

$$\lim_{n \to \infty} \sup_x |f_n(x) - f(x)| = 0$$

with probability 1 *is the uniform convergence of the series* (2).

Let

(6) $$\sum_{j=1}^{\infty} |a_j| j^{\alpha} < \infty ,$$

where $\alpha > 0$ is a given constant, and

(7) $$\sum_{n=1}^{\infty} q(n) e^{-\frac{n\beta}{q^{2+2\alpha}(n)}} < \infty .$$

Theorem 2. *Under conditions* (3), (6) *and* (7)

$$\lim_{n \to \infty} \sup_x q^{\alpha}(n) |f_n(x) - f(x)| = 0$$

holds with probability 1.

Let us denote by A a not necessarily bounded interval of R. $E_k(A)$ stands for the class of functions having k continuous and bounded derivatives on A.

Let

$$\varphi_j(x) \in E_k(A). \tag{8}$$

Let us define the norm on $E_k(A)$ by

$$\|f\|_k^A = \sup_{0 \leq l \leq k} \sup_{x \in A} |f^{(l)}(x)|.$$

Denote

$$\alpha_{j,l} = \sup_{x \in A} |\varphi_j^{(l)}(x)|,$$

$$\mu_{n,l} = \sup_{0 \leq j \leq n} \alpha_{j,l},$$

$$\beta_{n,k} = \sup_{0 \leq l \leq k} \mu_{n,l}.$$

Suppose that for each $\lambda > 0$

$$\sum_{n=1}^{\infty} q(n) \exp\left\{-\frac{\lambda n}{\beta_{n,k}^2 \beta_{n,0}^2 q^2(n)}\right\} < \infty \tag{9}$$

and that $\sum_{j=0}^{n} a_j \varphi_j^{(l)}(x)$ is convergent uniformly to $f^{(l)}(x)$ for $n \to \infty$, then

$$\sum_{j=0}^{n} a_j \varphi_j^{(l)}(x) \Rightarrow f^{(l)}(x) \qquad \text{for any} \qquad l \leq k. \tag{10}$$

Theorem 3. *Suppose that conditions* (8), (9) *hold. Then*

$$\lim_{n \to \infty} \|f_n(x) - f(x)\|_k^A = 0 \tag{11}$$

is valid with probability 1 *if and only if* (10) *holds.*

Conditions of Theorem 3 proved to be not only sufficient as it was shown in [10] but necessary too. Moreover condition (2°) of Theorem 2 of [10] is unnecessary. Observe that if for any $\alpha > 0$

(12) $$\sum_{n=1}^{\infty} q(n) \exp\left\{ -\frac{\lambda n}{\beta_{n,k}^2 \beta_{n,0}^2 q^{2+2\alpha}(n)} \right\} < \infty$$

then under conditions (6), (8), (12)

$$\lim_{n \to \infty} q^{\alpha}(n) \| f_n(x) - f(x) \|_k^A = 0$$

with probability 1.

2. One can consider the distance of $f_n(x)$ from $f(x)$ in the metric of the space $L_2(-\infty, \infty)$ that is the convergence of the random variables:

$$V_n = \int_{-\infty}^{\infty} [f_n(x) - f(x)]^2 dx \, .$$

Suppose that

(13) $$\sum_{j=0}^{\infty} a_j^2 < \infty \, .$$

Theorem 4. *Under conditions* (3), (4) *and* (13)

$$\lim_{n \to \infty} V_n = 0$$

with probability 1.

3. Let $(x_1, y_1), (x_2, y_2), \ldots, (x_n, y_n)$ be a sample of independent observations of the two-dimensional random vector (ξ, η) having density function $f(x, y)$. Denote the density function of ξ by $f(x)$ and the regression curve of η with respect to ξ by $y(x)$ $f(x, y)$ has the following expansion:

$$f(x, y) = \sum_{i,j=0}^{\infty} a_{ij} \varphi_{ij}(x, y)$$

where $\varphi_{ij}(x, y)$ is a system of two-dimensional orthonormal functions.

Suppose that the following conditions hold:

(14) $$\sum_{i,j=0}^{m} a_{ij}\varphi_{ij}(x, y) \Rightarrow f(x, y) ,$$

(15) $$|\varphi_{ij}(x, y)| \leqslant C_1$$

(C_1 does not depend on x, y and i, j),

(16) $$\min_x f(x) = \mu > 0 ,$$

(17) $$\int_{-\infty}^{\infty} y \sum_{i,j=0}^{\infty} a_{ij}\varphi_{ij}(x, y)dy = \sum_{i,j=0}^{\infty} a_{ij} \int_{-\infty}^{\infty} y\varphi_{ij}(x, y)dy ,$$

(18) $$\sum_{n=1}^{\infty} q_1(n)q_2(n) \exp\left\{-\frac{\beta_1 n}{q_1^2(n)q_2^2(n)}\right\} < \infty$$

for each $\beta_1 > 0$, where $q_1(n)$, $q_2(n)$ converge to infinity as $n \to \infty$.

As the estimation of $y(x)$ consider the statistics

$$y_n(x) = \frac{\Phi_n(x)}{f_n(x)}$$

where $f_n(x)$ defined by (5),

$$\Phi_n(x) = \sum_{i,j=0}^{q_1(n), q_2(n)} a_{ij}K_{ij}(x) ,$$

$$K_{ij}(x) = \int_{-\infty}^{\infty} y\varphi_{ij}(x, y)dy ,$$

$$a_{ijn} = \frac{1}{n} \sum_{k=1}^{n} \psi_{ij}(x_k, y_k)$$

where $\psi_{ij}(x, y)$-s are orthonormal functions to $\varphi_{ij}(x, y)$ in the sense of [1].

Theorem 5. *Suppose that conditions* (3), (4), (14)-(18) *hold. Then for* $n \to \infty$

$$\sup_x |y_n(x) - y(x)| = 0$$

with probability 1.

Proof of the theorems.

Proof of Theorem 1. Obviously

$$\mathsf{E}a_{jn} = a_j .$$

We need the following lemmas.

Lemma 1. *For each* $\epsilon > 0$

$$\mathsf{P}\left\{\sum_{j=0}^{q(n)} |a_{jn} - a_j| \geqslant \epsilon\right\} \geqslant 2(1 + q(n))e^{-\frac{n\epsilon^2}{2C^2(1+q(n))^2}} .$$

Using the theorem of Hoeffding [12] we have

$$\mathsf{P}\left\{\sum_{j=0}^{q(n)} |a_{jn} - a_j| \geqslant \epsilon\right\} \leqslant$$

$$\leqslant \sum_{j=0}^{q(n)} \mathsf{P}\left\{\frac{1}{n}\sum_{i=1}^{n} |\varphi_j(x_i) - \mathsf{E}\varphi_j(x_i)| \geqslant \frac{\epsilon}{n(1 + q(n))}\right\} \leqslant$$

$$\leqslant 2(1 + q(n))e^{-\frac{n\epsilon^2}{2C^2(1+q(n))^2}}$$

Lemma 2.

$$\lim_{n\to\infty} \sup_x |f_n(x) - \mathsf{E}f_n(x)| = 0$$

with probability 1.

Proof.

$$\mathsf{P}\{\sup_x |f_n(x) - \mathsf{E}f_n(x)| \geqslant \epsilon\} =$$

$$= \mathsf{P}\left\{\sup_x \left|\sum_{j=0}^{q(n)} a_{jn}\varphi_j(x) - \mathsf{E}\sum_{j=0}^{q(n)} a_{jn}\varphi_j(x)\right| \geqslant \epsilon\right\} \leqslant$$

$$\leqslant \mathsf{P}\left\{\sum_{j=0}^{q(n)} |a_{jn} - a_j| \geqslant \frac{\epsilon}{c}\right\} \leqslant 2(1 + q(n))e^{-\frac{n\epsilon^2}{2C^4(1+q(n))^2}} .$$

The assertion follows from the Borel – Cantelli lemma and (4).

Lemma 3. *for the validity of*

$$\lim_{n\to\infty} \sup_x |f_n(x) - h(x)| = 0$$

with probability 1 *is necessary and sufficient that*

$$\lim_{n\approx\infty} \sup_x |\mathsf{E}f_n(x) - h(x)| = 0$$

where $h(x)$ *is an arbitrary function.*

Proof. Necessity. Let

$$\lim_{n\to\infty} \sup_x |f_n(x) - h(x)| = 0$$

with probability 1, then the necessity follows from the following inequality and Lemma 2

$$\sup_x |\mathsf{E}f_n(x) - h(x)| \leqslant \sup_x |\mathsf{E}f_n(x) - f_n(x)| + \sup_x |f_n(x) - h(x)| .$$

Sufficiency. Let

$$\lim_{n\to\infty} \sup_x |\mathsf{E}f_n(x) - h(x)| = 0 ,$$

then the sufficiency follows from the following inequality and Lemma 2

$$\sup_x |f_n(x) - h(x)| \leqslant \sup_x |f_n(x) - \mathsf{E}f_n(x)| + \sup_x |\mathsf{E}f_n(x) - h(x)| .$$

Now we are ready for the proof of Theorem 1.

Sufficiency. From the statement of Lemma 3 immediately follows the sufficiency of Theorem 1. Indeed

$$\lim_{n\to\infty} \sup_x |\mathsf{E}f_n(x) - f(x)| = \lim_{n\to\infty} \sup_n \Big| \sum_{i=1+q(n)}^{\infty} a_j\varphi_j(x)\Big| = 0 .$$

Necessity of the conditions of Theorem 1. Let

$$\lim_{n\to\infty} \sup_x |f_n(x) - f(x)| = 0$$

with probability 1. Then the necessity immediately follows from the statement of Lemma 3. Indeed

$$\sup_x \left| \sum_{j=1+q(n)}^{\infty} a_j\varphi_j(x) \right| = \sup_x |f_n(x) - \mathsf{E}f_n(x)| .$$

This proves Theorem 1. To the proof of Theorem 2 we need the following Lemmas.

Lemma 4. *Assuming that* (7) *holds*

$$\lim_{n\to\infty} \sup_x q^\alpha(n) |f_n(x) - \mathsf{E}f_n(x)| = 0$$

with probability 1.

Indeed,

$$\mathsf{P}\{\sup_x q^\alpha(n) |f_n(x) - \mathsf{E}f_n(x)| \geqslant \epsilon\} \leqslant$$

$$\leqslant \sum_{j=0}^{q(n)} \mathsf{P}\left\{ |a_{jn} - a_j| \geqslant \frac{\epsilon}{C(1+q(n))^{1+\alpha}} \right\} \leqslant$$

$$\leqslant 2(1+q(n)) e^{-\frac{n\epsilon^2}{2C^4(1+q(n))^{2+2\alpha}}}$$

Thus Lemma 4 follows from the Borel – Cantelli lemma.

Lemma 5. *For the validity of*

$$\lim_{n\to\infty} \sup_x q^\alpha(n) |f_n(x) - h(x)| = 0$$

with probability 1 *is necessary and sufficient that*

$$\lim_{n\to\infty} \sup_x q^\alpha(n) |\mathsf{E}f_n(x) - h(x)| = 0$$

where $h(x)$ *is an arbitrary function.*

The proof of this Lemma is the same as that of Lemma 3.

Proof of Theorem 2. The proof follows from the statement of Lemma 5 and conditions (3) and (6).

Clearly

$$\lim_{n\to\infty} \sup_x q^{\alpha}(n) | \mathsf{E} f_n(x) - f(x) | =$$

$$= \lim_{n\to\infty} \sup_x q^{\alpha}(n) \Big| \sum_{j=1+q(n)}^{\infty} a_j \varphi_j(x) \Big| \leqslant$$

$$\leqslant \lim_{n\to\infty} C q^{\alpha}(n) \sum_{j=1+q(n)}^{\infty} |a_j| < \lim_{n\to\infty} C \sum_{j=1+q(n)}^{\infty} |a_j| j^{\alpha} = 0 ,$$

this proves Theorem 2.

The proof of Theorem 3 is the same as that of Theorem 1.

Proof of Theorem 4. Put

$$V_n = \int_{-\infty}^{\infty} [f_n(x) - f(x)]^2 dx = \int_{-\infty}^{\infty} [f_n(x) - \mathsf{E} f_n(x)]^2 dx +$$

$$+ \int_{-\infty}^{\infty} [\mathsf{E} f_n(x) - f(x)]^2 dx = V_n^* + V_n^{**} .$$

Consider V_n^*

$$V_n^* = \int_{-\infty}^{\infty} [f_n(x) - \mathsf{E} f_n(x)]^2 dx = \int_{-\infty}^{\infty} \Big[\sum_{j=0}^{q(n)} (a_{jn} - a_j)\varphi_j(x) \Big]^2 dx =$$

$$= \sum_{j=0}^{q(n)} (a_{jn} - a_j)^2 .$$

By (3) we have

$$V_n^* \leqslant 2C \sum_{j=0}^{q(n)} |a_{jn} - a_j| .$$

Now we show that if (4) holds then with probability 1

(19) $\qquad \lim V_n^* = 0 .$

Indeed by Lemma 1 we have

$$\mathsf{P}\{V_n^* \geqslant \epsilon\} \leqslant 2(1 + q(n)) e^{-\frac{n\epsilon^2}{8C^4(1+q(n))^2}} .$$

From the Borel – Cantelli lemma follows (19). By condition (13) follows that

$$\lim_{n \to \infty} V_n^{**} = 0 .$$

This proves Theorem 4.

The proof of Theorem 5 based on the inequality of [3] and a theorem of Hoeffding [12].

REFERENCES

[1] S.H. Siraždinov, To the theory of multivariate Hermite polynomials, *Trudy Inst. Mat. i Meh. AN. UzSSR*, 5 (1949), 70-95.

[2] E.A. Nadaraja, On non-parametric estimation of probability density and regression, *Teorija Verojatn. i Primenen.*, 10 (1965), 199-203.

[3] E.A. Nadaraja, A note an non-parametric estimation of probability density and regression curve, *Teorija Verojatn. i Primenen.*, 15 (1970), 139-142.

[4] N.N. Čencov, Estimation for the density function on the base of observations, *Dokl. Akad. Nauk. SSSR*, 147 (1962), 45-48.

[5] E. Parzen, On estimation of probability density and mode, *Ann. Math. Statist.*, 33 (1962), 1065-1076.

[6] I. Văduva, Contribuţii la teoria estimaţiilar Statistice ale densităţilor de repartiţie şi aplicaţii, *Studii şi cercetări matematice*, 20 (1968), 1207-1276.

[7] P.K. Bhattacharya, Estimation of a probability density function and its derivatives, *Sankhyā*, 29 (1967), 373-382.

[8] E.F. Schuster, Estimation of a probability density function and its derivatives, *Ann. Math. Statist.*, 40 (1969), 1187-1195.

[9] S.G. Schwartz, Estimation of probability density by an orthogonal series, *Ann. Math. Statist.*, 38 (1967), 1261-1265.

[10] D. Bosq, Estimation non paramétrique de la densité et de ses dérivés, *C. R. Acad. Sci. Paris,* 269 (1969), 1010-1012.

[11] M. Rosenblatt, Curve estimates, *Ann. Math. Statist.*, 42 (1971), 1815-1842.

[12] W. Hoeffding, Probability inequalities for sums of bounded random variables, *J. Amer. Statist. Assoc.*, 301 (1963), 13-30.

COLLOQUIA MATHEMATICA SOCIETATIS JÁNOS BOLYA I
9. EUROPEAN MEETING OF STATISTICIANS, BUDAPEST (HUNGARY), 1972.

ON SOME DISTRIBUTIONS OF A GENERALIZED RESTRICTED RANDOM WALK

S.G. MOHANTY – S. VELLORE

1. INTROUCTION

This note gives an extension of results in [3] and [5]. Using the usual vector notation $x = (x_1, x_2, \ldots, x_n)$, let $x \cdot y$ represent the dot product of x and y. We define a restricted random walk in the $(n+1)$-dimensional space by a sequence $(t_1, t_2, \ldots, t_{\alpha + (m+1) \cdot a})$ of random vectors $t_i = (t_{i0}, t_{i1}, \ldots, t_{in})$ such that

(i) $m_i > 0$, $a_i > 0$ and $\alpha \geqslant 0$ are integers $(i = 1, 2, \ldots, n)$,

(ii) $t_{ij} = 0$ for $j < 0$ or $j > n$,

(iii) $$\begin{cases} t_{i0} = 0 \text{ or } 1 \\ t_{ij} = 0 \text{ or } -m_j \quad (j = 1, 2, \ldots, n) \end{cases}$$

in such a way that

$$
\text{(iv)} \begin{cases} \sum_{j=0}^{n} t_{ij} = 1 \text{ or } -m_h \text{ for some } h \ (1 \leqslant h \leqslant n) \\ \sum_{i=1}^{\alpha+(m+1)\cdot a} t_{i0} = \alpha + m \cdot a \\ \sum_{i=1}^{\alpha+(m+1)\cdot a} t_{ij} = -a_j m_j \ (j = 1, 2, \ldots, n). \end{cases}
$$

Letting $s_0 = 0$, $s_j = \sum_{u=1}^{j} \sum_{v=0}^{n} t_{uv}$ $(j = 1, 2, \ldots, \alpha + (m+1) \cdot a)$ we note that the sequence $(s_0, s_1, \ldots, s_{\alpha+(m+1)\cdot a})$ defines a restricted random walk on a line in the usual sense where a particle either moves one unit to the right or m_j $(j = 1, \ldots, n)$ units to the left and s_j represents the position of the particle at time j, given that there are exactly $\alpha + {} + m \cdot a$ steps to the right and a_j steps of m_j units to the left. Also one can interpret this formulation in terms of several candidates ballot problem: there are $n + 1$ candidates numbered $0, 1, \ldots, n$ and candidate 0 has $\alpha + m \cdot a$ votes whereas condidate j has a_j votes $(j = {} = 1, \ldots, n)$. That the random walk reaches the point α for the first time can be viewed as the ballot problem where the votes of candidate 0 is always greater than the sum (over all candidates) of the votes for candidate j times m_j.

The following statistics which have been defined in [3] and [5] for $n = 1$, are given below for the sake of completeness.

$$\gamma = |\{j\colon s_{j-1} > 0,\ s_j = 0,\ s_{j+1} = 1\}|$$

$$\lambda = |\{j\colon s_j = 0,\ s_{j-1}s_{j+1} < 0\}|$$

$$\lambda' = |\{j\colon s_{j-1} = 0,\ s_j = 1\}|$$

$$\pi = |\{j\colon s_j > 0\}|$$

$$\chi = \max\{s_j\}$$

$$\sigma = \min\{j\colon s_j = \chi\}$$

$$\psi = \min\{s_j\}$$

$$\varphi = \min\{j\colon s_j = \psi \text{ and for some } (j_0, j_1, \ldots, j_{-\psi}),$$

$$0 = j_0 < j_1 < \ldots < j_{-\psi} = \psi,$$

$$s_{j_{t-1}+1} > -t,\ s_{j_{t-1}+2} > -t, \ldots, s_{j_t} = -t,$$

$$t = 1, 2, \ldots, -\psi \} .$$

Here cardinality of a set A is denoted by $|A|$.

When a sequence satisfies, $s_j s_{j+1} \geqslant 0$ for all j, we say that it satisfies "Condition c".

The main purpose of this paper is to show that the results in [3] and [5] on the above statistics when $n = 1$, can be extended to a general n, which involve the evaluation of distributions (also joint distributions) of the above statistics and the establishment of some equivalence relations. While the basic method does not change from the previous papers, the results are obtained by the use of the following identity [2].

Let

$$(\alpha | b) = \frac{\alpha(\alpha - 1) \ldots \ \alpha - \sum_i b_i + 1}{b_1 ! b_2 ! \ldots b_n !} .$$

Then

$$(1) \qquad \sum_{j=0}^{m} \frac{\alpha}{\alpha + b \cdot j} (\alpha + b \cdot j | j) \frac{\gamma}{\gamma + b \cdot (m - j)} (\gamma + b \cdot (m - j) | m - j) = \\ = \frac{\alpha + \gamma}{\alpha + \gamma + b \cdot m} (\alpha + \gamma + b \cdot m | m) .$$

Another point to note is that the results on distributions though have an appearance similar to those for $n = 1$, are harder to guess, until one completes the computation. Thus we illustrate these by considering only some distributions.

It is easy to verify the one to one correspondence between the random walk and a set of lattice paths from the origin to $(\alpha + \sum_i m_i a_i, a_1, \ldots$ $\ldots, a_n)$ in $(n + 1)$-dimensional space, when we represent a movement of one unit to the right by a unit along the x_0 axis and a movement of m_j

units to the left by a unit on x_j axis. The total number of such paths is $(\alpha + (m+1) \cdot a | a)$ and the probability distribution $\mathrm{P}(\cdot)$ of any characteristic $(\cdot)$ of the restricted random walk is given by $\dfrac{N^{\alpha}(\cdot)}{(\alpha + (m+1) \cdot a | a)}$ where $N^{\alpha}(\cdot)$ is the number of paths corresponding to $(\cdot)$. We write $N(\cdot)$ for $N^{0}(\cdot)$ and $N_c(\cdot)$ for $N(\cdot)$ under "Condition c". Because of the above remark, we present the results in the form of $N^{\alpha}(\cdot)$, $N(\cdot)$ and $N_c(\cdot)$.

A result on lattice path counting that is helpful subsequently is as follows [4], Lemma 2.

The number of paths from the origin to $(\alpha + m \cdot a,\ a_1, a_2, \ldots, a_n)$ not touching the hyperplane $x_0 = m \cdot a$ except at the origin is

$$\frac{\alpha}{\alpha + (m+1) \cdot a} (\alpha + (m+1) \cdot a | a) \tag{2}$$

and therefore the number of paths not crossing the same hyperplane is

$$\frac{\alpha + 1}{\alpha + 1 + (m+1) \cdot a} (\alpha + 1 + (m+1) \cdot a | a)\,. \tag{3}$$

2. DISTRIBUTIONS AND EQUIVALENCE RELATIONS

The first theorem gives the distribution of $\gamma, \lambda, \lambda'$ and χ whereas the second on the joint distributions of (χ, ρ) and (λ', l).

Theorem 1. *Let* $C(l) = \{l\colon l_i \geq 0,\ \sum l_i = l\}$ *and* $l^- = (l_1, \ldots, l_{n-1})$. *Then*

(a) $N^{\alpha}(\gamma = l) =$

$$= \begin{cases} \displaystyle\sum_{C(l)} (l | l^-) \frac{\alpha + l \cdot m}{\alpha + l \cdot m + (m+1) \cdot (a - l)} \times \\ \times\ (\alpha + l \cdot m + (m+1) \cdot (a - l) | a - l) \\ \textit{for } l = 0, 1, \ldots, \max_i \{a_i\} \\ 0 \qquad \textit{otherwise;} \end{cases}$$

(b) $N_c(\lambda = l) =$

$$= \begin{cases} 2 \sum\limits_{C(l+1)} (l+1 \mid l^-) \dfrac{l \cdot (m+1)}{a \cdot (m+1)} (a \cdot (m+1) \mid a - l) \\ for \;\; l = 0, 1, \ldots, \max\limits_i \{a_i - 1\} \\ 0 \qquad\qquad otherwise; \end{cases}$$

(c) $N_c(\lambda' = l) =$

$$= \begin{cases} \sum\limits_{C(l)} (l \mid l^-) \dfrac{1 + l \cdot (m+1)}{1 + a \cdot (m+1)} (1 + a \cdot (m+1) \mid a - l) \\ for \;\; l = 0, 1, \ldots, \max\limits_i \{a_i\} \\ 0 \qquad\qquad otherwise; \end{cases}$$

(d) $N(\chi = l) =$

$$= \begin{cases} \dfrac{1}{1 + a \cdot (m+1)} (1 + a \cdot (m+1) \mid a) \quad for \;\; l = 0; \\ \sum\limits_{A(l)} \dfrac{l}{l + j \cdot (m+1)} \times \\ \times (l + j \cdot (m+1) \mid j)(m+1) \cdot (a - j) - l \mid a - j) - \\ - \sum\limits_{B(l,a)} \dfrac{l+1}{l + 1 + (m+1) \cdot (a-k)} \times \\ \times (l + 1 + (m+1) \cdot (a-k) \mid a - k)((m+1) \cdot k - l - 1 \mid k) \\ for \;\; l = 1, 2, \ldots, m \cdot a \\ where \;\; A(l) = \{j \colon 0 \leqslant j_i \;\; for\ every \;\; i \;\; and \;\; m \cdot (a - j) \geqslant l\} \\ and \;\; B(l, a) = \{k \colon l < m \cdot k \;\; and \;\; k_i \leqslant a_i \;\; for\ every \;\; i\}; \\ 0 \qquad\qquad otherwise. \end{cases}$$

We will discuss the case $n = 2$, as the general case is similar.

Proof. (a) It suffices to show that the summand in (a) for $n = 2$ represents the number of paths from the origin to $(\alpha + \sum m_i a_i, a_1, a_2)$ having exactly l_i touches in x_i direction $i = 1, 2$, such that $l_1 + l_2 = l$. For $l = 0$, it represents the number of paths with no touches which by (2) is equal to

$$\frac{\alpha}{\alpha + \sum_i (m_i + 1)a_i} \binom{\alpha + \sum_i (m_i + 1)a_i}{a_1, a_2}.$$

Also it can be checked that the result is true for $l = 1$. Using induction, we assume the result to be true for $l - 1$ touches with l_i touches in x_i direction $(i = 1, 2)$ such that $l_1 + l_2 = l - 1$. Any path with l touches of which exactly l_1 are in x_1 direction has two segments, the first segment ending at the first point of touch in either x_1 or x_2 direction and the second segment starting from there. Thus by induction hypothesis the required number is

$$\sum_{j_1=0}^{a_1-l_1} \sum_{j_2=0}^{a_2-l+l_1} \frac{m_1}{m_1 + \sum_i (m_i + 1)j_i} \binom{m_1 + \sum_i (m_i+1)j_i}{j_1, j_2} \binom{l-1}{l_1 - 1} \times$$

$$\times \frac{\alpha + (l_1 - 1)m_1 + (l - l_1)m_2}{\alpha + (l_1 - 1)m_1 + (l - l_1)m_2 + (m_1 + 1)(a_1 - 1 - j_1 - l_1 + 1) + (m_2 + 1)(a_2 - j_2 - l + l_1)} \times$$

$$\times \binom{\alpha + (l_1 - 1)m_1 + (l - l_1)m_2 + (m_1 + 1)(a_1 - 1 - j_1 - l_1 + 1) + (m_2 + 1)(a_2 - j_2 - l + l_1)}{a_1 - j_1 - l_1, a_2 - j_2 - l + l_1} +$$

$$+ \sum_{j_1=0}^{a_1-l_1} \sum_{j_2=0}^{a_2-l+l_1} \frac{m_2}{m_2 + \sum_i (m_i + 1)j_i} \binom{m_2 + \sum_i (m_i + 1)j_i}{j_1, j_2} \binom{l-1}{l_1} \times$$

$$\times \frac{\alpha + l_1 m_1 + (l - l_1 - 1)m_2}{\alpha + l_1 m_1 + (l - l_1 - 1)m_2 + (m_1 + 1)(a_1 - j_1 - l_1) + (m_2 + 1)(a_2 - j_2 - l + l_1)} \times$$

$$\times \binom{\alpha + l_1 m_1 + (l - l_1 - 1)m_2 + (m_1 + 1)(a_1 - j_1 - l_1) + (m_2 + 1)(a_2 - j_2 - l + l_1)}{a_1 - j_1 - l_1, a_2 - j_2 - l + l_1}$$

which on using the identity (1) simplifies to

$$\binom{l}{l_1} \frac{\alpha + l_1 m_1 + (l - l_1) m_2}{\alpha + l_1 m_1 + (l - l_1) m_2 + (m_1 + 1)(a_1 - l_1) + (m_2 + 1)(a_2 - l + l_1)} \times$$

$$\times \binom{\alpha + l_1 m_1 + (l - l_1) m_2 + (m_1 + 1)(a_1 - l_1) + (m_2 + 1)(a_2 - l + l_1)}{a_1 - l_1,\ a_2 - l + l_1}.$$

This completes the proof.

We omit the proofs of others with the following remarks.

(i) From the above we may note that the main feature of the proof for $n > 1$, consists of considering a touch from different directions. This fact introduces the first factor and the summation in (a), (b) and (c).

(ii) In proving (b), (c) and (d), we need (3).

(iii) The methods of proof are similar to those in [5].

As in [5], the relation $N_c(\lambda' = l) = \frac{1}{2}[N_c(\lambda = l - 1) + N_c(\lambda = l)]$ for all l is also true here, which can be established from (b) and (c).

Theorem 2. (a) *For* $l = 1, \dots, \boldsymbol{m} \cdot \boldsymbol{a}$ *and* $l \leqslant r \leqslant (\boldsymbol{m} + 1) \cdot \boldsymbol{g} + l$, g_i *satisfying* $l \leqslant \boldsymbol{m} \cdot (\boldsymbol{a} - \boldsymbol{g})$,

$$N(\chi = l,\ \rho = r) = \sum_{C(r,l)} \frac{l}{l + (\boldsymbol{m} + 1) \cdot \boldsymbol{j}} (l + (\boldsymbol{m} + 1) \cdot \boldsymbol{j} \mid \boldsymbol{j}) \times$$

$$\times\ ((\boldsymbol{m} + 1) \cdot (\boldsymbol{a} - \boldsymbol{j}) - l \mid \boldsymbol{a} - \boldsymbol{j}) -$$

$$- \sum_{B(l, \boldsymbol{a} - \boldsymbol{j})} \frac{1}{1 + (\boldsymbol{m} + 1) \cdot (\boldsymbol{a} - \boldsymbol{j} - \boldsymbol{k})} \times$$

$$\times (1 + (\boldsymbol{m} + 1) \cdot (\boldsymbol{a} - \boldsymbol{j} - \boldsymbol{k}) \mid \boldsymbol{a} - \boldsymbol{j} - \boldsymbol{k}) \times$$

$$\times ((\boldsymbol{m} + 1) \cdot \boldsymbol{k} - l - 1 \mid \boldsymbol{k})$$

where $C(r, l) = \{(j_1, \dots, j_n) : (\boldsymbol{m} + 1) \cdot \boldsymbol{j} + l = r\}$.

(b) *For* $l = 1, \dots, \max_i \{a_i\}$, $r = (\boldsymbol{m} + 1) \cdot \boldsymbol{j} - l$ *such that* $\boldsymbol{j}$ *satisfies* $l \leqslant \sum j_i$ *and* $j_i \leqslant a_i$ *for every* i,

$$N_c(\lambda' = l,\ \pi = r) = \sum_{D(r,l)} \frac{l+1}{l+1+(m+1)\circ(a-j)} \times$$

$$\times\ ((m+1)\cdot(a-j)+l+1 \mid a-j) \times$$

$$\times \sum_{C(l)} (l \mid l^-) \frac{l\cdot m}{l\cdot m+(m+1)\cdot(j-l)} \times$$

$$\times\ (l\cdot m+(m+1)\cdot(j-l) \mid j-l)$$

where $D(r,l) = \{(j_1, \ldots, j_n)\colon (m+1)\cdot j = r\}$ *and* $C(l)$ *is as defined in Theorem 1.*

Proofs are similar to that in $n = 1$. Note that the first summation in (a) and (b) does not exist for $n = 1$.

We conclude by mentioning the equivalence relation

$$N_c(\lambda' = l,\ \pi = r) = N(\psi = -l,\ \varphi = r)\,.$$

One way of establishing this is by finding the right hand side (ref. [5]) and equating with (b) of Theorem 2. A simpler wayto check is to follow the construction method given in [1], by ignoring the change of sign suggested in that method.

Acknowledgement. Appretiation is extended to National Research Council of Canada for financial support of this work.

REFERENCES

[1] E. Csáki – I. Vincze, On some combinatorial relations concerning the symetric random walk, *Acta Sci. Math. Hungar.*, 25 (1963), 231-235.

[2] S.G. Mohanty, Some convolutions with multinomial coefficients and related probability distribution, *SIAM Review,* 8 (1966), 501-509.

[3] S.G. Mohanty, On some generalizations of a restricted random walk, *Studia Sci. Math. Hungar.*, 3 (1968), 225-241.

[4] S.G. Mohanty, On queues involving batches, *J. Appl. Prob.*, 9 (1972), 430-435.

[5] B.R. Handa – S.G. Mohanty, On some distribution concerning a restricted random walk, *Studia Sci. Math. Hungar.*, 4 (1969), 99-108.

COLLOQUIA MATHEMATICA SOCIETATIS JÁNOS BOLYAI
9. EUROPEAN MEETING OF STATISTICIANS, BUDAPEST (HUNGARY), 1972.

ON STEPWISE REGRESSION IN ORTHOGONAL MODELS

E. MOHN – I. HOLME – R. VOLDEN

1. THE MOTIVE FOR THE WORK ON STEPWISE REGRESSION PROCEDURES

Stepwise regression analysis is one of the most common techniques for building up linear models for the purpose of prediction. Today, most of the larger computer installations are provided with programs for stepwise regression. The use of such programs in practical statistical work has increased during the last years. Unfortunately the misuse of the programs has probably also increased in the same period. The reason can be that when using these programs one does not know the probability of making erroneous inclusions of the possible independent variables.

Now, the theoretical statistician can say that the misuse is not his problem. However, in our opinion such an attitude can not be recommended. Therefore, we have considered one of the most common procedures of today, namely that of Efroymson [1] and have tried to describe its main features in an analytical way. As a first attack we look at the simplest possible case, with orthogonal independent variables.

Our motive for doing this work has been to get information about the "good" and "bad" properties of the procedure. Special interest has been attached to the bad properties since such information will give us:

(i) ideas of how to improve the procedure if possible and

(ii) background for giving warnings to the users for special unfavourable cases of use.

2. MODEL, PROBLEM AND METHOD

The model to be considered is the following:

$$y^{n\times 1} = X^{*}\beta^{(p+1)\times 1} + e$$

where $\beta^{*} = (\beta_0, \beta_1, \ldots, \beta_p)$, $XX^{*} = I$ and $e \sim N(0, \sigma^2 I)$.

The *problem* is to decide which of $\beta_1, \ldots, \beta_p$, if any, are different from zero. If we assert that a certain β_i is different from zero, the corresponding variable is included in the model.

The *method* should satisfy the conditions

$$\mathsf{P}_0(\text{asserting exactly } j \text{ of the } \beta_i\text{'s} \neq 0) = \alpha_j \qquad (j = 1, \ldots, p)$$

where $\{\alpha_j\}$ are small predetermined numbers and P_0 means probability computed under the condition that β_1 up to β_p equal zero.

Several different methods may be used in this multiple decision problem. We shall here consider one of them, namely the procedure of Efroymson and modify this procedure such that the conditions given above are satisfied.

3. THE PROCEDURE OF EFROYMSON AND THE MODIFIED EFROYMSON'S PROCEDURE

The procedure of Efroymson uses a forward selection technique for choosing variables to the regression equation with the additional possibility of making exclusions of variables. At all steps, except the first, we try to exclude a variable. If we succeed, we continue to the next step. In the opposite case we try to include a new variable. If we succeed, we continue to the next step, otherwise we stop the procedure. At the first step we start to include a variable, after having forced the constant term into the regression.

In addition, the procedure is supplemented by:

(i) a selection rule choosing at each step a candidate for exclusion or inclusion

(ii) a criterion deciding if exclusion (inclusion) should be carried out.

This was the procedure for the general non-orthogonal case. In our case, where the independent variables are orthogonal, the procedure is easy to describe since only inclusions are possible. An inclusion of a variable corresponds to asserting that the corresponding β_i is different from zero.

Before I describe the procedure in a more formal way I shall introduce some notations. Let

$$\nu = n - p - 1$$

$$\hat{\beta} = Xy$$

$$s^2 = \frac{1}{\nu}(y - X^*\hat{\beta})^*(y - X^*\hat{\beta})$$

$$S = \nu s^2 .$$

Further we define $S_1, \ldots, S_p$ to be $\hat{\beta}_1^2, \ldots, \hat{\beta}_p^2$ arranged in increasing order and $x_{(1)}, \ldots, x_{(p)}$ the corresponding variables. Now, let us for a moment assume that m variables $x_{j_1}, \ldots, x_{j_m}$ are included in the regression equation in addition to the constant x_0 in some way or another. Then the residual sum of squares will be:

$$SS = y'y - \hat{\beta}_0^2 - \hat{\beta}_{j_1}^2 - \ldots - \hat{\beta}_{j_m}^2$$

where SS is expressed by the estimates of the regression coefficients corresponding to $x_{j_1}, \ldots, x_{j_m}$.

In the Efroymson procedure the selection rule for choosing inclusion candidate at a certain step is the following: Choose that variable not in the regression which give the greatest reduction in the residual sum of squares when included. If m variables have been included by the Efroymson's procedure these m variables would be $x_{(p)}, \ldots, x_{(p-m+1)}$ included in the same order. The residual sum of squares would be:

$$SS = y'y - \hat{\beta}_0^2 - S_p - S_{p-1} - \ldots - S_{p-m+1} .$$

The next inclusion candidate would be the one among the remaining $p - m$ variables which had the greatest square on the estimated regression coefficient, i.e. $x_{(p-m)}$ with the square S_{p-m}. Before $x_{(p-m)}$ is included we have:

$$SS_{\text{old}} = y'y - \hat{\beta}_0^2 - S_p - S_{p-1} - \ldots - S_{p-m+1} ,$$

The reduction $= S_{p-m}$.

After the inclusion of $x_{(p-m)}$ we have

$$SS_{\text{new}} = y'y - \hat{\beta}_0^2 - S_p - \ldots - S_{p-m+1} - S_{p-m} .$$

The inclusion takes place if the reduction divided by the new residual sum of squares is greater than a constant. Thus

$$\text{include } x_{(p-m)} \text{ iff } \frac{S_{p-m}}{y'y - \hat{\beta}_0^2 - S_p - \ldots - S_{p-m}} > \text{constant} .$$

If the opposite inequality holds $x_{(p-m)}$ is not included and the procedure is stopped.

We now define the statistics:

$$R_1^{(p)} = \nu \frac{S_1}{S}, \; R_j^{(p)} = (\nu + j - 1) \frac{S_j}{S + S_1 + \ldots + S_{j-1}}$$

$$(j = 2, \ldots, p) .$$

Then an equivalent criterion to that above is:

$$\text{include } x_{(p-m)} \text{ iff } R_{p-m}^{(p)} > \text{constant} .$$

The whole procedure of Efroymson may thus be written:

For $j = 1, 2, \ldots, p-1$ include the variables

$x_{(p)}, \ldots, x_{(p-j+1)}$ iff

$$(R^{(p)}_{p-j} \leq c_{p-j}) \cap \bigcap_{p-j+1}^{p} (R^{(p)}_i > c_i)$$

include all variables iff $\bigcap_{1}^{p} (R^{(p)}_i > c_i)$

where $\{c_i\}$ are predetermined constants.

In the original version of the procedure, the constants were chosen equal to a constant k which could be specified by the user. However, in computer programs of the procedure the constants $\{c_i\}$ are often chosen as percentage points in the Fisher distribution. In all cases, the choice of the constants determines the properties of the procedure. We shall here choose them such that the conditions mentioned earlier – that the probability is α_j of having exactly j false inclusions of variables in the case when $\beta_1, \ldots, \beta_p$ equal zero – is fulfilled.

From the description above we see that at least one variable is included if $R^{(p)}_p > c_p$. On the other hand

$$\mathsf{P}_0(\text{at least one variable is included}) =$$

$$= \sum_{j=1}^{p} \mathsf{P}_0(\text{exactly } j \text{ variables are included}) =$$

$$= \sum_{j=1}^{p} \alpha_j = \alpha, \text{ say}.$$

Thus the constant c_p is determined by

$$\mathsf{P}_0(R^{(p)}_p > c_p) = \alpha .$$

That is, c_p is the upper α point in the distribution of $R^{(p)}_p$ when $\beta_1 = \ldots = \beta_p = 0$. This distribution has been considered by Draper, Guttman & Kanemasu [2].

In order to determine the other constants, we use:

$$\mathsf{P}_0\left\{(R^{(p)}_{p-j} \leqslant c_{p-j}) \cap \bigcap_{p-j+1}^{p} (R^{(p)}_i > c_i)\right\} = \alpha_j$$

$$(j = 1, \ldots, p-1)\,.$$

We first determine c_{p-1} when c_p is known, then c_{p-2} when c_p and c_{p-1} are known, and so on until all constants are determined.

To compute these constant we need the joint distribution of $R^{(p)}_1, \ldots, R^{(p)}_p$ when $\beta_1 = \ldots = \beta_p = 0$. This distribution is found to be

$$p! \prod_{i=1}^{p} f(r_i;\, 1,\, \nu + i - 1)$$

where $r_1 > 0$, $r_{i+1} > \dfrac{(\nu + i)r_i}{\nu + i - 1 + r_i}$ $(i = 1, \ldots, p-1)$ and $f(z; 1, \nu)$ is the F-density with 1 and ν degrees of freedom. From this result the joint distribution of $R^{(p)}_i, \ldots, R^{(p)}_p$ is deduced for $i = 2, \ldots, p$. This distribution can be written in a special product form which under certain conditions makes the computation of the constants relatively easy. It may be shown that the constant c_j is then the upper γ_j-point in the distribution of $R^{(j)}_j$, where γ_j is a function of the α_j's and of c_p, $c_{p-1}, \ldots$ $\ldots, c_{j+1}$. We therefore may determine c_p from the distribution of $R^{(p)}_p$, then c_{p-1} from the distribution of $R^{(p-1)}_{p-1}$, and so on until the last constant c_1 is determined from the distribution of $R^{(1)}_1$.

4. THE PERFORMANCE OF THE PROCEDURE

The measure of performance which is used is the following: Suppose $\beta_1 = \ldots = \beta_m = 0$, $\beta_i \neq 0$ for $i = m+1, \ldots, p$. Then we have studied

$$\Delta(\beta) = \mathsf{P}\,\{(\text{do not include variables } x_i,\ i = 1, \ldots, m) \cap$$

$$\cap\ (\text{include variables } x_i,\ i = m+1, \ldots, p)\}\,.$$

For the procedure of Efroymson we have not been able to find an expression for $\Delta(\beta)$ suitable for numerical treatment. We have therefore estimated $\Delta(\beta)$ by Monte Carlo experiments and compared the performance of this procedure with that of a method based on the largest F-ratio.

CONCLUSIONS

The performance of the method based on the largest F-ratio on the whole has a somewhat more favourable behaviour than that of the modified Efroymson procedure. This is quite marked if one considers the probability of at least one correct inclusion and no false as a measure of preference. Efroymson's procedure has a tendency of making more erroneous inclusions.

In the cases $p = 2$, $\beta_1 = \beta_2 \neq 0$ and $p = 3$, $\beta_1 = \beta_2 = \beta_3 \neq 0$ where no erroneous conclusions are possible the Efroymson's procedure is preferable when ν is not too small. For small ν the Efroymson's procedure has the unfavourable property of not including variables at all when $\beta_1 = \ldots = \beta_p = b \to \infty$.

REFERENCES

[1] M.A. Efroymson, Multiple Regression Analysis, *Mathematical Methods for Digital Computers,* ed. by A. Ralston and H.S. Wilf, 191-203. Wiley, New York, 1962.

[2] N.R. Draper – I. Guttman – H. Kanemasu, The Distribution of certain Regression Statistics, *Biometrika,* 58 (1971), 295-298.

COLLOQUIA MATHEMATICA SOCIETATIS JÁNOS BOLYAI
9. EUROPEAN MEETING OF STATISTICIANS, BUDAPEST (HUNGARY), 1972.

SYSTEMATIZATION AND CODIFICATION OF THE EXPERIMENTAL DESIGNS – THE SURVEY OF THE WORKS OF SOVIET STATISTICIANS

V.V. NALIMOV

1. INTRODUCTION

We think that now in the experimental design there exist at least two basic problems. The first is the logic regulation of the whole diversity of the designs in practical use and the second is the creation of sufficiently universal numerical methods for constructing new designs. A great number of approaches to the problem of experimental design have called to life a lot of designs which are difficult to regard in the frame-work of some universal concept. It is clearly understood by all those who have to give consultations, to teach students, to write books... In our country a group of scientists connected with Laboratory of Statistical Methods of Moscow State University have begun to work on codification and systematization of the experimental designs.

But first of all I would like to make some retrospective remarks. Twelve years ago in our country we made first attempts to apply Box's methods (the designing of response surface) to the problems of chemistry, technical physics, metallurgy. The results proved to be excellent – the experimenters perfectly understood the logic of designing and from their own experience became aware of the high efficiency of those methods. Since that time experimental design has been widely used in different fields of experimental activities. But it proved to be too difficult to render Box's main ideas to critically-minded statisticians. And soon together with them we had to recognize certain logical inconsistency of the method.

In this method everything is safe enough for the linear designs. Indeed, if we deal with k independent variables, then in the case of the condition $(k + 1) \equiv 0 \pmod 4$ the linear designs are given by the Hadamard matrices – they are quadratic matrices X of the dimension $N \times N$, which consist of $+1$ and -1, satisfying the condition $X^*X = NI$. Hadamard matrices geometrically give the regular simplex, the vertices of which are a subset of the vertices of the cube, limiting the region of experiment. The Hadamard matrices can also be regarded as designs built on a sphere circumscribed around the cube under consideration. It is intuitively clear that the space of independent variables selected for an experiment is used here in the most optimal way. It is possible to prove rigorously that the regression coefficients θ_i are estimated here with the equal and minimum possible variances. The designs given by Hadamard matrices have some other pleasant properties. They are orthogonal: $\operatorname{cov}(\hat{\theta}_i,\hat{\theta}_j) = 0$; rotatable: the matrix X^*X proves to be invariant to the orthogonal rotation of the coordinate axes etc. But the things were much worse with regard to optimality of the second-order designs. Box and Hunter (1957) suggested that rotatability should be used as a criterion of optimality for second-order designs. It is not too easy to explain to a faultfinding opponent why it is good. Actually, rotatable second-order designs should be built on a ball inscribed into the cube, by which the boundaries of the experiment are set on. The angles of this cube remain unused and the regression coefficients are no longer estimated with minimum possible variance.

Concurrent with Box's approach, the American mathematician J. Kiefer developed the general theory of the experimental designs. In his works the D-optimum criterion plays the central role: the matrix of indedependent variables should be constructed so that the determinant of the information matrix $\frac{1}{N}X^*X$ was the maximum amongst the whole possible set of designs. In this case the ellipsoid of the variance of the estimators of the regression coefficients will have its minimal shape. This approach to the experimental design is natural generalization of the criterion of the jointly efficient estimates, introduced into statistics by R. Fisher. The linear designs given by Hadamard matrices proved D-optimal. However, D-optimal second-order designs find no practical application. Their realization demands too many experiments (for $k = 5$ more than 1000).

2. THE CONSTRUCTION OF OUASI-*D*-OPTIMUM DESIGNS BY ROUND-OFF OF THE CONTINUOUS DESIGNS

In our laboratory an attempt was made to construct almost saturated quasi-D-optimal designs. Proceeding from some geometric considerations there was made the round-off (to an integer) of the weights of the experimental points in Kiefer's continuous designs*. After that with the help of computer the value of the determinant of the matrix $\frac{1}{N}X^*X$ was calculated and the best design was chosen out to the design built by rounding-off. Here is one of the results: for $k = 5$ (the number of estimated parameters is 21) a quasi-D-optimal design Ki_{75} was found which can be compared with a D-optimal design and Box's design in the value of the determinant.

The design	The number of experimental points N	$\frac{1}{N}X^*X$
D-optimal design	–	$0{,}63 \cdot 10^{-6}$
Ki_{75}	52	$0{,}22 \cdot 10^{-6}$
Box's rotatable design	52	$0{,}28 \cdot 10^{-28}$
Box's rotatable design with semireplica	52	$0{,}55 \cdot 10^{-28}$

*In theoretical works the continuous designs ξ_k are built:

$$\begin{Bmatrix} p_1, \dots, p_k \\ x_1, \dots, x_k \end{Bmatrix}$$

where $\{x_1, \dots, x_k\}$ – the spector of designs, p_i – the weight of the ith point of design. $\sum p_i = 1$, p_iN is not obligatory an integer.

In the sense of D-optimality this design is much more useful than a corresponding Box's rotatable design. The efficiency increases here at the expense of the use of the angles of a multidimensional cube which gives the region of experiment.

Perhaps it is more rightful to consider Box's second-order rotatable designs as the designs constructed on a ball inscribed into the cube under consideration. Then we must estimate the value of the determinant of their information matrix as compared with the respective D-optimum design on a ball. It was found that when $\rho < 0{,}7$ the Box's design behaves better than the D-optimal one, but when $\rho \geqslant 0{,}7$ the situation becomes much worse. It is rather typical of minimax problems.

All that was mentioned above can be found in our article [1].

3. NEW DISCRETE D-OPTIMAL DESIGNS

The work on construction of the D-optimal and quasi-D-optimal second-order designs on a cube and on a ball was continued by L.L. Pesotchinsky [2].

The problem of constructing continuous D-optimal designs of the second order in a pure theoretical sense cannot be considered solved, for the number of points in the spectrum of the known D-optimal designs often exceeds the upper boundary H found by Kiefer and his colleagues. In this case it is reasonable to try to construct D-optimal design with the lesser number of points in the spectrum. In the paper by L.L. Pesotchinsky a new class of continuous D-optimal second-order designs on a cube is suggested which contains D-optimum designs with the number of points in the spectrum lesser than H for the dimensions $k \leqslant 10$. For the dimensions $k = 4, 5, 6$ continuous D-optimum designs were found with the minimal number of points in the spectrum. The results obtained are also used in construction of D-optimal continuous designs on a ball.

Further in this work the author tries to solve the problem of constructing quasi-D-optimal designs on a cube. These are designs containing a small number of observation and nearly D-optimal in their properties.

The designs obtained (DP-designs) have a little lesser number of observations than, for example, the previously built quasi D-optimal designs (Ki_{75}). Besides, they also have a great advantage as compared with previously built (by means both of numerical methods and simple round-off) quasi-D-optimal designs – these designs are symmetrical, i.e. the respective covariance matrix has a block structure and contains a great number of "zero" covariances of the estimators of parameters. Besides, the designs of such kind are asymptotically D-optimal, i.e. there characteristics converge to characteristics of continuous D-optimal designs with the increase of k.

4. OTHER CRITERIA

In spite of all its attraction, D-optimality criterion is not the only possible one. Below we give 22 criteria for designing the experiments (only in case of surface design)

1. D-optimality: $\max |\frac{1}{N} X^*X|$.

2. G-optimality: the minimum of maximal in the region of designing standardized variance of the model estimate.

3. The minimum of the average in the region of designing standardized variance of model estimate.

4. The minimum of maximal variance of the estimate of regression coefficients: $\min \max \sigma^2 \{\hat{\theta}_i\}$.

5. A-optimality: the minimum of the sum of squares of the principal semi-axes of the ellipsoid of the variance of the estimates of the regression coefficients (minimum of the trace of the covariance matrix of parameters estimates $(\frac{1}{N} X^*X)^{-1}$.

6. The minimum of the maximal axis of the ellipsoid of variances of the estimates of the regression coefficients.

7. Maximal precision of estimation of the coordinates of the extremum.

8. The minimal error under the condition that the response surface of degree $(d+1)$th is approximated by the polynomial of degree d.

9. The best possibility of estimating lack of fit by the presentation of the results of observation by the polynomial of a given degree.

10. The closeness of the number of observations N to the number of the estimated parameters m.

11. Rotatability: the invariance of the matrix X^*X to the orthogonal transformations of the coordinates.

12. Orthogonality: cov $\{\hat{\theta}_i \hat{\theta}_j\} = 0$.

13. The possibility of performing non-linear transformations of certain kind of independent variables preserving the optimality of the design.

14. The possibility of splitting the design into orthogonal blocks in order to eliminate uncontrolled time drift.

15. The compositivity of the design; the possibility of using the points of the design, built for the presentation of the results by the polynomial of degree d as a subset of points for an optimal design of the degree $d+1$; this problem may arise when the polynomial of the degree d inadequately represents the results of observations.

16. Insensitivity to rough errors in the results of observations.

17. Simplicity of calculation.

18. Insensitivity to errors in independent variables.

19. The possibility of estimating lack of fit at each step.

20. Clearness of the presentation of results.

21. Uniformity: variance of model estimate $d(x, \xi)$ close to the centre of the experiment should not depend on the distance when measured from the centre of experiment.

22. The minimum of sums of relative errors in the estimation of regression coefficients: $\min \sum_{i=1}^{m} \frac{\sigma(\hat{\theta}_i)}{\theta_i}$.

These criteria or at least part of them can be regarded as axioms and the respective designs as theorems. Here in most cases axioms prove incompatible. They do not form mathematical structures in the sense of Bourbaki. Instead of laconically formulated structures rich in their logical corollaries, here we deal with the mosaic axioms. And in this lies the difference between pure and applied mathematics. The only thing we can do is to perform numeric comparison of the whole possible variaty of designs.

5. CATALOGUES OF SECOND-ORDER DESIGNS

Our colleagues Mrs. T.I. Golikova, Mrs. L.A. Pantchenko, M.Z. Freedman [3] have performed such a comparison. It has been made on the basis of the catalogue of the second-order designs which they have composed. Almost for each dimension a design may be found which is symmetrical and has a small number of observations near to an optimal one in its characteristics; such a design also has a covariance matrix of blocked structure where most covariances are equal to zero (e.g., it is the design B for dimensions 3 and 4 on a cube, Hartley's design – which is also called "the design B with semireplica" – for dimension 5 on a cube. It is clear that it is not always preferable to use asymmetrical saturated designs close to D-optimal; it is reasonable only under very strict limitations imposed upon the number of observations.

The catalogue contains all known designs which are used to obtain the estimates of models which are second-order polynomials:

$$\mathrm{E}y = f^*(x)\theta \ ,$$

where $f^*(x) = (1, x_1, \ldots, x_k, x_1^2, \ldots, x_k^2, x_1, x_2, \ldots, x_{k-1}x_k)$, $\theta^* = (\theta_1, \ldots, \theta_m)$ – is the vector of unknown parameters. The regions where the observations may be taken are as follows:

1) a hypercube: $-1 \leqslant x_i \leqslant 1, \ i = 1, \ldots, k;$

2) a ball: $\sum_{i=1}^{k} x_i^2 = 1$.

For a cube: $k = 2 - 7$,

for a ball: $k = 2 - 8$.

For each design entering the catalogue the following characteristics are given.

1. Matrix of the design. If possible, a design is put down in a reduced form; e.g., if a design contains the complete factor experiment 2^k in the vertices of a cube, the star points $\alpha = 1$ and a zero point is written down as follows*

$$\left.\begin{matrix}1\\ \vdots\\ 8\end{matrix}\right\} \quad \text{CFE } 2^k; \quad b = 1,$$

$$\left.\begin{matrix}9\\ \vdots\\ 14\end{matrix}\right\} \quad \text{star points;} \quad \alpha = 1,$$

$$15 \quad \text{zero point.}$$

Some designs are put down in their complete form.

2. The determinant of a normed information matrix

$$|D(\xi)| = \left|\frac{1}{N}(X^*X)\right|$$

where X is the matrix of independent variables with the elements

$$x_{ij} = \| f_j(x_l) \| .$$

3. The average in the region normed variance of the model estimate (under the condition that the variance of one observation is $\sigma^2 = 1$)

$$d_{av} = \frac{\int_X d(x\ \xi)dx}{V_X} = \frac{\int_X f^x(x)D^{-1}(\xi)f(x)dx}{V_X}$$

where V_X is the volume of the region where the design is performed.

*CFE stands for "complete factor experiments".

4. The maximal eigenvalue λ_{max} of the normed covariance matrix $D^{-1}(\xi)$.

5. The trace of the normed covariance matrix.

6. For symmetrical designs – the maximal in the region normed variance of the model estimator

$$\max_{X} d(x, \xi) .$$

7. The value of variance $d(x, \xi)$ in the points of experiment.

8. Normed variances of the estimates of the model coefficients

$$N\sigma^2(\hat{\theta}_i) \qquad (i = 1, \ldots, k) .$$

9. Correlation matrix of parameters estimates. If possible, the matrix is put down in a reduced (abbreviated) form.

10. Some parameters which are necessary for calculating the parameters of a model without the help of computer.

11. For asymmetric designs – matrix $L = (X^*X)^{-1}X^*$ which can be used for calculating parameters of a model without the help of computer: $\hat{\theta} = LY$.

12. The projections of the function $d(x, \xi)$ into some two-dimensional planes.

13. For rotatable designs – the diagram of the function $d(x, \xi) = f(R, \xi)$ where R is the distance to the center of experiment.

14. For non-rotatable designs – the diagram of the functions $d(x, \xi)$ in the directions to the centers of cube faces of various dimensions; in this case we take the centers of those faces which include only zeroes and "plus ones" as coordinates. (Of course, it is not exhaustive characteristics for asymmetric designs).

The designs included in the catalogue:

1. D-optimal continuous designs.

2. Saturated precise *D*-optimal designs obtained by numerical methods (Box – Draper, Mrs. Dubova – Fedorov) (for dimension 2, 3, 4 on a cube).

3. Designs close to *D*-optimal in characteristics containing a small number of points (for dimensions 4, 5, 6 on a cube).

4. Box's rotatable designs.

5. Orthogonal designs.

6. Rechtshafner's saturated designs.

7. Designs which contain a complete factor experiment (or fractional replica) in the vertices of a cube or star points with an arm $\alpha = 1$ (designs of *B*-type).

8. Simplex-sum rotatable designs (symmetrical and asymmetrical ones.

9. Saturated simplex-sum designs.

10. Box – Benkens' three-levelled designs which are rotatable for dimensions 4 and 7.

11. Hartley's designs.

12. Westlake's designs.

13. Some other designs.

The data included in the catalogue give the opportunity to compare various designs from the viewpoint of various criteria of optimality and to estimate the advantages and the losses which we have when applying this or that design. As an example Table 1 unites some results for second-order design for $k = 5$.

There follow some conclusions drawn from the initial consideration of the catalogue data:

1. For designs on a cube characteristics differ considerably. E.g., on a cube it is evidently inconvenient to use Box's rotatable designs, simplex-

sum rotatable designs and orthogonal designs (for these designs the values of $|D(\xi)|$ are several orders less than those of the optimal design).

2. For designs on a ball characteristics of various designs are rather close to each other (with small exclusions).

3. Both for a cube and a ball symmetrical designs can be found close to optimal ones in all characteristics considered. Practically it seems more rational to oriente oneself in many cases at finding such compromise solutions [7].

6. SEQUENTIAL NUMERICAL METHODS OF CONSTRUCTING QUASI-*D*-OPTIMAL DESIGNS

Another problem which our colleagues tried to solve is the investigation of the sequential numerical methods of designing of saturated quasi-*D*-optimal designs. This work was made by Mrs. G.S. Dubova under the supervision of Dr. V.V. Fedorov [5].

It was already mentioned above that continuous *D*-optimal designs which are widely represented in literature, in practical use are to be rounded-off to the discrete ones which are close to precise *D*-optimal design in their characteristics. Such round-off is feasible when $N > m$ (m is the number of parameters, N is the total number of observations); however, it is difficult or even impossible when $N = m$.

A numerical method of constructing precise *D*-optimal designs was suggested and saturated $N = m$ *D*-optimal designs for a number of polynomial models have been built.

The principal idea of the method is based upon the fact that if two designs ξ_0 and ξ_1 differ only by one point of their spectra, then the determinants of their variance matrices are connected as follows:

$$|D^{-1}(\xi_1)| = \frac{|D^{-1}(\xi_0)|}{1 + \Delta(\overset{0}{x}_i, \overset{1}{x}_i)} \,.$$

Evidently, this plan may be improved if the value $\Delta(\overset{0}{x}_i, \overset{1}{x}_i) > 0$. From this an iteration procedure follows.

1. There is an arbitrary non-singular design ξ_s.

2. Points $\overset{s}{x}_i$ and $\overset{s+1}{x}_i$ are found satisfying

$$\max_i \max_{x \in X} \Delta(\overset{s}{x}_i, \overset{s+1}{x}_i) .$$

3. If the maximum obtained is a positive value, substitute $\overset{s+1}{x}_i$ for $\overset{s}{x}_i$ and obtain the design ξ_{s+1} instead of ξ_s. Then the procedure is to be iterated. The process of improving the design comes to an end when $\Delta > 0$ cannot be found. The detailed description of the algorithm can be found in the book by V. V. Fedorov [4]. Algol program and peculiarities of the operation of the algorithm and also the designs constructed and their characteristics can be found in the preprint of the Laboratory of Statistical methods of Moscow State University [5].

Some results are given as an example in Table 2. Geometric distribution of points for the case $k = 2$ is given in Fig. 1.

Neglectable modification of the program allows constructing compositional design. The necessity of it arises when we find that the polynomial of degree d proves inadequate. Then we must build a design of degree $d + 1$, using the design of degree d as a nucleus.

7. THE CONSTRUCTION OF WEIGHING *D*-OPTIMAL DESIGNS

The designs constructed are the designs of weighing on spring balance. *D*-optimal designs with the minimum number of observations are found in [6]. For even and odd dimensions k a type of information matrix of *D*-optimal designs is found. The following necessary conditions have been obtained for N observations in *D*-optimal designs.

For dimension $k = 3 \pmod 4$ $\quad N > k$.
For dimension $k = 1 \pmod 4$ $\quad N \geqslant 2k \ (k > 1)$.
For dimension $k = 0 \pmod 4$ $\quad N \geqslant 2k + 2$.
For dimension $k = 2 \pmod 4$ $\quad N \geqslant k + 1$.

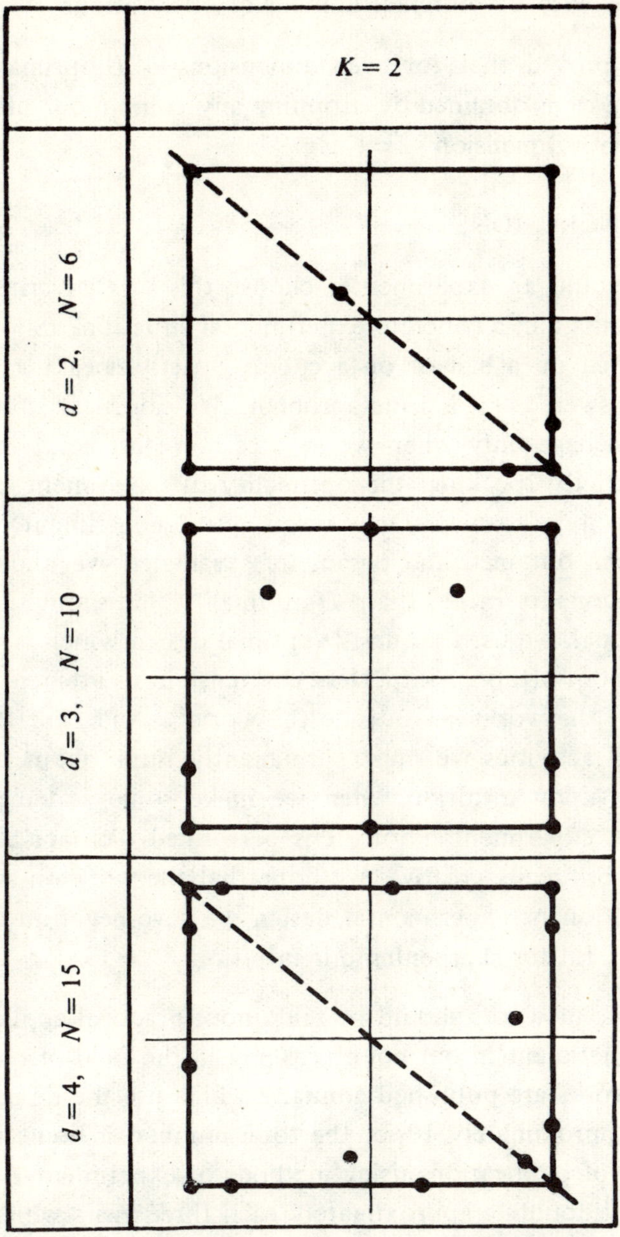

Figure 1.
D-optimal designs by Dubova

For odd dimensions the result is the algorithm of constructing D-optimal designs with the minimal number of observations N.

It was proved that for even dimensions a D-optimal design (with dimension $2k$) is obtained by dropping any column out of the matrix of the design with dimension $2k + 1$.

8. CONCLUSIONS

How should an experimenter choose this or that criterion for constructing a design in a concrete experimental situation? Should the design be constructed on a ball or on a cube? Strictly speaking, this question cannot be answered – it is a metaproblem. We can speak about something in a metalanguage only when we have certain statements in the objective language. Strictly speaking, the optimality of experiment can be judged only after it is performed. Once in an actual experiment only after the data had been obtained and parameters evaluated we understood that it was orthogonality rather than D-optimality that should have been our concern, though we used a quasi-D-optimal design which is very bad from the viewpoint of orthogonality. Here Wittgenstein's statement is appropriate: "The goal of the world lies outside the world". And nevertheless in our experimental activities we must permanently think about metaproblems relying only upon intuition when we make some statement about the results of an experiment before it is performed. Perhaps it is this that makes our work really creative. We hope that the work on systematization and codification of experimental design we have begun in our laboratory will prove useful for sharpening our intuition.

A couple of words should be said about practical application of the design of experiment in our country. Only in the field of chemical research about 500 works are published annually which use the design of experiment. It is approximately 1% of the total number of home publications. The number of publications using methods of experimental design grows exponentially doubling approximately each three years while the total number of publications doubles during 12-15 years.

TABLE 1.

CHARACTERISTICS OF SOME SECOND-ORDER DESIGNS THE DIMENSION OF THE SPACE OF INDEPENDENT VARIABLES BEING $k = 5$

Design	Dimension k	Number of observ. N	$\|D(\xi)\|$	d_{av}	λ_{max}	tr	d_{max}
D-optimal	5	–	$0.64 \cdot 10^{-6}$	15.5	20.96	81.38	21
B_5 (the design by Golikova)*	5	42	$0.68 \cdot 10^{-7}$	14.1	17.14	111.70	34.2
Ha_5 (Hartley's design)	5	27	$0.17 \cdot 10^{-7}$	11.0	11.04	83.33	27.4
Ki_{75} (rounded-off Kiefer's design)	5	52	$0.22 \cdot 10^{-6}$	18.1	24.21	96.03	34.5
DP_5 (Pesotchinsky's design)	5	50	$0.40 \cdot 10^{-6}$	14.4	27.0	74.1	24.1

* These designs have one observation in each vertex and in the centers of $(k - 1)$-dimensional facets

TABLE 2.

COMPLETE POLYNOMIAL REGRESSION ON A CUBE.
THE DETERMINANTS OF SATURATED DESIGNS COMPARED TO THOSE OF D-OPTIMAL CONTINUOUS DESIGNS
(known in literature)
(THE DETERMINANTS OF VARIANCE MATRICES ARE NORMED)

Dimension	The degree of polynomial	saturated $\|D^{-1}(\xi)\|$	continuous $\|D^{-1}(\xi)\|$
$k = 2$	$d = 2$	$0.174 \cdot 10^3$	$0.875 \cdot 10^2$
	$d = 3$	$0.167 \cdot 10^8$	
	$d = 4$	$0.532 \cdot 10^{17}$	
$k = 3$	$d = 2$	$0.539 \cdot 10^4$	$0.173 \cdot 10^4$
$k = 4$	$d = 2$	$0.297 \cdot 10^6$	$0.457 \cdot 10^5$

REFERENCES

[1] V.V. Nalimov – T.I. Golikova – N.G. Mikeshina, On practical use of the concept of *D*-optimality, *Technometrics,* 12 (1970), 799-815.

[2] L.L. Pesotchinsky, *D*-optimal and quasi-*D*-optimal precise designs for quadratic regression on a cube and on a ball, Preprint of Moscow University, No. 42, 1972.

[3] T.I. Golikova – L.A. Pantchenko – M.Z. Freedman, *Catalogue of second-order designs,* Moscow University Press (to be published in Russian); see also *New Ideas in Experimental Design,* "Nauka", 1969 (in Russian).

[4] V.V. Fedorov, *Theory of Optimal Experimentation,* Academic Press, 1972.

[5] I.S. Dubova – V.V. Fedorov, The Tables of Optimal Designs (Saturated *D*-optimal Designs on a Cube), Preprint of Moscow University (to be published, in Russian).

[6] V.Z. Brodsky – T.I. Golikova, The construction of *D*-optimal weighting designs with minimal number of observations, *Teorija Verojatn. i Primenen.*, 18 (1972), 578.

[7] T.I. Golikova – L.A. Panchenko, Systematization of desigs for estimating polynomial second-order models in optimal experimented design, (in Russian), Preprint of Moscow University, 1974.

COLLOQUIA MATHEMATICA SOCIETATIS JÁNOS BOLYAI
9. EUROPEAN MEETING OF STATISTICIANS, BUDAPEST (HUNGARY), 1972.

A k-FOLD QUASI-POISSON PROCESS

D. OAKES

1. INTRODUCTION

Let k be a positive integer. A point process P on the real line R will be said to satisfy the condition $C(k)$ if

(i) the respective counts (numbers of events) N_i $(i = 1, 2, \dots, k)$ of P in any k contiguous intervals $[a_{i-1}, a_i)$ of R are independently distributed, having a Poisson distribution with mean $a_i - a_{i-1}$, but

(ii) P is not a Poisson process.

It can be shown that no point process can simultaneously satisfy $C(k)$ for all k (cf. Daley and Vere-Jones [1], Theorem 2.5). Moran [3] and Goldman [2], quoting an example of L. Shepp, each exhibit processes which satisfy $C(1)$. Shepp modifies the property of the Poisson process which states that conditionally on the number of events occurring in a Borel subset A of R, their positions are independently and uniformly distributed over R. He does not consider the properties of the sequence of intervals (times between successive events) of his process. In contrast, Moran constructs an interval sequence which can be imbedded in R to give a stationary point process satisfying $C(1)$.

By generalizing Shepp's example, D. Szász [4] has obtained processes which satisfy $C(k)$ for an arbitrary fixed k. In this note a generalization of Moran's construction is outlined. This yields a new process P_k which also satisfies $C(k)$.

2. CONSTRUCTION OF THE PROCESS

Our construction proceeds in three steps. Details of these, and the proof that the process defined does satisfy $C(k)$ will be given elsewhere.

The first step is to construct a $(k+1)$-tuple $(X_1, X_2, \ldots, X_{k+1})$ of random variables such that

(a) each X_i $(i = 1, 2, \ldots, k+1)$ is exponentially distributed with unit mean,

(b) $(X_1, X_2, \ldots, X_k)$ are mutually independent, as are $(X_2, X_3, \ldots, X_{k+1})$,

(c) for each i $(1 \leqslant i \leqslant k)$ the k-tuples $(X_1, \ldots, X_{i-1}, X_i + X_{i+1}, X_{i+2}, \ldots, X_{k+1})$ have the joint distributions that they would have if $(X_1, X_2, \ldots, X_{k+1})$ were independent,

(d) the random variables $(X_1, X_2, \ldots, X_{k+1})$ are not mutually independent.

The second step is to construct an infinite sequence $\{X_i\colon i = 0, \pm 1, \pm 2, \ldots\}$ of random variables such that, for each integer r, the $(k+1)$-tuple $(X_{r(k+1)+1}, X_{r(k+1)+2}, \ldots, X_{r(k+2)})$ satisfies (a), (b), (c) and (d) above, and such that $(k+1)$-tuples corresponding to different values of r are mutually independent. This is to be the interval sequence of our process.

The final step is to imbed the interval sequence in R in a suitable way. This is achieved by constructing the stationary renewal process with interval sequence $\{Y_r\}$, where

$$Y_r = \sum_{i=1}^{k+1} X_{r(k+1)+i} \qquad (r = 0, \pm 1, \pm 2, \dots)$$

and interpolating.

It can be shown that the resulting point process P_k does satisfy $C(k)$ and thus may be called a "k-fold Quasi-Poisson process". Finally, we note that if $k \geqslant 2m - 1$ then the joint distributions of the counts of P_k. in any class of m intervals of R, not necessarily contiguous or disjoint, are the same as those of a Poisson process with unit rate.

REFERENCES

[1] D.J. Daley – D. Vere-Jones, A summary of the theory of point processes, *Stochastic Point Processes: Statistical Analysis, Theory and Application,* Wiley, New York, 1972.

[2] J.R. Goldman, Stochastic Point Processes and Limit Theorems, *Ann. Math. Statist.*, 38 (1967), 771-779.

[3] P.A.P. Moran, A non-Markovian Quasi-Poisson Process, *Studia Sci. Math. Hungar.*, 2 (1967), 425-429.

[4] D.O.H. Szász, Once more on the Poisson process, *Studia Sci. Math. Hungar.*, 5 (1970), 441-444.

COLLOQUIA MATHEMATICA SOCIETATIS JÁNOS BOLYAI
9. EUROPEAN MEETING OF STATISTICIANS, BUDAPEST (HUNGARY), 1972.

AN ESTIMATION FOR THE RENEWAL FUNCTION OF AN IFR-DISTRIBUTION

A. OBRETENOV

The distribution functions $F(x)$ for which the failure rate

$$r(t, x) = \frac{F(t + x) - F(t)}{1 - F(t)}$$

is an increasing function of t for any fixed $x \geq 0$ are called IFR distribution functions. They play an important role in many problems of the theory of reliability [1].

The paper deals with the estimation of the difference between the renewal function of an IFR distribution function and that of the exponential one. The following theorem is proved:

Theorem. *If* $F(x)$ *is an IFR distribution function and* $H(t)$ *is the renewal function of* $F(x)$, *then for all* $t \geq 0$

$$(1) \qquad 0 \leq \frac{t}{\mu_1} - H(t) \leq 1 - \frac{\alpha}{\mu_1}$$

where

$$\alpha = \lim_{n\to\infty} \frac{\mu_{n+1}}{(n+1)\mu_n}, \qquad \mu_n = \int_0^\infty x^n \, dF(x) .$$

The equality in (1) is reached by the exponential and the degenerate distributions.

In order to prove this theorem we shall establish a lemma.

Lemma. *If* $F(x)$ *is an IFR distribution function, then for every* $t \geqslant 0$ *and* $n = 1, 2, \ldots$ *we have the inequality*

$$\text{(2)} \qquad \lambda_n(t) = \frac{\bar{F}_{n-1}(t)}{\int_t^\infty \bar{F}_{n-1}(x)dx} \leqslant \alpha^{-1}$$

where $F_0 = F$, $\bar{F}_{n-1} = 1 - F_{n-1}$, μ_n *is the* n*-th moment of* $F(x)$ *and*

$$\text{(3)} \qquad F_n(x) = \frac{1}{\alpha_1^{(n-1)}} \int_0^x \bar{F}_{n-1}(t)dt, \qquad \alpha_1^{(n-1)} = \int_0^\infty \bar{F}_{n-1}(x)dx .$$

Proof. We know, if $F(x)$ is an IFR distribution function, then the function $F_1(x)$ defined by (3) and $n = 1$ is an IFR distribution function, too, further $\lambda_2 \geqslant \lambda_1$ and for the same reason we have $\lambda_3 \geqslant \lambda_2$, etc.

In general

$$\lambda_n(t) \geqslant \lambda_{n-1}(t) \geqslant \ldots \geqslant \lambda_1(t) ,$$

where every $\lambda_k(t)$ is a decreasing function of t, as the failure function of the IFR distribution function $F_{k-1}(t)$. The function $\lambda_n(t)$ represented by the corresponding iterated distribution is

$$\text{(2')} \qquad \lambda_n(t) = \frac{\bar{F}_{n-1}(t)}{\int_t^\infty \bar{F}_{n-1}(x)dx} .$$

According to a theorem (Theorem 6.2, [2]) the function $C_n(t) = F_n(\alpha_n t)$ tends to $1 - e^{-t}$, when $n \to \infty$. Beside that, if $\beta_n \to \beta$ then

$C_n(\beta_n t) \to 1 - e^{-\beta t}$. As $\alpha_n \geqslant \alpha_{n+1}$, then $\dfrac{\alpha_n t}{\alpha_{n+1}} \geqslant t$ and because of the monotony of $\lambda_n(t)$, we have

(4) $$\lambda_n\left(\frac{\alpha_n t}{\alpha_{n+1}}\right) \geqslant \lambda_n(t) \geqslant \lambda_1(t)\,.$$

Let us put $\beta_n = \dfrac{1}{\alpha_{n+1}}$, then $\bar{C}_n(\beta_n t) = \bar{F}_n\left(\dfrac{\alpha_n t}{\alpha_{n+1}}\right) \to e^{-\frac{t}{\alpha}}$ and as to the integral $\int\limits_{\frac{\alpha_n t}{\alpha_{n+1}}}^{\infty} \bar{F}_n(x)dx$ we have the limit

(5) $$\lim_{n\to\infty} \int_t^{\infty} \bar{C}_n(\beta_n u)dn = \int_t^{\infty} e^{-\frac{u}{\alpha}} du = \frac{1}{\alpha} e^{-\frac{t}{\alpha}}\,.$$

In (5) we have used a limit under the integral and now we must verify that. Being $F_n(t)$ an IFR distribution function $\bar{F}_n\left(\dfrac{\alpha_n t}{\alpha_{n+1}}\right) = 1 - C_n(\beta_n t)$ will be an IFR distribution function, too. According to a well-known property of the IFR distributions

(6) $$\bar{C}_n(\beta_n u) \leqslant [\bar{C}_n(\beta_n t)]^{\frac{u}{t}}$$

for every $u > t$.

Let t be fixed. Let us choose a number $\epsilon > 0$, such that $q = \epsilon + e^{-\frac{t}{u}} < 1$, then for every great enough n, we have $\bar{C}_n(\beta_n t) < e^{-\frac{t}{\alpha}} + \epsilon$ because of the limit relation. The inequality (6) implies

(7) $$\bar{C}_n(\beta_n u) < q^{\frac{u}{t}}$$

for every $n > n_0$ and for every u greater than t. As $q < 1$ (7) shows us that $\bar{C}_n(\beta_n u)$ are uniformly bounded in relation to n by an integrable function in $(0, \infty)$.

From (2′) and (5) it follows

$$\lim_{n\to\infty} \lambda_n \left(\frac{\alpha_n t}{\alpha_{n+1}}\right) = \alpha^{-1}$$

and if in the inequality (4) $n \to \infty$, we get $\lambda_1(t) \leqslant \frac{1}{\alpha}$. When we apply in (4) the inequality $\lambda_n \geqslant \lambda_k$ $(n > k)$, then the same inequalities remain true for every k.

The exponential distribution $1 - e^{-\frac{t}{\mu_1}}$ belongs to the IFR distributions as well as to the DFR distributions, because its failure function $\lambda(t) = \mu_1^{-1}$ is constant. In this sense it appears as a border between these two types of the distributions.

Now we prove the theorem.

Proof. As $F(x)$ is an IFR distribution function, the function $r(t, x)$ is increasing in t for fixed $x \geqslant 0$. Let N_t be the number of renewals in $[0, t)$ for a renewal process defined by $F(x)$. The random variable $\eta_t = t_{N_t+1} - t$, where t_{N_t+1} is the first failure after t, has the distribution function

$$\text{(8)} \qquad \mathrm{P}(\eta_t \leqslant x) = \int_{-0}^{t} [F(t-u+x) - F(t-u)]dH(u) + F(t+x) - F(t) .$$

From (8) and from the monotonicity of $r(t, x)$ in t, we receive

$$\text{(9)} \qquad \begin{aligned} &\mathrm{P}(\eta_t \leqslant x) \leqslant \\ &\leqslant r(t,x) \int_{-0}^{t} \bar{F}(t-u)dH(u) + r(t,x)\bar{F}(t) = r(t,x) . \end{aligned}$$

From the definition of the random variable η_t and from the Wald's equality we have

$$\mathrm{E}\eta_t = \mu_1[\mathrm{E}N_t + 1] - t$$

or

$$\text{(10)} \qquad \frac{t}{\mu_1} - H(t) = 1 - \mu_1^{-1}\mathrm{E}\eta_t, \qquad \mathrm{E}N_t = H(t) .$$

Now we estimate the mean value of η_t. Using the inequality (9) for $E\eta_t$, we receive an underbound

$$E\eta_t = \int_0^\infty 1 - P(\eta_t \leqslant x)dx \geqslant \int_0^\infty 1 - r(t, x)dx .$$

The integral on the right side of this inequality is

$$\int_0^\infty \bar{r}(t, x)dx = \frac{1}{\bar{F}(t)} \int_0^\infty \bar{F}(t) - F(t + x) + F(t)dx =$$

$$= \frac{1}{\bar{F}(t)} \int_t^\infty \bar{F}(x)dx ,$$

hence

(11) $$E\eta_t \geqslant \frac{\int_t^\infty \bar{F}(x)dx}{\bar{F}(t)}$$

and according to the proved inequality for the failure function of the iterated IFR distributions of F, the quotient on the right side of (11) surpasses α. Consequently

(12) $$E\eta_t \geqslant \alpha .$$

The inequality (12), used in (10) gives us the upper estimation we look for. That the under estimation is zero is seen from (10), because $E\eta_t \leqslant \mu_1$ as the mean of the rest lifetime after the moment t (see [1]).

REFERENCES

[1] R.E. Barlow – F. Proschan, *Mathematical Theory of Reliability,* John Wiley and Sons, I uc., New York, London, Sydney, 1965.

[2] W.L. Harkness – R. Shantaram, Convergence of a sequence of transformations of distribution functions, *Pacif. Journ. of Math.,* 31 (1969), 403-415.

COLLOQUIA MATHEMATICA SOCIETATIS JÁNOS BOLYAI
9. EUROPEAN MEETING OF STATISTICIANS, BUDAPEST (HUNGARY), 1972.

AN INFORMATION-TYPE MEASURE OF DIFFERENCE OF PROBABILITY DISTRIBUTIONS BASED ON TESTING STATISTICAL HYPOTHESES

F. ÖSTERREICHER

§1. INTRODUCTION

Let P_0, P_1 be two probability measures on a measurable space $(\Omega, \mathfrak{A})$. The fundamental lemma of Neyman – Pearson implies, that for each $\alpha \in [0, 1]$ there exists a best test φ_α according to the test-problem (α, P_0, P_1).

The function $i_{(P_0, P_1)}$ defined by

$$i_{(P_0, P_1)}(\alpha) := \int_\Omega \varphi_\alpha \, dP_1$$

will play an important part in the following considerations. Some properties of this function are summarized in

Lemma 1.1. $i_{(P_0, P_1)}$ *is concave, increasing and continuous and it holds*

$$\alpha \leqslant i_{(P_0, P_1)}(\alpha) \leqslant 1 \quad \forall \alpha \in [0, 1]$$

where

$$\alpha = i_{(P_0, P_1)}(\alpha) \ \forall \alpha \in [0, 1] \quad \textit{iff} \quad P_0 = P_1$$

and

$$1 = i_{(P_0, P_1)}(\alpha) \ \forall \alpha \in [0, 1] \quad \textit{iff} \quad P_0 \perp P_1 .$$

Because of these properties one can suppose, that the function

$$i_{(P_0, P_1)}(\alpha) - \alpha$$

is suitable to define measures of difference of probability distributions. One of those is

$$\max_{\alpha \in [0, 1]} (i_{(P_0, P_1)}(\alpha) - \alpha)$$

which may be regarded as f-divergence* with non-strict convex function $f(u) = \frac{1}{2} | 1 - u |$ because of

Lemma 1.2.

$$\max_{\alpha \in [0, 1]} (i_{(P_0, P_1)}(\alpha) - \alpha) = \frac{1}{2} | P_1 - P_0 |$$

where $| P_1 - P_0 |$ *denotes the variation distance of* P_0 *and* P_1.

Proof. Because of

$$\max_{\alpha \in [0, 1]} (i_{(P_0, P_1)}(\alpha) - \alpha) =$$

$$= i_{(P_0, P_1)}(P_0 \{p_1 > p_0\}) - P_0 \{p_1 > p_0\}$$

*f-divergences had been introduced by I. Csiszár in [6]. The following definition is unessentially modified.

Definition. The quantity $I_f(P_0, P_1) := \int_\Omega p_1 f \frac{p_0}{p_1} d\lambda - f(1)$ is called f-divergence of P_0 and P_1, where f: $[0, \infty] \to R$ is a convex function.

it holds

$$\max_{\alpha\in[0,1]} (i_{(P_0,P_1)}(\alpha) - \alpha) =$$

$$= \int_{\{p_1>p_0\}} dP_1 - \int_{\{p_1>p_0\}} dP_0 = \frac{1}{2}\int_{\Omega} |p_1 - p_0|\, d\lambda$$

where p_i $(i = 0,1)$ is the Radon – Nykodym-derivative of P_i with respect to a probability measure λ, which is dominating P_0 and P_1.

But because strict convexity of f has to be assumed for "good" comparison of experiments (using f-divergences), $\max_{\alpha\in[0,1]} (i_{(P_0,P_1)}(\alpha) - \alpha)$ is not in every relation an appropriate measure of difference of probability distributions.

§2. DEFINITION AND PROPERTIES OF TEST-DIVERGENCE

Definition 2.1. The quantity

$$I_T(P_0, P_1) := \int_0^1 (i_{(P_0,P_1)}(\alpha) - \alpha)\,d\alpha$$

is called Test-divergence of the probability distributions P_0 and P_1.

Remark. The intuitive background for the meaning of the Test-divergence lies in the fact, that the Test-divergence is the half volume of the Risk-set of the decision-problem given by the special case of testing simple hypotheses.

Some of the properties of the Test-divergence are summarized in

Theorem 2.1.

1) $0 \leq I_T(P_0, P_1) \leq \frac{1}{2}$

where

$$0 = I_T(P_0, P_1) \quad \textit{iff} \quad P_0 = P_1$$

and

$$\frac{1}{2} = I_T(\mathsf{P}_0, \mathsf{P}_1) \quad \textit{iff} \quad \mathsf{P}_0 \perp \mathsf{P}_1 .$$

2) I_T *is symmetrical*

3) $\frac{1}{4}|\mathsf{P}_1 - \mathsf{P}_0| \leq I_T(\mathsf{P}_0, \mathsf{P}_1) \leq \frac{1}{2}|\mathsf{P}_1 - \mathsf{P}_0|$.

Proof. All points can be verified by properties of $i_{(\mathsf{P}_0, \mathsf{P}_1)}$. For 3) concavity is essential:*

$$\alpha \cdot \frac{\mathsf{P}_1\{p_1 > p_0\}}{\mathsf{P}_0\{p_1 > p_0\}} 1_{[0, \mathsf{P}_0\{p_1 > p_0\})}(\alpha) + (\mathsf{P}_1\{p_1 > p_0\} +$$

$$+ \frac{\mathsf{P}_1\{p_1 \leq p_0\}}{\mathsf{P}_0\{p_1 \leq p_0\}}(\alpha - \mathsf{P}_0\{p_1 > p_0\}))1_{[\mathsf{P}_0\{p_1 > p_0\}, 1]}(\alpha) -$$

$$- \alpha \leq i_{(\mathsf{P}_0, \mathsf{P}_1)}(\alpha) - \alpha \leq \mathsf{P}_1\{p_1 > p_0\} - \mathsf{P}_0\{p_1 > p_0\} .$$

A relation between f-divergences and Test-divergence, which follows from Theorem 3.1 in [4], Theorem 2 in [5] and Theorem 2.1 (3), is established in

Theorem 2.2. *Let* $f\colon [0, \infty] \to \bar{R}$ *be a convex function, such that* f *is strict convex at the point* 1 *and* $f(0)$ *and* $\lim_{u \to \infty} \frac{f(u)}{u}$ *are finite. Then there exist a real* $C_0 > 0$, *depending only on* f, *and a non-negative function* $\psi\colon [0, \infty) \to [0, \infty]$, *also depending only on* f, *with* $\psi(0) = \lim_{u \to 0} \psi(u) = 0$, *such that*

$$C_0(I_f(\mathsf{P}_0, \mathsf{P}_1))^2 \leq I_T(\mathsf{P}_0, \mathsf{P}_1) \leq \psi(I_f(\mathsf{P}_0, \mathsf{P}_1)) .$$

*1_A denotes the indicator function of the set A.

§3. ANALOGIES BETWEEN f-DIVERGENCES AND TEST-DIVERGENCE

A well-known result is

Theorem 3.1. *Let* $(\mathfrak{B}_n,\ n \in N)$ *be an increasing sequence of sub-sigmaalgebras of* $\mathfrak{A}$*, such that* $\mathfrak{A}$ *is generated by* $(\mathfrak{B}_n,\ n \in N)$ *and let* $f\colon [0, \infty] \to \bar{R}$ *be a convex function, such that* $I_f(\mathsf{P}_0, \mathsf{P}_1) < \infty$. *Then*

$$I_f(\mathsf{P}_0^{\mathfrak{B}_n}, \mathsf{P}_1^{\mathfrak{B}_n}) \uparrow I_f(\mathsf{P}_0, \mathsf{P}_1)$$

(at which $\mathsf{P}_i^{\mathfrak{B}_n}$ *is the restriction of* P_i *to* $\mathfrak{B}_n$; $i \in \{0, 1\}$, $n \in N$*).*

The analogue statement for the Test-divergence is

Theorem 3.2. *Let* $(\mathfrak{B}_n,\ n \in N)$ *be an increasing sequence of sub-sigmaalgebras of* $\mathfrak{A}$*, such that* $\mathfrak{A}$ *is generated by* $(\mathfrak{B}_n,\ n \in N)$. *Then*

$$I_T(\mathsf{P}_0^{\mathfrak{B}_n}, \mathsf{P}_1^{\mathfrak{B}_n}) \uparrow I_T(\mathsf{P}_0, \mathsf{P}_1)\,.$$

Proof. Let be $\alpha \in (0, 1)$. Because of the fundamental lemma of Neyman – Pearson there exist best tests φ_α; $(\varphi^{\mathfrak{B}_n})_\alpha$, $n \in N$ for the test-problems $(\alpha, \mathsf{P}_0, \mathsf{P}_1)$; $(\alpha, \mathsf{P}_0^{\mathfrak{B}_n}, \mathsf{P}_1^{\mathfrak{B}_n})$, $n \in N$. Consider those versions $\varphi_n := \mathsf{E}_\lambda^{\mathfrak{B}_n} \varphi_\alpha$ of the conditional expectation of φ_α with respect to a probability measure λ, which is equivalent to the pair $(\mathsf{P}_0, \mathsf{P}_1)$, for which

(1) $\quad 0 \leqslant \varphi_n \leqslant 1 \qquad \forall n \in N\,.$

The martingale theorem implies, that

(2) $\quad \varphi_n \to \varphi_\alpha \quad [\lambda]\text{-a.s.}$

Because of (1), (2) and the convergence theorem of Lebesgue it holds

1) $$\lim_{n\to\infty} \mathsf{E}_{\mathsf{P}_0}\varphi_n = \lim_{n\to\infty} \int \varphi_n p_0 \, d\lambda = \int \lim_{n\to\infty} \varphi_n p_0 \, d\lambda =$$
$$= \varphi_\alpha p_0 \, d\lambda = \mathsf{E}_{\mathsf{P}_0}\varphi_\alpha = \alpha$$

2) $$\lim_{n\to\infty} \mathsf{E}_{\mathsf{P}_1}\varphi_n = \mathsf{E}_{\mathsf{P}_1}\varphi_\alpha = i_{(\mathsf{P}_0, \mathsf{P}_1)}(\alpha)$$

and with the notations

$$\alpha_n := \mathsf{E}_{\mathsf{P}_0} \varphi_n \quad \text{and} \quad i_n(\alpha) := \mathsf{E}_{\mathsf{P}_1} \varphi_n$$

therefore

(3) $\alpha_n \to \alpha$

(4) $i_n(\alpha) \to i_{(\mathsf{P}_0, \mathsf{P}_1)}(\alpha)$.

From (1) and $\mathsf{E}_{\mathsf{P}_0} \varphi_n = \mathsf{E}_{\mathsf{P}_0}^{\mathfrak{B}_n} \varphi_n = \alpha_n$ it follows, that φ_n is a test for the test-problem $(\alpha_n, \mathsf{P}_0^{\mathfrak{B}_n}, \mathsf{P}_1^{\mathfrak{B}_n})$. Therefore the following relations hold

(5) $i_n(\alpha) \leqslant i_{(\mathsf{P}_0^{\mathfrak{B}_n}, \mathsf{P}_1^{\mathfrak{B}_n})}(\alpha_n) \quad \forall n \in N$.

It can be easily shown, that

(6) $i_{(\mathsf{P}_0^{\mathfrak{B}_n}, \mathsf{P}_1^{\mathfrak{B}_n})}(a) \leqslant i_{(\mathsf{P}_0^{\mathfrak{B}_{n+1}}, \mathsf{P}_1^{\mathfrak{B}_{n+1}})}(a) \leqslant i_{(\mathsf{P}_0, \mathsf{P}_1)}(\alpha) \quad \forall n \in N$.

(3), (4), (5), (6) and the continuity of $i_{(\mathsf{P}_0, \mathsf{P}_1)}$ imply

(7) $i_{(\mathsf{P}_0^{\mathfrak{B}_n}, \mathsf{P}_1^{\mathfrak{B}_n})}(a_n) \to i_{(\mathsf{P}_0, \mathsf{P}_1)}(a)$.

With (3), (7) and Lemma 2.1 in [3] it follows

$$i_{(\mathsf{P}_0^{\mathfrak{B}_n}, \mathsf{P}_1^{\mathfrak{B}_n})}(a) \to i_{(\mathsf{P}_0, \mathsf{P}_1)}(a)$$

and because of (6) even

$$i_{(\mathsf{P}_0^{\mathfrak{B}_n}, \mathsf{P}_1^{\mathfrak{B}_n})}(a) \uparrow i_{(\mathsf{P}_0, \mathsf{P}_1)}(a) .$$

Another application of the convergence theorem of Lebesgue completes the proof.

Another wellknown theorem is

Theorem 3.3.* *Let* $\mathfrak{B}$ *be a subsigmaalgebra of* $\mathfrak{A}$ *and* f *a strict convex function, such that* $I_f(\mathsf{P}_0, \mathsf{P}_1) < \infty$. *Then it holds*

*See e.g. Theorem 1 in [6] and Lemma 2.3 in [4].

$$I_f(\mathsf{P}_0, \mathsf{P}_1) \geq I_f(\mathsf{P}_0^{\mathfrak{B}}, \mathsf{P}_1^{\mathfrak{B}})$$

with equality iff $\mathfrak{B}$ *is sufficient with respect to the pair* $(\mathsf{P}_0, \mathsf{P}_1)$.

The proof of the analogue theorem for the Test-divergence is not quite trivial.

Theorem 3.4. *Let* $\mathfrak{B}$ *be a subsigmaalgebra of* $\mathfrak{A}$. *Then it holds*

$$I_T(\mathsf{P}_0, \mathsf{P}_1) \geq I_T(\mathsf{P}_0^{\mathfrak{B}}, \mathsf{P}_1^{\mathfrak{B}})$$

with equality iff $\mathfrak{B}$ *is sufficient with respect to the pair* $(\mathsf{P}_0, \mathsf{P}_1)$.

Because of these large analogies between f-divergences and Test-divergence one could suppose, that there exists a convex function f, such that

$$I_f(\mathsf{P}_0, \mathsf{P}_1) = I_T(\mathsf{P}_0, \mathsf{P}_1) \qquad \forall (\Omega, \mathfrak{A});\ \mathsf{P}_0, \mathsf{P}_1$$

but the contrary can be shown.

Theorem 3.5. *There exists no convex function* f, *such that*

$$I_f(\mathsf{P}_0, \mathsf{P}_1) = I_T(P_0, \mathsf{P}_1)$$

holds for all $(\Omega, \mathfrak{A})$; $\mathsf{P}_0, \mathsf{P}_1$.

REFERENCES

[1] J. Neveu, *Mathematical Foundations of the Calculus of Probability*, Holden – Day, Inc., San Francisco – London – Amsterdam, 1965.

[2] L. Schmetterer, *Mathematische Statistik*, Springer Verlag, Wien – New York, 1966.

[3] W. Sendler, Einige masstheoretische Sätze bei der Behandlung trennscharfer Tests, *Z. Wahrscheinlichkeitstheorie verw. Geb.*, 18 (1971), 183-196.

[4] I. Csiszár, Information-type measures of difference of probability distributions and indirect observations, *Studia Sci. Math. Hungar.*, 2 (1967), 299-318.

[5] I. Csiszár, On topological properties of *f*-divergences, *Studia Sci. Math. Hungar.*, 2 (1967), 329-339.

[6] I. Csiszár, Eine Informationstheoretische Ungleichung und ihre Anwendung auf den Beweis der Ergodizität von Markoffschen Ketten, *Publ. Math. Inst. Hung. Acad.*, 8 (1963), 85-107.

[7] T. Ferguson, *Mathematical Statistics. A decision theoretical approach,* Academic Press, New York – London, 1967.

COLLOQUIA MATHEMATICA SOCIETATIS JÁNOS BOLYAI
9. EUROPEAN MEETING OF STATISTICIANS, BUDAPEST (HUNGARY), 1972.

STATISTICAL STUDIES OF THE COSTS OF SIX-MAN VERSUS TWELVE-MAN JURIES

W.R. PABST, JR.

During the past year, the six-man jury has come to replace the traditional twelve-man jury at least in civil trials in the United States District Courts as a result of the landmark descision of the United States Supreme Court in Williams v. Florida, 399 U. S. 58 (1970). This paper deals with the possible saving in jury time (and costs) as a result of this reduction. Reducing jury size from twelve to six does not mean a corresponding reduction in total juror time because of the heavy overhead time involved in jury selection.

Many statistical studies are concerned with other aspects of the reduction in jury size. By using traditional binomial sampling theory, D.F. Walbert* concludes that the probability of conviction may be higher

*David F. Walbert, "The Effect to Jury Size on the Probability of Conviction: An Evaluation of Williams v. Florida," 22 *Case Western Reserve Law Review* 529, 1971.

for "weak" cases and lower for "strong" cases with the six-man jury. H. Friedman* also resorts to the sampling operating characteristic curves in showing the possible effect of the change in jury size and also the possible lessening of the unanimity requirement. In bitterly opposing the reduction, Zeisel** cites the lower probability of including minority groups on the six-man jury. All these studies assume that there will be some monetary and manpower savings as a result of this change. This paper studies this aspect of the change.

Information is available from other court studies*** on the amount of time involved in those civil cases just before and just after the change took place in the actual cases of the United States District Court for the District of Columbia. During the first half of 1971, some 69 civil cases were tried with a twelve-man jury, while during the latter half, 78 cases with a six-man jury. The mix of cases, about half involving suits arising from automobile accidents and the rest mostly personal injury suits, was about the same during the two periods of time.

Two steps are involved in the selection of a jury: the first is the holding of the voir dire panel ("the time to see and to speak out") during which baised or unqualified jurors may be disqualified by the attorneys of either side; the second is the selection of the jury from the unchallenged panelists.

The appended charts show the frequency distribution of the times required for the voir dire panels and for the trials, and the numbers of people required in the panels, the numbers challenged, and those not used. From these, it is possible to make some estimates of the relative costs involving the two situations.

*Herbert Friedman, "Trial by Jury: Criteria for Convictions, Jury Size and Type I and Type II Errors," *The American Statistician,* Vol. 26, No. 2, April 1972.

**Hans Zeisel, "The Waning of the American Jury," *American Bar Association Journal,* April 1972.

***William R. Pabst, Jr., "A sutdy of Juror Waiting Time Reduction," prepared for the Law Enforcement Assistance Administration, United States Department of Justice, May 31, 1971; and "An End to Juror Waiting," *Judicature,* Vol. 55, No. 7, March 1972.

The information on times shows that there was almost no difference between the voir dire panel and trial times as between the six-man and the twelve-man juries, as follows:

Type of Jury	Number of Panels	Average Voir Dire Panel Times	Number of Trials	Average Trial Times
Six-Man	78	52.0 min.	71	7.80 hours
Twelve-Man	69	51.1 min.	66	7.80 hours

Several long lasting panels and long lasting trials, outliers from the distributions, were excluded from these averages to avoid distortion. Obviously, the difference in times between the two types of juries is not significant.

The information on the number of people used in the voir dire panels is given below:

Type of Jury	Number of Panels	Average Number		
		Per Panel	Challenged	Not Used
Six-Man	78	21.67	6.46	6.94
Twelve-Man	69	27.54	7.36	6.68

The average number per panel changed from 27.54 to 21.67, a reduction of only 21 %, despite the fact that only about half as many were to be selected. In practice, the number selected for the six-man jury was either seven or eight (six, plus one or two alternates), divided about half and half, and the number selected for the twelve-man jury was almost always 14 (twelve, plus two alternates). The alternates hear the case with the members of the jury, but do not participate in the jury deliberation unless they replace a regular juror during the course of the trial.

The surprising thing is that the number of challenges is almost as large for the six-man jury as for the twelve-man jury. Making challenges of prospective jurors is part of the initial maneuvering of the trial lawyer, and it does appear that the number of challenges is independent of the percentage of people to be selected. The number not used on the average in the panels was also about the same for both types of juries. The safety margin is thus apparently geared to the number of challenges rather than to the percentage of jurors to be selected from the panel.

The overall direct saving in manhours for the six-man jury can thus be structured by multiplying the average panel size by panel times plus the jury size by trial time, and comparing the two situations, as follows:

Six-Man:

$$21.67 \text{ men (panel)} \times 52.0 \text{ min.} +$$
$$+ 7.5 \text{ men (jury)} \times 7.80 \text{ hours} = 18.8 + 58.5 = 77.3 \text{ hours}$$

Twelve-Man:

$$27.54 \text{ men (panel)} \times 52.1 \text{ min.} +$$
$$+ 14.0 \text{ men (jury)} \times 7.80 \text{ hours} = 23.9 + 109.2 =$$
$$= 133.1 \text{ hours}$$

This shows a saving in direct manhours per trial of 55.8 hours, or about 41.9%.

This saving of nearly 42% in direct labor is substantial, but the nature of the juror selection system is such that the overall savings may be more directly related to the size of the panels than to the reduction in direct juror hours. The possible explanation of this is that the size of the juror call-in from day to day is dictated largely by the size of the panels and by the daily peaks generated when several panels are called simultaneously. Were these 147 civil cases the only ones heard by the court, this indirect factor could be calculated and total savings in manpower made precise. But, in addition to the 147 civil cases, the court also heard 500 criminal cases in which the same judges and the same jurors participated. Thus it is not possible to sort out completely the effect on overhead that the potential savings from the reduction to six-man civil juries might have had.

Since the change to the six-man civil juries occurred in June, general information comparing the first and second halves of the year throws some light on what happened. The following table collects pertinent information.

Average Numbers	1st Half 1971	2nd Half 1971	Reduction	Percentage Reduction
Daily Call-In	120	105	15	12.5
Carry Over (in trials)	70	60	10	14.1
Jurors Available	190	165	25	13.4
Daily Peaks Usage	132	111	21	16.1
Peak Usage/ Jurors Available	67%	61%		

Since the first half of 1971 used twelve-man civil juries, and the second half used the six-man juries, the reductions between the halves shown above are in the right direction, but are larger than might be expected. Since the 147 civil cases is about 22.7% of the total number of civil and criminal cases, and since the direct saving potential in civil cases was found above to be 41.9%, the overall saving that might be expected for the court from a full measure of civil direct savings would be about 9.5%.

The fact that the reduction in the number of jurors called in, and the reduction in the daily peak usage, were both much larger than 9.5%, suggests that other management changes were taking place in the court between the first and second halves of 1971. The reduction in daily peak usage of 16.1% suggests that it was able to get almost as much reduction in juror needs from better scheduling as from the switch to the six-man jury in civil cases.

The only blemish on the general record was that the Court was able to reduce the number of jurors available by only 13.4%, not as much as the percentage reduction of peak usage. Thus the utilization ratio, that is, the peak usage to jurors available, decreased from 0.67 to 0.61. The supply of jurors called in was, on the average, not decreased as much as the demand.

CONCLUSIONS

The change to a six-man jury in civil cases made possible a direct saving in juror time (and costs) of about 42%, despite the fact that the reduction in panel size was only 21%.

Although hidden by noncomparable elements, the direct savings possible appeared to be realized in the United States District Court for the District of Columbia. In fact, the larger than expected savings in overall juror time for both civil and criminal cases reflect the contention of many that better management can be more effective in reducing juror costs than in reducing the size of juries. Better management practices were certainly instituted during this time.

The reduction in juror time with the six-man jury does not reduce the judge time, the lawyer or witness time, for the time involved in the voir dire panels and trials was almost exactly the same whether the six-man or twelve-man jury was used.

The reduction in jury size could therefore have little effect on court delay.

Whether the legal consequences of a six-man jury are compensated

Whether the legal consequences of a six-man jury are compensated by the reduction in the number of jurors, as the references strongly doubt, the direct potential savings of this small part of overall court costs are appreciable.

VOIR DIRE TIMES

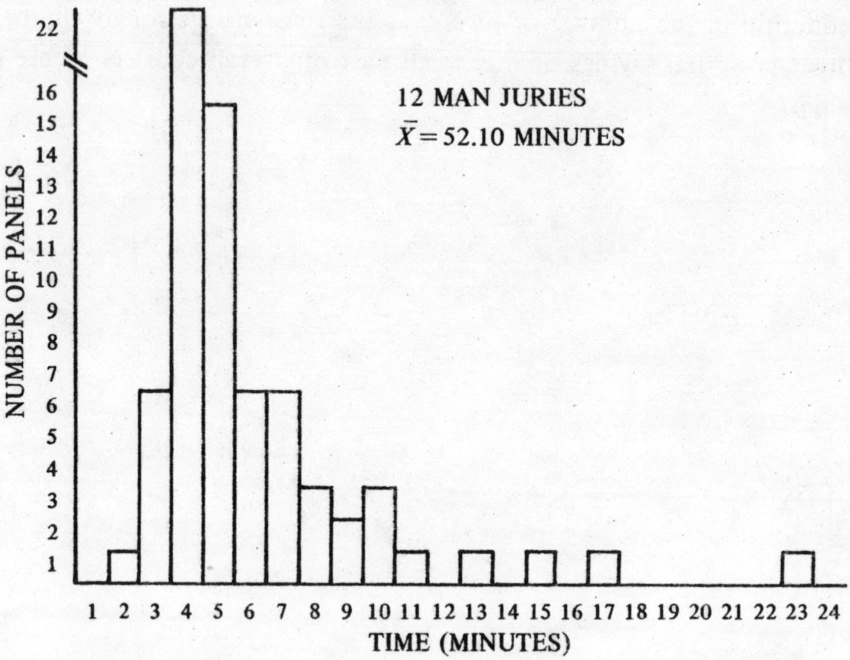

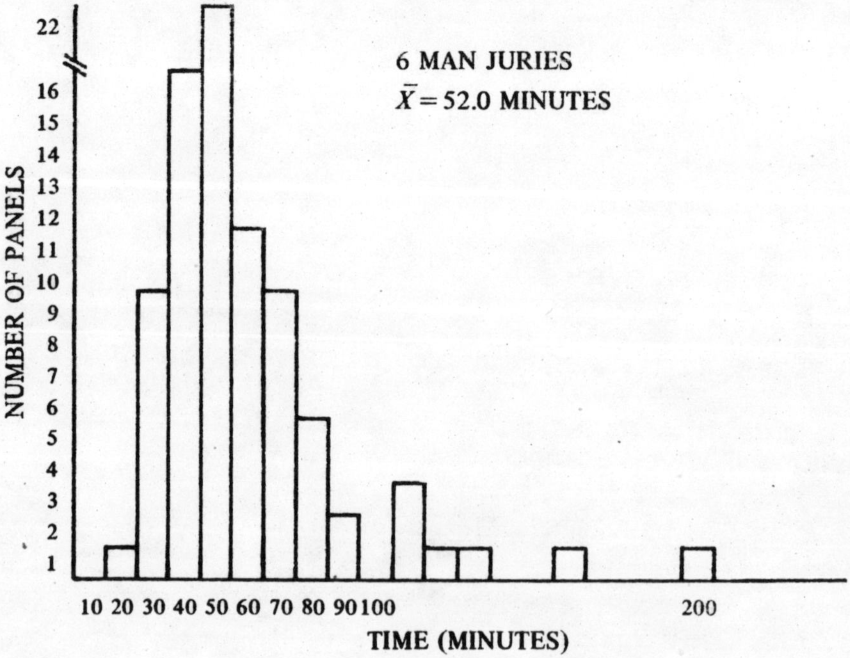

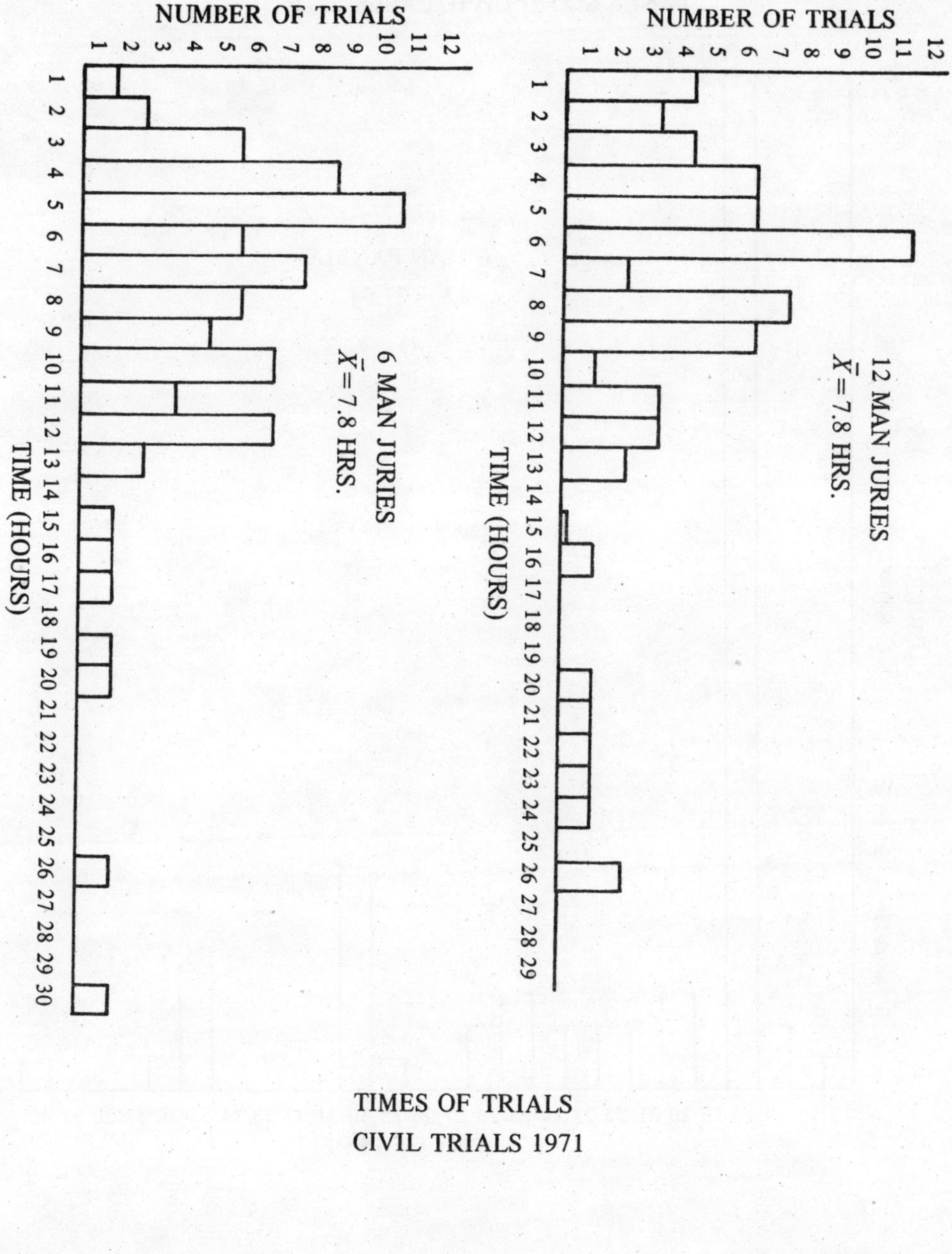

TIMES OF TRIALS
CIVIL TRIALS 1971

PANEL SIZES CIVIL CASES 1971

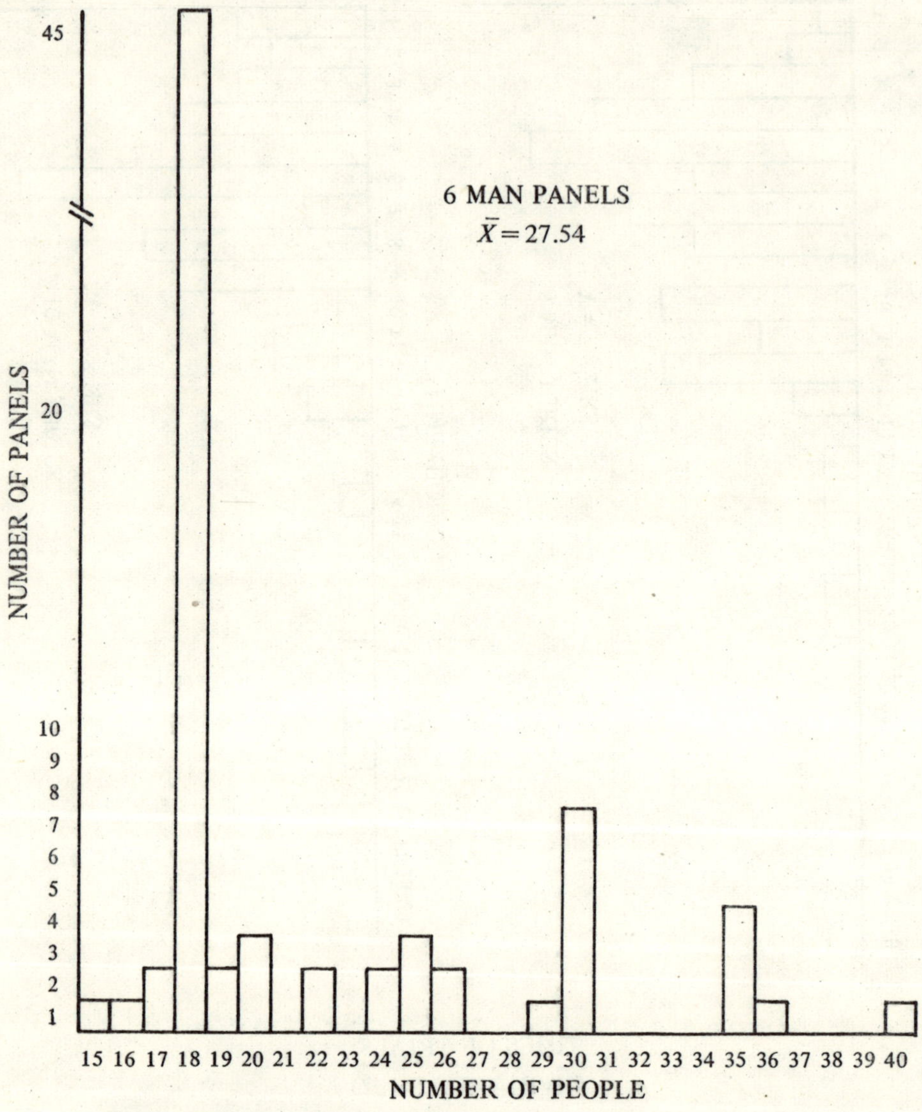

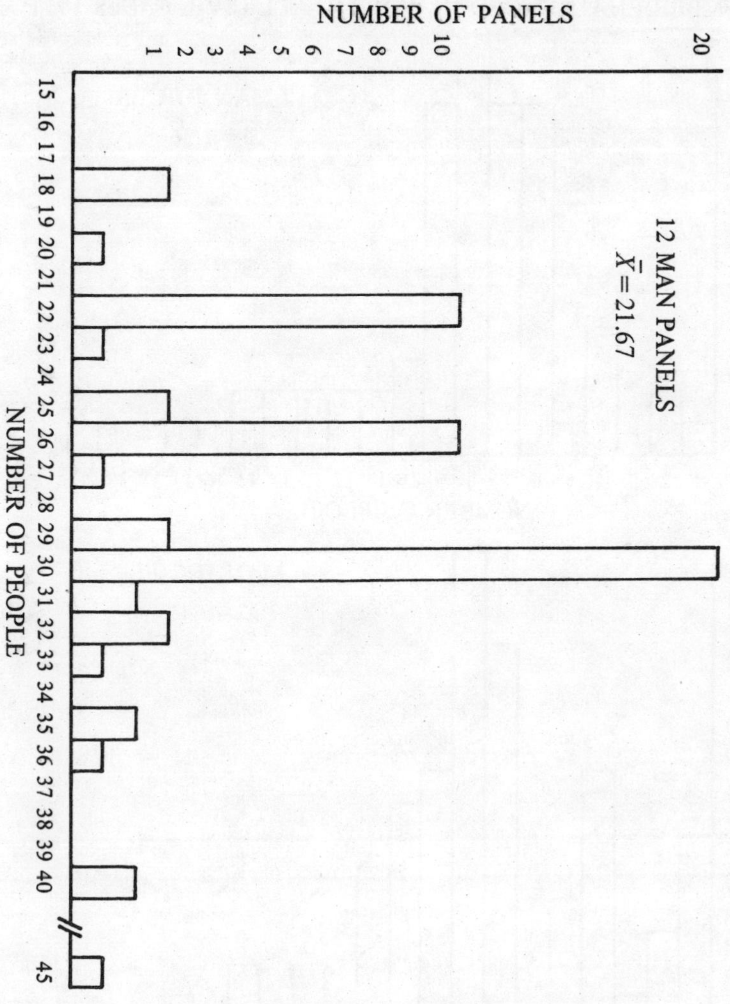

PANEL SIZES CIVIL CASES 1971

NUMBER CHALLENGED IN PANELS ALL CIVIL CASES 1971

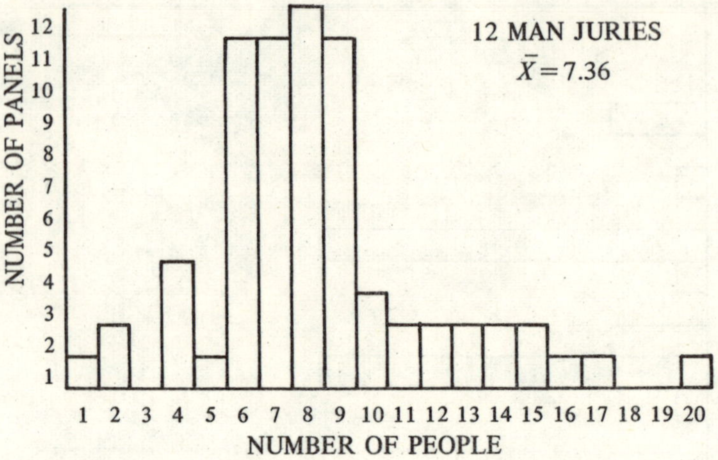

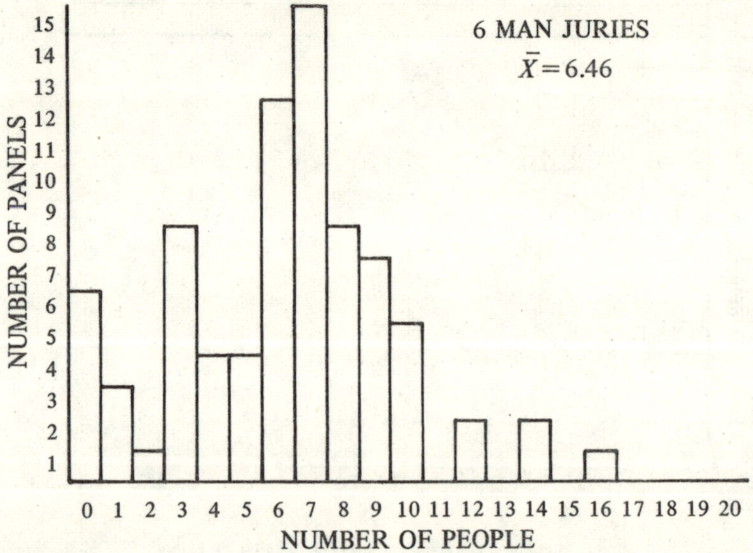

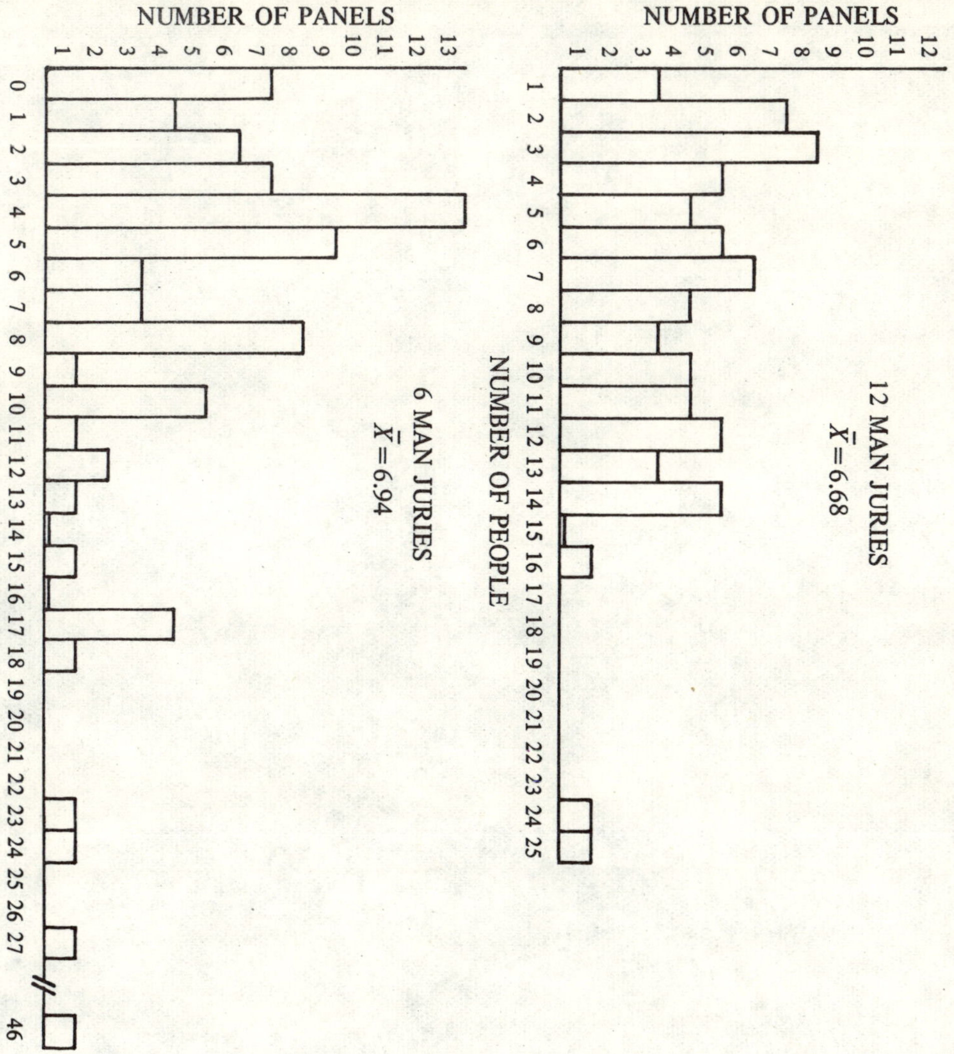

NUMBER NOT USED IN PANELS
ALL CIVIL TRIALS 1971

COLLOQUIA MATHEMATICA SOCIETATIS JÁNOS BOLYAI
9. EUROPEAN MEETING OF STATISTICIANS, BUDAPEST (HUNGARY), 1972.

ON SUMS OF A RANDOM NUMBER OF RANDOM MULTI-DIMENSIONAL VECTORS

V. PAULAUSKAS

It is well-known the fundamental importance of various central limit theorems and laws of large numbers in the theory of the mathematical statistics. Many methods of estimation of unknown parameters reduce, when the sample size is large, to the averaging of some function of the observations. The first works on limit theorems for sums of a random number of random variables due to Anscombe (1952) were inspired by needs of statistics too. At present the theory of sums of a random number of random variables is highly developed and is applied for solving many problems of mathematical statistics, theory of service and so on.

We shall point out some aspects of such theory in multi-dimensional case.

Let $x = (x_1, x_2, \dots, x_k)$, $y = (y_1, \dots, y_k)$ be vectors from R_k. We shall write $x \cdot y = (x_1 y_1, \dots, x_k y_k)$, $\frac{x}{y} = \left(\frac{x_1}{y_1}, \dots, \frac{x_k}{y_k}\right)$, $(x, y) = \sum_{i=1}^{k} x_i y_i$, $|x| = (x, x)^{\frac{1}{2}}$. If $x \in R_k$ and $A \subset R_k$ then $xA =$

$= \{z: z = x \cdot y,\ y \in A\}$. Let $\xi_1, \xi_2, \ldots, \xi_n = (\xi_{n1}, \ldots, \xi_{nk})$ be independent identically distributed random vectors (r.v.) with vector of means $a = (a_1, \ldots, a_k)$ and non-singular convariance matrix. Let $F(A)$ be the distribution of the r.v. $\xi_1 - a$, B be it's covariance matrix and $b_i^2 = \mathsf{E}(\xi_{1i} - a_i)^2$. Let $\omega = \omega(t)$ be an integer valued random variable dependent on some parameter t and independent, for all t, of all r.v. ξ_j, $(j = 1, 2, \ldots)$. Let us denote

$$\alpha \equiv \alpha(t) = \mathsf{E}\omega, \quad \gamma \equiv \gamma(t) = \mathsf{E}|\omega - \alpha|,$$

$$\beta^2 = \mathsf{E}(\omega - \alpha)^2, \quad \omega_r = \mathsf{P}\{\omega = r\}.$$

We shall consider the r.v. $Z_\omega = \dfrac{\sum_{i=1}^{\omega} \xi_i - \alpha a}{\sigma}$ where $\sigma = (\sigma_1, \ldots, \sigma_k)$, $\sigma_i^2 = \alpha b_i^2 + a_i \beta^2$ and the limit distribution of Z_ω $F_\omega(A) = \mathsf{P}\{Z_\omega \in A\}$, as $t \to \infty$ and $\alpha \to \infty$. Let $I_k(d, A)$, $d \in R_1$, $A \subset R_k$ stands for degenerate k-dimensional distribution with unit mass at point $(d, d, \ldots, d) \in R_k$ and $H_\omega(A) = \sum_{i=1}^{\infty} \omega_r I_k\left(\dfrac{r - \alpha}{\beta}, A\right)$. Let $\Phi_B(A)$ denote the k-dimensional normal distribution with zero means and covariance matrix B and let $\mathscr{E}$ denote the class of all universally measurable convex sets of R_k.

Theorem 1. *Let $\alpha \to \infty$, $\dfrac{\gamma}{\alpha} \to 0$ as $t \to \infty$. Then*

$$\Delta = \sup_{A \in \mathscr{E}} |F_\omega(A) - \widetilde{H}_\omega * \Phi_{\alpha B_1}(A)| \to 0, \qquad t \to \infty,$$

where $*$ denotes the convolution, B_1 is the covariance matrix of the r.v. $\dfrac{\xi_1 - a}{\sigma}$ and $\widetilde{H}_\omega(A) = H_\omega\left(\dfrac{\sigma}{\beta\alpha} A\right)$.

If we want to estimate the speed of convergence in this limit theorem we must require some conditions on moments of ξ_1 that we shall do by the help of so-called pseudomoments. Let us denote $\nu_3 = \sup_{|t|=1} \mathsf{E}^{-\frac{3}{2}}(\xi_1, t)^2 \int_{R_k} |(x, t)|^3 |(F - \Phi_B)(dx)|$, $\beta(A) = \inf_{x \in \delta A} (x, x)^{\frac{1}{2}}$,

$\beta_\Lambda(A) = \inf_{x \in \delta A} (\Lambda^{-1} x, x)^{\frac{1}{2}}$, where δA denotes the boundary of A.

Theorem 2. *If the r.v.'s ξ_i in Theorem* 1 *satisfy the condition* $\nu_3 < < \infty$, *then*

$$\Delta \leqslant C \cdot k^4 \alpha^{-\frac{1}{2}} \max(\nu_3, \nu_3^{\frac{1}{4}}) + Ck\gamma\alpha^{-1},$$

(C is an absolute constant not the same in different places).

Theorem 3. *Let* ξ_i, $(i = 1, 2, \dots)$ *be independent identically distributed r.v.-s with distribution* $G(A)$, $\mathsf{E}\xi_1 = 0$ *and nonsingular covariance-correlation matrix* Λ. *Then for all* $A \in \mathscr{E}$

$$\left| P\left\{\frac{1}{\sqrt{\alpha}} \sum_{i=1}^{\omega} \xi_i \in A\right\} - \Phi_\Lambda(A)\right| \leqslant$$

$$\leqslant C \cdot k^4 \frac{\max(\bar{\nu}_3, \bar{\nu}_3^{\frac{1}{4}})}{(1 + \beta_\Lambda^3(A))\sqrt{\alpha}} + \frac{1}{1 + \beta^3(A)} \left(\frac{C \cdot k^4 \gamma}{\alpha} + \frac{C \cdot k^2 \beta^2}{\alpha^2}\right),$$

where

$$\bar{\nu}_3 = \int_{R_k} (\Lambda^{-1}x, x)^{\frac{3}{2}} |(G - \Phi_\Lambda)(dx)|.$$

The obtained theorems generalize and strengthen results of [1]. The proofs are given in [2], and there are a little better estimates in Theorems 2 and 3 if we deal with distribution functions or in the case $k = 1$ ([2]).

If ω_t is a stopping time for the sequence $\xi_i = (\xi_{i1}, \dots, \xi_{ik})$, $i = 1, 2, 3, \dots$ of independent identically distributed random vectors with zero means and non-singular correlation matrix Λ, then we can give the following generalization of a result due to J. Tomkó [3].

Let N_t, A_t, B_t and C_t be functions such that all of them increase unboundedly as $t \to \infty$, $(A_t + B_t)N_t^{-1} \to 0$ as $t \to \infty$ and

$$\mathsf{P}\{N_t - A_t \leqslant \omega_t \leqslant N_t + B_t\} \geqslant 1 - \frac{C}{C_t}.$$

Without loss of generality we can regard N_t, A_t and B_t taking only integer values. Let λ_i $(i = 1, 2, \dots, k)$ be eigenvalues of Λ. We shall consider the randomly stopped sum

$$S_t = \frac{\sum_{i=1}^{\omega_t} \xi_i}{\bar{\sigma} \, N_t}, \qquad \text{where} \quad \bar{\sigma} = (\bar{\sigma}_1, \ldots, \bar{\sigma}_k), \quad \bar{\sigma}_j^2 = \mathsf{E}\xi_{1j}^2 ,$$

and its distribution $F_t(A) = \mathsf{P}\{S_t \in A\}$.

Theorem 4. *If* $\nu_3 < \infty$, *then*

$$\sup_{A \in \mathscr{E}} |F_t(A) - \Phi_\Lambda(A)| \leqslant C \cdot k^4 \cdot N_t^{-\frac{1}{2}} \max(\nu_3, \nu_3^{\frac{1}{4}}) +$$

$$+ 3 \sqrt[3]{\frac{4}{\pi}} \left(\sum_{i=1}^{k} \frac{1}{\sqrt{\lambda_i}} \right)^{\frac{2}{3}} \cdot k \left(\frac{A_t + B_t}{N_t} \right)^{\frac{1}{3}} + \frac{C}{C_t} .$$

REFERENCES

[1] S.H. Sirazdinov – M. Mamatov – S.K. Formanov, Uniform estimation in limit theorems for sums of a random number of random variables, *Izvestija AN. UzSSR.*, (1970), 28-34.

[2] V. Paulauskas, On sums of a random number of random multidimensional vectors, *Litov. Mat. Sb.*, 12 (1972), 109-131.

[3] J. Tomkó, The rate of convergence in limit theorems for service systems with finite queue capacity, *J. Appl. Prob.*, 9 (1972), 87-102.

COLLOQUIA MATHEMATICA SOCIETATIS JÁNOS BOLYAI
9. EUROPEAN MEETING OF STATISTICIANS, BUDAPEST (HUNGARY), 1972.

GENERALIZATION OF CHERNOFF'S RESULT ON THE ASYMPTOTIC DISCERNIBILITY OF TWO RANDOM PROCESSES

A. PEREZ

Based on the generalized Shannon – McMillan's limit theorem for entropy densities, the present paper extends to the general stationary case Chernoff's well-known result on the asymptotic discernibility of two random stationary processes of the independent type.

1. INTRODUCTION

Let $\{\xi_n\}$ $(n = 0, \pm 1, \pm 2, \ldots)$ be a sequence of abstract valued random variables on the same measurable space $(X_0, \mathscr{X}_0)$ and denote by $(X, \mathscr{X})$ the corresponding infinite product space and by $(X_{r,s}, \mathscr{X}_{r,s})$ that corresponding to the coordinates from r to s.

Let either P or Q be the probability measure induced by the above sequence on $(X, \mathscr{X})$ and denote by H_{P} and H_{Q} the respective statistical hypotheses occuring with a priori probabilities p and q, $p + q = 1$. Provided that p and q are both positive, their exact values are irrelevant for the asymptotic behaviour of the probability of error in discriminating

H_P and H_Q on the base of a growing number of observed random variables of the above sequence; thus, it is possible for the sake of simplicity to take in the sequel $p = q = \frac{1}{2}$ and restrict us to the study of the maximum likelihood error probabilities. We shall denote by $e_{Pn}(P, Q)$ and $e_{Qn}(P, Q)$ the corresponding conditional error probabilities relative to n successive observations and by $e_n(P, Q)$ their mean value.

In his 1952 paper [1] H. Chernoff has determined the asymptotic rate of convergence to zero of the above error probabilities for the case of a sequence of mutually independent and identically distributed random variables, i.e. under the hypothesis that P and Q are stationary *independent* random processes. If P_0 and Q_0 are their one-dimensional restrictions, it namely holds

$$(1.1) \qquad \lim_{n\to\infty} \frac{1}{n} \log e_n(P, Q) = \log H_{\alpha_0}(P_0, Q_0) ,$$

where

$$(1.2) \qquad H_{\alpha_0}(P_0, Q_0) = \min_{0 \leqslant \alpha \leqslant 1} H_\alpha(P_0, Q_0)$$

and

$$H_\alpha(P_0, Q_0) = \text{alpha-entropy of } P_0 \text{ with respect to } Q_0 =$$

$$(1.3) \qquad = \int \left(\frac{dP_0}{dW}\right)^{\alpha} \left(\frac{dQ_0}{dW}\right)^{1-\alpha} dW ,$$

W being a measure dominating P_0 and Q_0*.

Our generalization of the Chernoff's result (1.1) to the case of P and Q stationary *dependent* random processes may be stated as follows (throughout $e_n(P, Q)$ may be replaced by the sum $e_{Pn}(P, Q) + e_{Qn}(P, Q)$ of the corresponding conditional error probabilities):

$$(1.4) \qquad \lim_{n\to\infty} \frac{1}{n} \log e_n(P, Q) = \lim_{n\to\infty} \frac{1}{n} \log H_{\alpha_n}(P_{0,n-1}, Q_{0,n-1}) ,$$

*In the case $\alpha = 0$ (resp. $\alpha = 1$) it is necessary to consider the definition (1.3) as the limit for $\alpha \downarrow 0$ (resp. for $\alpha \uparrow 1$).

where $\mathsf{P}_{0,n-1}$ and $\mathsf{Q}_{0,n-1}$ are the n-dimensional restrictions on $(X_{0,n-1}, \mathscr{X}_{0,n-1})$ of P and Q, respectively, and

$$H_{\alpha_n}(\mathsf{P}_{0,n-1}, \mathsf{Q}_{0,n-1}) = \min_{0 \leqslant \alpha \leqslant 1} H_\alpha(\mathsf{P}_{0,n-1}, \mathsf{Q}_{0,n-1}), \tag{1.5}$$
$$(n = 1, 2, \ldots).$$

The right-hand limit figuring in (1.4) is what we have called in [2] minimal alpha-entropy rate of the random process P with respect to the random process Q. In the Chernoff case this rate is, of course, equal to $\log H_{\alpha_0}(\mathsf{P}_0, \mathsf{Q}_0)$ and, thus, the equality (1.4) reduces to the equality (1.1).

The method of proving (1.4) in the sense $\leqslant$ is similar to that of Chernoff and is based on the inequality

$$e_n(\mathsf{P}, \mathsf{Q}) \leqslant e_{\mathsf{P}n}(\mathsf{P}, \mathsf{Q}) + e_{\mathsf{Q}n}(\mathsf{P}, \mathsf{Q}) \leqslant H_{\alpha_n}(P_{0,n-1}, \mathsf{Q}_{0,n-1}), \tag{1.6}$$
$$(n = 1, 2, \ldots).$$

However, our method of proving (1.4) in the opposite sense $\geqslant$, i.e. of estimating the left-side member from below, differs essentially from the Chernoff's method and is based on our generalized Shannon – McMillan's limit theorem for entropy densities (incompletely proved in [3] because of Lemma 2.2 which is replaced here by Lemma 2.1) and on the inequality (cf. [4])

$$\frac{1}{n}\log e_n(\mathsf{P}, \mathsf{Q}) \geqslant \int \min\left\{-\frac{1}{n}\log\frac{d\mathsf{R}_{0,n-1}}{d\mathsf{P}_{0,n-1}},\ -\frac{1}{n}\log\frac{d\mathsf{R}_{0,n-1}}{d\mathsf{Q}_{0,n-1}}\right\} d\mathsf{R},$$

valid for every probability measure R on $(X, \mathscr{X})$ such that $\mathsf{R}_{0,n-1}$ is dominated by $\mathsf{P}_{0,n-1}$ and $\mathsf{Q}_{0,n-1}$ for every finite n.

The Shannon – McMillan's limit theorem above states that in the stationary case under some general condition GC (which will be considered in detail below) the quantities figuring in { } of the second member of this inequality converge in the R-mean for $n \to \infty$. If, moreover, R is ergodic, then the corresponding limits are equal [R] to the respective generalized Shannon's entropy rates, $h(\mathsf{R}, \mathsf{P})$, of R with respect to P, and $h(\mathsf{R}, \mathsf{Q})$, of R with respect to Q, up to the sign.

Using the estimate* $|\min(x, y) - \min(u, v)| \leqslant |x - u| + |y - v|$ and the Shannon – McMillan's theorem we obtain from the preceding inequality the following basic inequality

$$\liminf_{n\to\infty} \frac{1}{n} \log e_n(\mathsf{P}, \mathsf{Q}) \geqslant \min\{-h(\mathsf{R}, \mathsf{P}), -h(\mathsf{R}, \mathsf{Q})\}. \tag{1.7}$$

Let us recall that $h(\mathsf{R}, \mathsf{P})$ is defined by

$$h(\mathsf{R}, \mathsf{P}) = \lim_{n\to\infty} \frac{1}{n} \int \log \frac{d\mathsf{R}_{0,n-1}}{d\mathsf{P}_{0,n-1}} d\mathsf{R} \tag{1.8}$$

and similarly $h(\mathsf{R}, \mathsf{Q})$. Under our conditions, the limits in question exist and are given by

$$\begin{aligned} h(\mathsf{R}, \mathsf{P}) &= H(\mathsf{R}_{-\infty,1}, \mathsf{RP}_{-\infty,0,1}), \\ h(\mathsf{R}, \mathsf{Q}) &= H(\mathsf{R}_{-\infty,1}, \mathsf{RQ}_{-\infty,0,1}), \end{aligned} \tag{1.9}$$

where the second members are the generalized Shannon's entropies of the probability measure $\mathsf{R}_{-\infty,1}$ (induced by R on $(X_{-\infty,1}, \mathscr{X}_{-\infty,1})$, the measurable space corresponding to the coordinates from $-\infty$ to 1) with respect to the probability measures $\mathsf{RP}_{-\infty,0,1}$ and $\mathsf{RQ}_{-\infty,0,1}$, respectively. The latter are defined as follows:

Let E be an arbitrary set of the σ-algebra $\mathscr{X}_{-\infty,0}$ and F an arbitrary set of the σ-algebra $\mathscr{X}_1 = \mathscr{X}_0$. Let $p_{\mathsf{P}}(F | x_{-\infty,0})$ and $p_{\mathsf{Q}}(F | x_{-\infty,0})$ be the conditional probabilities of F given $x_{-\infty,0} \in (X_{-\infty,}, \mathscr{X}_{-\infty,0})$, thus, measurable with respect to the σ-algebra $\mathscr{X}_{-\infty,0}$, corresponding to P and Q, respectively. Let us introduce the set functions

$$\mathsf{RP}_{-\infty,0,1}(E \times F) = \int_E p_{\mathsf{P}}(F | x_{-\infty,0}) d\mathsf{R}_{-\infty,0}, \tag{1.10}$$

$$\mathsf{RQ}_{-\infty,0,1}(E \times F) = \int_E p_{\mathsf{Q}}(F | x_{-\infty,0}) d\mathsf{R}_{-\infty,0}. \tag{1.11}$$

*For this and many other valuable remarks we are indebted to the referee.

Provided that the conditional probabilities introduced above are well-defined probability measures on $\mathscr{X}_1$ for every $x_{-\infty,0} \in X_{-\infty,0}[\mathsf{R}_{-\infty,0}]$, the set functions (1.10) and (1.11) can be uniquely extended on the whole σ-algebra $\mathscr{X}_{-\infty,1}$ and define the two probability measures $\mathsf{RP}_{-\infty,0,1}$ and $\mathsf{RQ}_{-\infty,0,1}$.

The general condition for the validity of the Shannon – McMillan limit theorem mentioned above, assumes throughout as in paper [3] (cf. page 553 below in [3]), the existence of $\mathsf{RP}_{-n,0,1}$ and $\mathsf{RQ}_{-n,0,1}$ (cf. (1.10) and (1.11) where in the place of ∞ we put n) as well-defined probability measures on $\mathscr{X}_{-n,1}$ for $n = 1, 2, \ldots, \infty$; this means well-defined conditional probabilities $p_{\mathsf{P}}(F \mid x_{-n,0})$ and $p_{\mathsf{Q}}(F \mid x_{-n,0})$ for every $n = 1, 2, \ldots, \infty$, and every $x_{-n,0} \in (X_{-n,0}, \mathscr{X}_{-n,0})[\mathsf{R}_{-n,0}]$. Our hypothesis that R is a stationary probability measure on $(X, \mathscr{X})$ such that $\mathsf{R}_{0,n-1}$ is dominated by $\mathsf{P}_{0,n-1}$ and $\mathsf{Q}_{0,n-1}$ for every positive n, permits easily to ensure the desired definition for every finite n since it is sufficient to suppose regular the corresponding conditional probability functions, defined almost surely $[\mathsf{P}_{-n,0}]$ and $[\mathsf{Q}_{-n,0}]$ respectively, and, thus, almost surely $[\mathsf{R}_{-n,0}]$ also. However, for $n = \infty$ the situation is not so simple due to the possibility that $\mathsf{R}_{-\infty,0}$ may be, at least partly, singular with respect to $\mathsf{P}_{-\infty,0}$ and $\mathsf{Q}_{-\infty,0}$, while the corresponding conditional probability functions are in general defined almost surely $[\mathsf{P}_{-\infty,0}]$ and $[\mathsf{Q}_{-\infty,0}]$, respectively. Given R as above, there is only one way to define $p_{\mathsf{P}}(F \mid x_{-\infty,0})$ and $p_{\mathsf{Q}}(F \mid x_{-\infty,0})$, $F \in \mathscr{X}_1$, namely,

$$(1.12) \qquad p_{\mathsf{P}}(F \mid x_{-\infty,0}) = \lim_{n \to \infty} p_{\mathsf{P}}(F \mid x_{-n,0}),$$

$$(1.13) \qquad p_{\mathsf{Q}}(F \mid x_{-\infty,0}) = \lim_{n \to \infty} p_{\mathsf{Q}}(F \mid x_{-n,0}),$$

almost surely $[\mathsf{R}_{-\infty,0}]$, what implies the assumption that these limits exist not only almost surely $[\mathsf{P}_{-\infty,0}]$ and $[\mathsf{Q}_{-\infty,0}]$, respectively, as in general, but for a larger set of $x_{-\infty,0}$'s having a probability one with respect to R.

In addition, the general condition in question assumes that the generalized Shannon entropies below are finite:

(1.14) $\quad H(\mathsf{R}_{-\infty,1}, \mathsf{RP}_{-\infty,0,1}) < \infty\,, \quad H(\mathsf{R}_{-\infty,1}, \mathsf{RQ}_{-\infty,0,1}) < \infty\,.$

In the sequel we shall say that P and Q satisfy the condition GC with respect to R if the above assumptions are fulfiled so that the relation

$$\lim_{n\to\infty} \int \left| \log \frac{dp_{\mathsf{P}}(x_1 \,|\, x_{-\infty,0})}{dp_{\mathsf{P}}(x_1 \,|\, x_{-n,0})} \right| d\mathsf{R}_{-\infty,0} = 0 \tag{1.15}$$

holds and a similar relation for Q.

Let, further, S_1 be the set of all ergodic probability measures R with respect to which P and Q satisfy the condition GC.

Let

$$r(\mathsf{P}, \mathsf{Q}) = \sup_{\mathsf{R}\in S_1} \min\{-h(\mathsf{R}, \mathsf{P}), -h(\mathsf{R}, \mathsf{Q})\}\,. \tag{1.16}$$

We shall prove that the right-hand limit in (1.4), i.e. the minimal alpha-entropy rate of P with respect to Q is equal to $r(\mathsf{P}, \mathsf{Q})$, under the same general condition for which the equality (1.4) is proved. In other words, it shall be proved that

$$\lim_{n\to\infty} \frac{1}{n} \log H_{\alpha_n}(\mathsf{P}_{0,n-1}, \mathsf{Q}_{0,n-1}) = r(\mathsf{P}, \mathsf{Q})\,. \tag{1.17}$$

2. PROOF OF THE EQUALITY (1.4)

In the sequel we shall assume that the processes P and Q are different and that there is no finite n such that $\mathsf{P}_{0,n-1}$ is singular with respect to $\mathsf{Q}_{0,n-1}$ because, otherwise, beginning from such an n the error probability $e_n(\mathsf{P}, \mathsf{Q})$ will be constantly equal to zero. Excluding the two cases above, the alpha-entropies (1.5) will satisfy

$$0 < H_{\alpha_n}(\mathsf{P}_{0,n-1}, \mathsf{Q}_{0,n-1}) < 1 \quad (n = 1, 2, \ldots)\,. \tag{2.1}$$

Under this condition, it is possible to introduce the probability measures $\mathsf{R}^{(n)}_{0,n-1}$ on $\mathscr{X}_{0,n-1}$ by the relation

$$(2.2) \qquad \frac{d\mathsf{R}^{(n)}_{0,n-1}}{d\mathsf{W}_{0,n-1}} = \frac{\left(\dfrac{d\mathsf{P}_{0,n-1}}{d\mathsf{W}_{0,n-1}}\right)^{\alpha_n}\left(\dfrac{d\mathsf{Q}_{0,n-1}}{d\mathsf{W}_{0,n-1}}\right)^{1-\alpha_n}}{H_{\alpha_n}(\mathsf{P}_{0,n-1}, \mathsf{Q}_{0,n-1})}$$

where $\mathsf{W}_{0,n-1}$ is a measure dominating $\mathsf{P}_{0,n-1}$ and $\mathsf{Q}_{0,n-1}$.*

It is possible to see that

$$(2.3) \qquad \begin{aligned} &H(\mathsf{R}^{(n)}_{0,n-1}, \mathsf{P}_{0,n-1}) = H(\mathsf{R}^{(n)}_{0,n-1}, \mathsf{Q}_{0,n-1}) = \\ &= -\log H_{\alpha_n}(\mathsf{P}_{0,n-1}, \mathsf{Q}_{0,n-1}) . \end{aligned}$$

Let $\mathsf{T}^{(n)}$ be the probability measure on $(X, \mathscr{X})$ which is defined as the product measure resulting from $\mathsf{R}^{(n)}_{0,n-1}$ applied successively on $\ldots \mathscr{X}_{0,n-1}, \mathscr{X}_{n,2n-1}, \ldots$. It is clear that $\mathsf{T}^{(n)}$ is n-periodic and n-ergodic, i.e. ergodic with respect to a shift transformation of n steps.

On the base of $\mathsf{T}^{(n)}$ it is possible to construct a stationary and ergodic (1-ergodic) probability measure $\mathsf{V}^{(n)}$ on $(X, \mathscr{X})$ by taking

$$(2.4) \qquad \mathsf{V}^{(n)}(E) = \frac{1}{n}\sum_{k=0}^{n-1} \mathsf{T}^{(n)}(S^{-k}E)$$

for every set $E \in \mathscr{X}$, where S is the one-step shift transformation. Indeed, on the base of the n-periodicity of $\mathsf{T}^{(n)}$ we have

$$(2.5) \qquad \begin{aligned} \mathsf{V}^{(n)}(SE) &= \frac{1}{n}\sum_{k=0}^{n-1} \mathsf{T}^{(n)}(S^{-k+1}E) = \frac{1}{n}\mathsf{T}^{(n)}(SE) + \\ &+ \frac{1}{n}\sum_{k=1}^{n-1} \mathsf{T}^{(n)}(S^{-k+1}E) = \frac{1}{n}\mathsf{T}^{(n)}(S^{-n}(SE)) + \\ &+ \frac{1}{n}\sum_{k=0}^{n-2} \mathsf{T}^{(n)}(S^{-k}E) = \frac{1}{n}\sum_{k=0}^{n-1} \mathsf{T}^{(n)}(S^{-k}E) = \mathsf{V}^{(n)}(E) . \end{aligned}$$

Thus, $\mathsf{V}^{(n)}$ is stationary. For proving that $\mathsf{V}^{(n)}$ is, moreover, ergodic, we proceed as follows.

*In the case $\alpha_n = 0$ (resp. 1) it is necessary to consider the above definition (2.2) as the limit for $\alpha \downarrow 0$ (resp. for $\alpha \uparrow 1$).

Let $E \in \mathscr{X}$ be such that (1) $SE = E$ (invariant set) and (2) $\mathsf{V}^{(n)}(E) > 0$. It is sufficient to prove that $\mathsf{V}^{(n)}(E) = 1$. Indeed, since $S^{-k}E = E$ for every integer k, it holds

$$\mathsf{V}^{(n)}(E) = \frac{1}{n}\sum_{k=0}^{n-1} \mathsf{T}^{(n)}(S^{-k}E) = \mathsf{T}^{(n)}(E) > 0\,, \tag{2.6}$$

the latter inequality resulting from our hypothesis (2). But $\mathsf{T}^{(n)}$ is by its construction n-ergodic due to the independence of the coordinate random variables distant more than n. As a consequence, $\mathsf{T}^{(n)}(E) = 1$ since $S^n E = E$ and $\mathsf{T}^{(n)}(E) > 0$. Hence $\mathsf{V}^{(n)}(E) = 1$, according to (2.6).

Q.E.D.

For the proof of (1.4) we shall make use of the preceding results as well as of the following lemma (cf. Lemma 2.2 of [3]) which serves, in particular, to ensure the validity of the Shannon – McMillan limit theorem mentioned in the Introduction.

Lemma 2.1. *Let* R *and* P *be two stationary processes with* P *satisfying the condition GC with respect to* R, *so that in particular (cf.* (1.14) *and* (1.15)*)*

(i) $\quad H(\mathsf{R}_{-\infty,1}, \mathsf{RP}_{-\infty,0,1}) < \infty$

(ii) $\quad \lim_{n\to\infty} \int \left| \log \frac{dp_{\mathsf{P}}(x_1 \mid x_{-\infty,0})}{dp_{\mathsf{P}}(x_1 \mid x_{-n,0})} \right| d\mathsf{R} = 0\,.$

Then the sequence $\left\{ \log \frac{d\mathsf{R}_{-n,1}}{d\mathsf{RP}_{-n,0,1}} \right\}_{n \geq 1}$ *converges in the* R*-mean to* $\log \frac{d\mathsf{R}_{-\infty,1}}{d\mathsf{RP}_{-\infty,0,1}}$.

In particular, it follows that

$$\lim_{n\to\infty} H(\mathsf{R}_{-n,1}, \mathsf{RP}_{-n,0,1}) = H(\mathsf{R}_{-\infty,1}, \mathsf{RP}_{-\infty,0,1})\,. \tag{2.7}$$

Proof. The desired relation holds iff

$$\lim_{n\to\infty}\int\left|\log\frac{dp_{\mathsf{R}}(x_1|x_{-\infty,0})}{dp_{\mathsf{R}}(x_1|x_{-n,0})}-\log\frac{dp_{\mathsf{P}}(x_1|x_{-\infty,0})}{dp_{\mathsf{P}}(x_1|x_{-n,0})}\right|d\mathsf{R}=0 \tag{2.8}$$

and, thus, by (ii), iff

$$\lim_{n\to\infty}\int\left|\log\frac{dp_{\mathsf{R}}(x_1|x_{-\infty,0})}{dp_{\mathsf{R}}(x_1|x_{-n,0})}\right|d\mathsf{R}=0\,.$$

The latter is equivalent to the relation

$$\lim_{n\to\infty}\int\log\frac{dp_{\mathsf{R}}(x_1|x_{-\infty,0})}{dp_{\mathsf{R}}(x_1|x_{-n,0})}d\mathsf{R}=0 \tag{2.9}$$

since the integral represents a generalized Shannon's entropy. By defining $H_n(x_{-\infty,0})$, for $n=1,2,\ldots,\infty$ by

$$H_n(x_{-\infty,0})=H(p_{\mathsf{R}}(\cdot|x_{-n,0}),p_{\mathsf{P}}(\cdot|x_{-\infty,0}))$$

we see that (2.9) may be written

$$\lim_{n\to\infty}\int H_n(x_{-\infty,0})d\mathsf{R}=H(\mathsf{R}_{-\infty,1},\mathsf{RP}_{-\infty,0,1})<\infty\,, \tag{2.10}$$

according to (i).

On the base of the equality

$$p_{\mathsf{R}}(\cdot|x_{-m,0})=\mathsf{E}_{\mathsf{R}}\{p_{\mathsf{R}}(\cdot|x_{-n,0})|\mathscr{X}_{-m,0}\}$$

holding a.s. $[\mathsf{R}_{-\infty,0}]$ for any $m<n$, we obtain

$$p_{\mathsf{R}}(\cdot|x_{-m,0})\log p_{\mathsf{R}}(\cdot|x_{-m,0})\leqslant$$
$$\leqslant\mathsf{E}_{\mathsf{R}}\{p_{\mathsf{R}}(\cdot|x_{-n,0})\log p_{\mathsf{R}}(\cdot|x_{-n,0})|\mathscr{X}_{-m,0}\}\,,$$

since the function $u\log u$ is convex. Here, by $\mathsf{E}_{\mathsf{R}}\{g(x_{-n,0})|\mathscr{X}_{-m,0}\}$ we denote the conditional expectation of $g(x_{-n,0})$ with respect to R and $\mathscr{X}_{-m,0}$-measurable.

From the above inequality it follows that the process $\mathsf{E}_{\mathsf{R}}\{H_n(x_{-\infty,0})|\mathscr{X}_{-n,0}\}$ $(n=1,2,\ldots)$ is an R-semi-martingale with respect to the growing system of σ-algebras $\{\mathscr{X}_{-n,0}\}_{n\geqslant 1}$, constituted of a sequence of nonnegative random variables. It holds, namely,

$$\int \mathsf{E}_{\mathsf{R}}\{H_m(x_{-\infty,0})|\mathscr{X}_{-m,0}\}d\mathsf{R} \leqslant \int \mathsf{E}_{\mathsf{R}}\{H_n(x_{-\infty,0})|\mathscr{X}_{-n,0}\}\,d\mathsf{R} \leqslant$$
$$\leqslant \int H_\infty(x_{-\infty,0})d\mathsf{R} = H(\mathsf{R}_{-\infty,1}, \mathsf{RP}_{-\infty,0,1}) < \infty$$

according to (i). Thus, according to a well-known theorem, it holds

$$\lim_{n\to\infty} \int |\mathsf{E}_{\mathsf{R}}\{H_n(x_{-\infty,0})|\mathscr{X}_{-n,0}\} -$$
$$- \mathsf{E}_{\mathsf{R}}\{H_\infty(x_{-\infty,0})|\mathscr{X}_{-\infty,0}\}|\,d\mathsf{R} = 0$$

from which (2.9) follows immediately.

Theorem 2.1. *Let* P *and* Q *be two stationary processes satisfying the condition GC uniformly with respect to all the stationary and ergodic processes* $\mathsf{V}^{(n)}$ $(n = 1, 2, \ldots,)$ *introduced above* (cf. (2.4)). *Then the following limits exist and it holds*

$$\lim_{n\to\infty} \frac{1}{n} \log e_n(\mathsf{P}, \mathsf{Q}) = \lim_{n\to\infty} \frac{1}{n} \log H_{\alpha_n}(\mathsf{P}_{0,n-1}, \mathsf{Q}_{0,n-1}) . \tag{2.11}$$

Proof. The conditions of the theorem ensure the validity of the generalized Shannon – McMillan's limit theorem for the pair $(\mathsf{V}^{(n)}, \mathsf{P})$ as well as for the pair $(\mathsf{V}^{(n)}, \mathsf{Q})$ $(n = 1, 2, \ldots)$. Thus, the relation (1.7) may be written

$$\begin{aligned} &\liminf_{k\to\infty} \frac{1}{k} \log e_k(\mathsf{P}, \mathsf{Q}) \geqslant \\ &\geqslant \min\{-h(\mathsf{V}^{(n)}, \mathsf{P}), \ -h(\mathsf{V}^{(n)}, \mathsf{Q})\} \qquad (n = 1, 2, \ldots) . \end{aligned} \tag{2.12}$$

Let $\mathsf{P}^{(n)}$ and $\mathsf{Q}^{(n)}$ be the product measures on $(X, \mathscr{X})$ generated by $\mathsf{P}_{0,n-1}$ and $\mathsf{Q}_{0,n-1}$, respectively, similarly as the measure $\mathsf{T}^{(n)}$ was generated by $\mathsf{R}^{(n)}_{0,n-1}$ above. On the base of the relations (1.9) and (2.3) we obtain for the entropy rates (taking namely account of the definition of $\mathsf{V}^{(n)}$ and the n-periodicity of $\mathsf{P}^{(n)}$ and $\mathsf{Q}^{(n)}$)

$$\begin{aligned} &-h(\mathsf{T}^{(n)}, \mathsf{P}^{(n)}) = -h(\mathsf{T}^{(n)}, \mathsf{Q}^{(n)}) = \\ &= \frac{1}{n} \log H_{\alpha_n}(\mathsf{P}_{0,n-1}, \mathsf{Q}_{0,n-1}) \qquad (n = 1, 2, \ldots) , \end{aligned} \tag{2.13}$$

$$
\begin{aligned}
&|h(\mathsf{T}^{(n)}, \mathsf{P}^{(n)}) - h(\mathsf{T}^{(n)}, \mathsf{P})| = \\
&= \Big|\frac{1}{n}\sum_{k=0}^{n-1} [H(\mathsf{T}^{(n)}S^k, \mathsf{T}^{(n)}S^k \mathsf{P}_{-k,0,1}) - \\
(2.14) \qquad &- H(\mathsf{T}^{(n)}S^k, \mathsf{T}^{(n)}S^k \mathsf{P}_{-\infty,0,1})]\Big| = \\
&= \Big|\frac{1}{n}\sum_{k=0}^{n-1} \int \log \frac{dp_{\mathsf{P}}(x_1 \mid x_{-\infty,0})}{dp_{\mathsf{P}}(x_1 \mid x_{-k,0})}\, d\mathsf{T}^{(n)}S^k\Big| \leqslant \\
&\leqslant \frac{1}{n}\sum_{k=0}^{n-1} \int \Big|\log \frac{dp_{\mathsf{P}}(x_1 \mid x_{-\infty,0})}{dp_{\mathsf{P}}(x_1 \mid x_{-k,0})}\Big|\, d\mathsf{T}^{(n)}S^k .
\end{aligned}
$$

According to the hypotheses of the theorem, the relation (ii) of the preceding lemma holds if R is replaced by $\mathsf{V}^{(n)}$ $(n = 1, 2, \ldots)$, uniformly in n, and thus also if R is replaced by $\mathsf{T}^{(n)}S^k$ uniformly in n and $k = 0, 1, \ldots, n-1$.

As a consequence, on the base of the mean-Cesaro limit theorem we obtain from (2.14)

$$
(2.15) \qquad \lim_{n\to\infty} [h(\mathsf{T}^{(n)}, \mathsf{P}^{(n)}) - h(\mathsf{T}^{(n)}, \mathsf{P})] = 0 .
$$

On the other hand, it holds due to the definition of $\mathsf{V}^{(n)}$ by $\mathsf{T}^{(n)}$ and the stationarity of P the following inequality

$$
(2.16) \qquad h(\mathsf{V}^{(n)}, \mathsf{P}) \leqslant h(\mathsf{T}^{(n)}, \mathsf{P}) .
$$

Analogous relations hold if we replace P and $\mathsf{P}^{(n)}$ by Q and $\mathsf{Q}^{(n)}$, respectively.

On the base of the above relations beginning from (2.12) we successively obtain

$$
\begin{aligned}
&\liminf_{k\to\infty} \frac{1}{k} \log e_k(\mathsf{P}, \mathsf{Q}) \geqslant \\
(2.17) \qquad &\geqslant \limsup_{n\to\infty} \min\{-h(\mathsf{V}^{(n)}, \mathsf{P}), \ -h(\mathsf{V}^{(n)}, \mathsf{Q})\} \geqslant \\
&\geqslant \limsup_{n\to\infty} \min\{-h(\mathsf{T}^{(n)}, \mathsf{P}), \ -h(\mathsf{T}^{(n)}, \mathsf{Q})\} =
\end{aligned}
$$

$$= \limsup_{n\to\infty} \min\{-h(\mathrm{T}^{(n)}, \mathrm{P}^{(n)}), -h(\mathrm{T}^{(n)}, \mathrm{Q}^{(n)})\} =$$

$$= \limsup_{n\to\infty} \frac{1}{n} \log H_{\alpha_n}(\mathrm{P}_{0,n-1}, \mathrm{Q}_{0,n-1}) .$$

Combining (2.17) with (1.6) we easily prove the existence and the equality of the limits in (2.11).

Q.E.D.

3. PROOF OF THE EQUALITY (1.17)

We shall prove the following

Theorem 3.1. *Let* P *and* Q *be two stationary processes satisfying the condition GC uniformly with respect to all the stationary and ergodic processes* $\mathrm{V}^{(n)}$ $(n = 1, 2, \dots)$ *introduced above (cf.* (2.4)*). Let* S_1 *be the set of ergodic probability measures used in the definition* (1.16) *of* $r(\mathrm{P}, \mathrm{Q})$.

Then the following equalities hold

$$r(\mathrm{P}, \mathrm{Q}) = \sup_{\mathrm{V}^{(n)}} \min\{-h(\mathrm{V}^{(n)}, \mathrm{P}), -h(\mathrm{V}^{(n)}, \mathrm{Q})\}, \tag{3.1}$$

$$\lim_{n\to\infty} \frac{1}{n} \log H_{\alpha_n}(\mathrm{P}_{0,n-1}, \mathrm{Q}_{0,n-1}) = r(\mathrm{P}, \mathrm{Q}) . \tag{3.2}$$

Proof. Let L be the limit in the left-hand side of (3.2), i.e. the minimal alpha-entropy rate of P with respect to Q, which exists according to Theorem 2.1 under the hypotheses of the theorem we are proving. Also the inequality below is valid here for $\epsilon > 0$ arbitrary provided that n is sufficiently large,

$$L - \epsilon \leqslant \min\{-h(\mathrm{V}^{(n)}, \mathrm{P}), \ -h(\mathrm{V}^{(n)}, \mathrm{Q})\} \leqslant L , \tag{3.3}$$

where the second inequality results from the inequality (2.12) since (2.11) holds. From (2.12) written with $\mathrm{R} \in S_1$ in the place of $\mathrm{V}^{(n)}$ it also follows that

$$r(\mathrm{P}, \mathrm{Q}) \leqslant L . \tag{3.4}$$

If, now, $r(\mathrm{P}, \mathrm{Q}) < L$, by taking in (3.3) $\epsilon < L - r(\mathrm{P}, \mathrm{Q})$ one would obtain

$$(3.5)\qquad \begin{aligned} &r(\mathsf{P}, \mathsf{Q}) < L - \epsilon \leqslant \min\{-h(\mathsf{V}^{(n)}, \mathsf{P}),\ -h(\mathsf{V}^{(n)}, \mathsf{Q})\} \leqslant \\ &\leqslant \sup_{R \in S_1} \min\{-h(\mathsf{R}, \mathsf{P}),\ -h(\mathsf{R}, \mathsf{Q})\} = r(\mathsf{P}, \mathsf{Q}), \end{aligned}$$

what is a contradiction. Thus, necessarily,

$$(3.6)\qquad r(\mathsf{P}, \mathsf{Q}) = L$$

so that (3.2) is proved. From (3.5) where we exclude the first inequality it follows also by taking the supremum of the min figuring in the first inequality with respect to all $\mathsf{V}^{(n)}$, $n = 1, 2, \ldots$, (note that $\mathsf{V}^{(n)} \in S_1$), that $r(\mathsf{P}, \mathsf{Q})$ satisfies (3.1) since ϵ is arbitrary.

Q.E.D.

In concluding this paper, let us remark that in the Markov case (i.e. for P and Q Markovian of arbitrary orders and stationary) the condition GC becomes extremely simple. The Markovian case was considered in particular by Koopmans [5] but the methods of proof used by him differ essentially from ours.

REFERENCES

[1] H. Chernoff, A measure of asymptotic efficiency for tests of a hypothesis based on the sum of observations, *Ann. Math. Statist.*, 23 (1952), 493-507.

[2] A. Perez, Discriminatory importance of the minimum α-entropy rate of one random process with respect to another, *Sixth Prague Conference on Information Theory,* 1971.

[3] A. Perez, Extensions of Shannon – McMillan's limit theorem to more general stochastic processes, *Trans. Third Prague Conference on Information Theory,* 1962, 545-574.

[4] A. Perez, Asymptotic discernibility of two stationary Markov chains, Second International Symposium of Information Theory in USSR, Tsahkadsor, 1971.

[5] L.H. Koopmans, Asymptotic rate of discrimination for Markov processes, *Ann. Math. Statist.*, 31 (1960), 982-994.

COLLOQUIA MATHEMATICA SOCIETATIS JÁNOS BOLYAI
9. EUROPEAN MEETING OF STATISTICIANS, BUDAPEST (HUNGARY), 1972.

ON THE STATISTICAL EXAMINATION OF POISSON PROCESSES WITH RANDOMLY CHANGING INTENSITY

J. PERGEL

In 1955 D.R. Cox examined the non-homogeneous Poisson processes $\xi(t)$ intensity of which is a stationary process $\Lambda(t)$. He computed the mean value and variance and other characteristics of the counting process N_t of $\xi(t)$. In 1963 D.P. Gaver studied the case where $\Lambda(t)$ is an alternating renewal process with the possible states λ_0, λ_1 and calculated the distribution function of the duration between two consecutive events of the process $\xi(t)$. In this note we consider the problem where $\Lambda(t)$ is a Markov process of two states λ_0, λ_1.

The stochastic behaviour of $\Lambda(t)$ can be described by two positive quantities μ_0 and μ_1 such that the staying of $\Lambda(t)$ in the states λ_0 and λ_1 will be exponentially distributed with parameters μ_0 and μ_1, resp. Our aim is to get estimations for the values $\lambda_0, \lambda_1, \mu_0, \mu_1$ observing only $\xi(t)$. Let the observed events of $\xi(t)$ take place at the points t_i $(i = \ldots, -1, 0, 1, 2, \ldots)$, then the values $x_i = t_i - t_{i-1}$ $(i = \ldots, -1, 0, 1, 2, \ldots)$ form a stationary time series.

The expectations $\mathsf{E}(x_i x_{i+k}) = r_k$ satisfy the formula

$$r_{k+1} = \pi_{00}^{(k)} m_0^2 + (\pi_{01}^{(k)} + \pi_{10}^{(k)}) m_0 m_1 + \pi_{11}^{(k)} m_1^2 , \tag{1}$$

where $\pi_{ij}^{(k)}$ can be determined recursively

$$\pi_{ij}^{(k)} = \pi_{i0}^{(k-1)} p_{0j} + \pi_{i1}^{(k-1)} p_{1j}, \qquad (k = 2, 3, \dots)$$

$$\pi_{ij}^{(1)} = p_i p_{ij} .$$

For the values m_0, p_i and p_{ij} we have the expressions

$$m_0 = \frac{\mu_0 + \mu_1 + \lambda_1}{\mu_1\lambda_0 + \mu_0\lambda_1 + \lambda_0\lambda_1} ,$$

$$p_0 = \frac{\mu_1\lambda_0}{\mu_0\lambda_1 + \mu_1\lambda_0} ,$$

$$p_{01} = \frac{\mu_0\lambda_1}{\mu_0\lambda_1 + \mu_1\lambda_0 + \lambda_0\lambda_1} ,$$

$$\mathsf{E}(x_i) = p_0 m_0 + p_1 m_1 .$$

The other values in the formulae can be get from these by an adequate permutation of the indices. The values $m_i, p_i, p_{ij}, \pi_{ij}$ have their intuitive meaning: p_{ij} are the transition probabilities of the Markov chain $\nu_k = \Lambda(t_k)$, p_i are the stationary probabilities of the states of this chain, $\pi_{ij}^{(k)}$ is the stationary probability

$$\pi_{ij}^{(k)} = \mathsf{P}(\nu_l = i, \nu_{l+k} = j) ,$$

and $m_i = \mathsf{E}(x_l \mid \Lambda(t_{l-1}) = i)$.

Knowing the expressions for the values r_k in principle we could use the method of least squares. However the equations which we get when trying to use it are very complicated. It seems to be more convenient to replace the values of r_k by their estimates

$$\tilde{r}_k = \sum_{l=k+1}^{n} x_l x_{l-k}$$

for some values of k and then to solve the equations (1). Here the question arises whether these equations have a unique solution converging to the real values of λ_i and μ_i. Using artificial data got from pseudorandom numbers some experiences are being made on the computer CDC 3300 concerning this problem.

Another variant of the problem is the case when we know the a priori joint distribution of the values $\lambda_0, \lambda_1, \mu_0, \mu_1$ and observing the process $\xi(t)$ we want to calculate the a posteriori joint distribution of these values. For this purpose let us denote by $p(t, \lambda_0, \lambda_1 \mu_0, \mu_1)$ the (non-complete) conditional joint density function of the random variables $\lambda_0, \lambda_1, \mu_0, \mu_1$ over the set where $\Lambda(t) = \lambda_0$ and by $q(t, \lambda_0, \lambda_1, \mu_0, \mu_1)$ the corresponding probability density function over the set where $\Lambda(t) = \lambda_1$ knowing the trajectory of $\xi(t)$ in the interval $[0, t]$. Then for $p(t, \lambda_0, \lambda_1, \mu_0, \mu_1)$ and $q(t, \lambda_0, \lambda_1, \mu_0, \mu_1)$ we have the integro-differential equation

$$-\frac{\partial p}{\partial t} = \left(\int_0^\infty \int_0^\infty \int_0^\infty \int_0^\infty [((s_0 + s_1)p(t, s_0, s_1, s_2, s_3) + \right.$$

$$+ (s_1 + s_3)q(t, s_0, s_1, s_2, s_3)]ds_0 ds_1 ds_2 ds_3 -$$

$$\left. - \lambda_0 - \mu_0 \right) p + \mu_1 q ,$$

and we obtain an analogous equation for $q(t, \lambda_0, \lambda_1, \mu_0, \mu_1)$ as well.

At the points t_i (i.e. at the events of $\xi(t)$) the functions $p(t, ., ., ., .)$ and $q(t, ., ., ., .)$ have a jump.

REFERENCES

[1] D.R. Cox, Some statistical methods concepted with series of events, *J. Roy. Statist. Soc. B,* 17 (1955), 129-164.

[2] D.R. Cox – P.A.W. Lewis, *The statistical analysis of series of events,* Methuen, London, 1966.

[3] D.P. Gaver, Random hazard in reliability problems, *Technometrics,* 5 (1963), 211-226.

COLLOQUIA MATHEMATICA SOCIETATIS JANOS BOLYAI
9. EUROPEAN MEETING OF STATISTICIANS, BUDAPEST (HUNGARY), 1972.

APPROXIMATE FORMULAS FOR PROBABILITY OF NONCROSSING OF LEVEL BY STATIONARY PROCESS

V.V. POZNJAKOV

Let $\xi(t)$ be a stationary Gauss process with zero mean function and unit variance and let us suppose that the sample functions are differentiable. The crossing level c is a constant, the time interval is $[0, t]$ and the correlation function $r(t)$ is n-times differentiable. It is well known thus the simple solution can be obtained when c or t tend to infinity and the distribution of the number of the crossings tends to the Poisson distribution or to the normal distribution. Such cases were investigated by many authors, among of them V.A. Volkonski, Ju.A. Rozanov, H. Cramér, Ju.K. Beljaev, M.P. Ersov.

On the other hand it is important to have simple formulas for cases of limited c and t. One can to do this as follows. Let $P(t)$ denote the probability of noncrossing of level c during the time interval $[0, t]$ and suppose

$$(1) \qquad P(t) = P(0) + \sum_{k=0}^{\infty} \frac{d^k P(t)}{dt^k}\bigg|_{t=0} \cdot \frac{t^k}{k!} \cdot$$

Let $[0, t]$ be divided by Δt to n equal parts and denote

$$A_k = \{\xi(k\Delta t) < c\}\,.$$

By stationarity and differentiability we get

$$\frac{d^k P}{dt^k}\Big|_{t=0} = \lim_{\Delta t \to 0} \frac{\sum_{l=0}^{k} (-1)^{k-l} C_k^l \mathrm{P}(A_0, A_1 \ldots A_l)}{\Delta t^k}\,. \tag{2}$$

The probability $\mathrm{P}(A_0, A_1 \ldots A_l)$ is an l-variate normal integral with correlation coefficients

$$r_{ij} = r(|i-j|\Delta t)\,. \tag{3}$$

The multivariate normal integral $F_k(c, r_{ij})$ can be expressed as follows

$$\begin{aligned} F_k(c, r_{ij}) &= \mathrm{P}(A_0 A_1 \ldots A_{k-1}) = F_k(c, R_0) + \\ &+ \int_L \sum_{p,q} \frac{\partial F_k(c, U_{ij})}{\partial U_{ij}}\, dU_{ij}\,; \end{aligned} \tag{4}$$

$$\frac{\partial F_k(c, U_{ij})}{\partial U_{pq}} = \frac{1}{2\pi} \frac{e^{-\frac{c^2}{1-U_{pq}}}}{\sqrt{1 - U_{pq}^2}} F_{k-2}(c^{pq}, U_{ij}^{pq}) \qquad ((i,j) \neq (p,q))\,,$$

$$c^{pq} = c \frac{(1 - U_{pq})(1 + U_{pq} - U_{pi} - U_{qi})}{\sqrt{(1 - U_{pq}^2 - U_{pi}^2 - U_{qi.}^2 + 2U_{pq} U_{qi} U_{pi})(1 - U_{pq}^2)}}\,,$$

$$U_{ij}^{pq} = \frac{B}{\sqrt{\Delta_{pqij}(1 - U_{pq}^2) - B^2}}\,,$$

$$B = U_{ij}(1 - U_{pq}^2) - U_{pi} U_{pj} - U_{gi} U_{gj} + U_{pq}(U_{pi} U_{qj} + U_{qi} U_{pj})$$

where

Δ_{pqij} denotes the determinant of the correlation matrix of the corresponding random variables, while

L is the integration curve in the space of the correlation coefficients from an arbitrary initial points R_0 to a final point $(r_{12}, r_{23}, \ldots, r_{k-1k})$.

The formula (4) was obtained in [1], [2]. By using the method described above we get

$$P'(0) = -\frac{dN(c, t)}{dt}\Big|_{t=0} \tag{5}$$

and

$$P''(0) = 0$$

where $N(c, t)$ is the mean rate of the crossings, thus

$$P(t) = P(0) - N(c, t) + O(t^3)\,, \qquad O(t^3) > 0\,.$$

An approximate evaluation of probability $P(t)$ can be achieved by a replacement of $P(t)$ by $\mathrm{P}(A_0 A_1 \ldots A_n)$. Obviously

$$P(t) - \mathrm{P}(A_0 A_1 \ldots A_n) < 0\,. \tag{6}$$

In order to estimate the differences (6) we carry out some transformations:

$$P(t) - \mathrm{P}(A_0 A_1 \ldots A_n) = P(t - \Delta t) - \mathrm{P}(A_0 A_1 \ldots A_{n-1}) -$$
$$- [P(t - \Delta t, \overline{\Delta t}) - \mathrm{P}(A_0 A_1 \ldots A_{n-1} \bar{A}_n)]\,,$$

$$\begin{aligned} &P(t - \Delta t, \overline{\Delta t}) - \mathrm{P}(A_0 A_1 \ldots A_{n-1} \bar{A}_n) = P(\overline{\Delta t}) - P(A_{n-1} \bar{A}_n) - \\ &- [P(\overline{t - \Delta t}) - P((\overline{A_0 A_1 A_2 \ldots A_{n-2}}) A_{n-1} \bar{A}_n)]\,, \end{aligned} \tag{7}$$

$$\mathrm{P}(A_{n-1}, \bar{A}_n) = \mathrm{P}(A_0 \bar{A}_1)\,.$$

$$P(\overline{t - \Delta t}, \overline{\Delta t}) - \mathrm{P}((\overline{A_0 A_1 \ldots A_{n-2}}) A_{n-1} \bar{A}_n) > 0\,,$$

$$P(\Delta t) < \mathrm{P}(A_0 A_1)$$

consequently

$$\mathrm{P}(A_0 A_1 \ldots A_n) - P(t) \leqslant n(\mathrm{P}(A_0 A_1) - P(\Delta t))\,.$$

It is easy to show

$$\frac{dP(A_0A_1)}{dt}\Big|_{t=0} = -\frac{dN(c,t)}{dt}\Big|_{t=0} \tag{8}$$

and

$$\frac{d^2P(A_0A_1)}{dt}\Big|_{t=0} = 0\,,$$

therefore, by (6) and (8), we obtain for $\Delta t \to 0$ that

$$|P(t) - P(A_0A_1A_2 \ldots A_n)| < nO(\Delta t^3)\,. \tag{9}$$

REFERENCES

[1] R. L. Plackett, A reduction formula for normal multivariate integral, *Biometrika*, 41 (1954), 351-360.

[2] V. V. Poznjakov, On a representation of the multivariate Gaussian integral, *Ukrain. Mat. Ž.*, 23 (1971), 562-566.

COLLOQUIA MATHEMATICA SOCIETATIS JÁNOS BOLYAI
9. EUROPEAN MEETING OF STATISTICIANS, BUDAPEST (HUNGARY), 1972.

THE CONVERGENCE OF AN AGE-DEPENDENT BRANCHING PROCESS ALLOWING IMMIGRATION

J. RADCLIFFE

1. INTRODUCTION

It is well-known that if $N(t)$ is the number of individuals in an age-dependent branching process at time t, and the process is supercritical, then $\mathsf{E}[N(t)] \sim n_1 e^{\alpha t}$. Further, if we let $V(t) = e^{-\alpha t}N(t)$, and provided suitable assumptions are made with regard to the lifetime distributions of individuals etc., then as $t \to \infty$, $V(t)$ converges in mean square and almost surely to a random variable V. In Section 2 analogous results are outlined for a process allowing immigration.

For a multitype process similar results can be obtained. Let $N(t) = (N_1(t), N_2(t), \dots, N_m(t))$, where $N_i(t)$ is the number of individuals of the ith type at time t in a process without immigration. Then $V(t) = e^{-\alpha t}N(t)$ converges in mean square and almost surely to a random vector V. If the process is irreducible then $V = \nu\eta$ a.s., where ν is a scalar random variable and $\eta = (\eta_1, \eta_2, \dots, \eta_m)$ is a constant vector. This implies that if the population does not become extinct then the proportion of the ith type, $\dfrac{N_i(t)}{N_1(t) + N_2(t) + \ldots + N_m(t)}$, converges almost surely to a fixed limit p_i. In Section 3 the corresponding results for a process allowing immigration are given. Let $Z(t) = (Z_1(t), Z_2(t), \ldots, \ldots, Z_m(t))$, where $Z_i(t)$ is the number of individuals of the ith type at time t. The asymptotic behaviour of the process is studied and it is shown that $W(t) = e^{-\alpha t} Z(t)$ converges in mean square and almost surely to a random variable W. Moreover, there exists a scalar random variable w such that $W = w\eta$ a.s. and $\dfrac{Z_i(t)}{Z_1(t) + Z_2(t) + \ldots + Z_m(t)}$ converges a.s. to p_i. This shows that the rate of increase of the population does not depend on the immigration, in the sense that the Malthusian parameter of the process allowing immigration is the same as that for the corresponding processs without an immigration component. Moreover, the asymptotic frequencies p_i are the same for the processes with and without an immigration component. In Section 4 the existence of a limit distribution is established for a multitype subcritical process allowing immigration.

2. A SINGLE TYPE PROCESS

We employ a model used by Jagers [3]. Individuals aged zero immigrate into the process in batches at times $\tau_1, \tau_2, \ldots, \tau_k$ which are the epochs of a renewal process, so that at time τ_k V_k individuals enter. The times $\delta_1 = \tau_1$, $\delta_k = \tau_k - \tau_{k-1}$ $(k = 2, 3, \ldots)$ between successive batches of immigrants and the number of immigrants V_k $(k = 1, 2, \ldots)$ are independent random variables such that

$$\mathrm{P}[\delta_k \leqslant t] = G_0(t) \qquad (k = 1, 2, \ldots)$$

$$\mathrm{P}[V_k = n] = p_{0n} \qquad (k = 1, 2, \ldots; \; n = 0, 1, 2, \ldots)$$

and $m_0 = \sum_{k=1}^{\infty} k p_{0k}$.

Let $Z(t)$ be the number of individuals in the process at time t, $M(t) = \mathrm{E}[Z(t)]$ and suppose $Z(0) = 0$. Once in the process each individual an age-dependent branching process as described by Mode [4], Chapter 3.

Let $N(t)$ be the number of individuals at time t after entry in one of these processes, and $\mu(t) = \mathsf{E}[N(t)]$. Suppose the lifetime of an individual is $G_1(t)$ and the mean number of offspring per individual is m_1.

We first look at the asymptotic behaviour of $M(t)$. By considering the point at which an immigrant first enters the process as a regeneration point of the process, an integral equation for the generating function of $Z(t)$ can be written down, from which a renewal type integral equation for $M(t)$ can be obtained by differentiation, which is as follows:

$$(2.1) \qquad M(t) = m_0 \int_0^t \mu(t-u)dG_0(u) + \int_0^t M(t-u)dG_0(u) \, .$$

Suppose the process is supercritical, then the Malthusian parameter $\alpha > 0$ is defined by

$$(2.2) \qquad m_1 \int_0^\infty e^{-\alpha u} dG_1(u) = 1 \, .$$

Under this condition it is well known e.g. Harris [2] that if $G_1(t)$ has a density $g_1(t) \in L_p$ for some $p > 1$, then

$$(2.3) \qquad \mu(t) = n_1 e^{\alpha t}[1 + O(e^{-\epsilon t})] \, .$$

Using this result together with some elementary analysis it follows that

$$(2.4) \qquad M(t) = \frac{m_0 n_1 a}{(1-a)} e^{\alpha t}[1 + O(e^{-\delta t})] \, ,$$

where $a = \int_0^\infty e^{-\alpha u} dG_0(u)$.

For details of the proof of this and other results of this Section the reader is referred to Radcliffe [5]. This describes the asymptotic behaviour of $M(t)$.

In order to prove the mean square and almost sure convergence of $W(t)$ we examine the asymptotic behaviour of $M_2(t,\tau) = \mathsf{E}[Z(t)Z(t+\tau)]$. By considering the time at which immigrants first enter as a regeneration point of the process the following integral equation for

$$F_2(s_1, s_2; t, \tau) = \mathsf{E}[s_1^{Z(t)} s_2^{Z(t+\tau)}]$$

is obtained:

$$F_2(s_1, s_2; t, \tau) = 1 - G_0(t + \tau) +$$

$$(2.5) \qquad + \int_t^{t+\tau} h_0[\Phi(s_2, t + \tau - u)]F(s_2, t + \tau - u)dG_0(u) +$$

$$+ \int_0^t h_0[\Phi_2(s_1, s_2; t - u, \tau)]F_2(s_1, s_2; t - u, \tau)dG_0(u) ,$$

where $\Phi(s, t) = \mathsf{E}[s^{N(t)}]$, $F(s, t) = \mathsf{E}[s^{Z(t)}]$,

$$\Phi_2(s_1, s_2; t, \tau) = \mathsf{E}[s_1^{N(t)} s_2^{N(t+\tau)}]$$

and $h_0(s)$ is the probability generating function of the number of offspring per individual. Differentiating with respect to s_1 and s_2 and letting $s_1 \uparrow 1$ and $s_2 \uparrow 1$, a renewal type integral equation for $M_2(t, \tau)$ is obtained which is of the form

$$(2.6) \qquad M_2(t, \tau) = f(t, \tau) + \int_0^t M_2(t - u, \tau)dG_0(u) ,$$

where $f(t, \tau)$ is an expression involving $\mu(t)$ and $\mu_2(t, \tau) = \mathsf{E}[N(t)N(t + \tau)]$.

If we use the result from Harris [2], that under suitable conditions

$$(2.7) \qquad \mu_2(t, \tau) = n_2 e^{\alpha(2t+\tau)}[1 + O(e^{-\epsilon_1 t})] ,$$

together with some elementary analysis, it can be shown that

$$(2.8) \qquad M_2(t, \tau) = de^{\alpha(2t+\tau)}[1 + O(e^{-ct})] .$$

From this result it can be shown by fairly standard methods such as those used in Harris [2] or Mode [4] that $W(t)$ converges in mean square to W.

In order to establish almost sure convergence, let

$$(2.9) \qquad Z(t) = B(t) - D(t) ,$$

where $B(t)$ is the number of individuals that have been born or have immigrated into the process by time t, and $D(t)$ is the number of individuals that have died by time t. Results can be obtained for $B(t)$ and $D(t)$ similar to those that have been mentioned converning $Z(t)$.

Let $B(t) = B^{(1)}(t) + B^{(2)}(t) + \ldots + B^{(k)}(t)$, where $B^{(r)}(t)$ is the number of descendents from the batch of immigrants that arrive at the rth event of the renewal process. Then if $G_0(0) = 0$, the expected number of renewals in $[0, t]$ is finite and hence k is a.s. finite (Feller [1]). Provided the process is regular $B^{(r)}(t)$ is a.s. finite. Hence $B(t)$ is a.s. finite. From the fact that $B(t)$ is a.s. a non-decreasing step function on $[0, \infty)$ and the result that

$$(2.10) \qquad M_2'(t, \tau) = \mathsf{E}[B(t)B(t+\tau)] = d' e^{\alpha(2t+\tau)}[1 + O(e^{-ct})]$$

the a.s. convergence of $W(t) = e^{-\alpha t}B(t)$ follows by the usual methods, e.g. Harris [2] or Mode [4]. Analogous results can be written down involving $D(t)$, and it follows that $W_D(t) = e^{-\alpha t}D(t)$ converges a.s. to a random variable W_D. Hence $W(t) = W_B(t) - W_D(t)$ converges a.s. to a random variable W.

3. A MULTITYPE PROCESS

Results similar to those of the previous section can be established for a multitype process. Suppose $Z(t) = (Z_1(t), Z_2(t), \ldots, Z_m(t))$ where $Z_i(t)$ is the number of individuals of the ith type in a process with immigration. At each epoch of the renewal process V_{ik} $(i = 1, 2, \ldots, m)$ individuals of type i enter and

$$\mathsf{P}[V_{ik} = n_1, \; V_{2k} = n_2, \ldots, V_{mk} = n_m] = p_{n_1, n_2, \ldots, n_m}$$
$$(k = 1, 2, \ldots; \; n_i = 0, 1, 2, \ldots; \; i = 1, 2, \ldots, m).$$

It can be shown (Radcliffe [6]) that if the process is irreducible then $W(t) = e^{-\alpha t}Z(t)$ converges in mean square and almost surely to a random variable W.

Theorem. *There exists a scalar random variable* w *such that*

(3.1) $W = w\eta$ a.s.

Proof. Let $W_i(t) = e^{-\alpha t}Z_i(t)$ and $M_{ij}(t,\tau) = \mathrm{E}[Z_i(t)Z_j(t+\tau)]$. Then

(3.2) $$\mathrm{E}[w_i(t)w_j(t+\tau)] = e^{-\alpha(2t+\tau)}M_{ij}(t,\tau) = d_{ij}[1 + O(e^{-ct})] .$$

Thus if $W_i(t)$ converges to W_i ,

(3.3) $$\mathrm{E}[W_iW_j] = d_{ij} .$$

The d_{ij} can be expressed in terms of the parameters of the branching process without immigration and it follows from Mode [4] that the d_{ij} can be expressed in the form

(3.4) $$d_{ij} = c\eta_i\eta_j .$$

So that

(3.5) $$\mathrm{E}[W_iW_j] = \theta\eta_i\eta_j .$$

From this it follows that

(3.6) $$\mathrm{E}[(\eta_k W_j - \eta_j W_k)^2] = 0 \qquad (j = 1, 2, \ldots, m;\ k = 1, 2, \ldots, m) .$$

All the 2×2 determinants of the random matrix

$$\begin{bmatrix} W_1 & W_2 & \cdots & W_m \\ \eta_1 & \eta_2 & \cdots & \eta_m \end{bmatrix}$$

are zero a.s. The matrix has rank one and $W = w\eta$ a.s.

It follows from this result that the proportion of the ith type in the population $\dfrac{Z_i(t)}{Z_1(t) + Z_2(t) + \ldots + Z_m(t)}$ converges a.s. to $\dfrac{\eta_i}{\eta_1 + \eta_2 + \ldots + \eta_m}$, which is the same limit to which $\dfrac{N_i(t)}{N_1(t) + N_2(t) + \ldots + N_m(t)}$ converges a.s. in a branching process without immigration conditional on the process not becoming extinct.

Reducible process can be treated by methods requiring only slight modification and results analogous to theorem 4.3 of Mode [4] established for processes allowing immigration.

4. THE LIMIT DISTRIBUTION IN THE SUBCRITICAL CASE

Jagers [3] has established the existence of a limit distribution for a subcritical branching process allowing immigration. In this section this result is extended to a generalised multitype process. Consider a branching process without immigration. Let $N_i(t)$ be the number of individuals of the ith type after time t, and $\mu_{ij}(t, \tau) = \mathsf{E}[N_j(t) | N(0) = \epsilon_i]$, where ϵ_i is a vector with 1 in the ith position and zero elsewhere. Let $m_{ij}(t)$ be the potential number of offspring of the jth type born to a type i individual after time t. For a more precise definition see Mode [4].

We shall need the following lemma which is a generalization of Lemma 4 of Jagers [3]:

Lemma 1.

$$\int_0^\infty \mu_{ij}(t)dt = a_{ij}\lambda_i , \tag{4.1}$$

where a_{ij} is the (i,j)th element of $(I-M)^{-1}$, M is the matrix whose (i,j)th element is $\int_0^\infty m_{ij}(dt)$ and $\lambda_i = \int_0^\infty tG_i(dt)$, $G_i(t)$ being the lifetime distribution of a type i individual.

Proof. $\mu_{ij}(t)$ satisfies the integral equation

$$\mu_{ij}(t) = \delta_{ij}(1 - G_i(t)) + \sum_{k=1}^{m} m_{ik}(ds)\mu_{kj}(t-s) , \tag{4.2}$$

(Mode (1971)), and hence

$$\Phi(\theta) = (I - H(\theta))^{-1}D(\theta) , \tag{4.3}$$

where $\Phi(\theta)$, $H(\theta)$ and $D(\theta)$ are the matrices whose (i,j)th elements are $\int_0^\infty \mu_{ij}(t)e^{-\theta t}dt$, $\int_0^\infty m_{ij}(dt)e^{-\theta t}$ and $\theta^{-1}\delta_{ij}(1 - G_i^*(\theta))$, where

$G_i^*(\theta) = \int_0^\infty e^{-\theta t} G_i(dt)$, respectively. Letting $\theta \to 0$ we obtain equation (4.1), which proves the Lemma.

Let

$$F(u_1, u_2, \ldots, u_m; t) = \sum_{n_1=0}^\infty \ldots \sum_{n_m=0}^\infty \mathsf{P}[Z_1(t) = n_1,\ Z_2(t) = n_2, \ldots, Z_m(t) = n_m] u_1^{n_1} u_2^{n_2} \ldots u_m^{n_m} .$$

In order to establish the existence of a limit distribution we first show that

$$F(u_1, u_2, \ldots, u_m) = \lim_{t \to \infty} F(u_1, u_2, \ldots, u_m; t)$$

exists. Let

$$h_0(u_1, u_2, \ldots, u_m) = \sum_{n_1=0}^\infty \ldots \sum_{n_m=0}^\infty p_{n_1, n_2, \ldots, n_m} u_1^{n_1} u_2^{n_2} \ldots u_m^{n_m}$$

and

$$\Phi_i(u_1, u_2, \ldots, u_m; t) = \mathsf{E}[u_1^{N_1(t)} u_2^{N_2(t)} \ldots u_m^{N_m(t)} \mid N(0) = \epsilon_i] .$$

$F(u_1, u_2, \ldots, u_m; t)$ satisfies the equation

$$F(u_1, u_2, \ldots, u_m; t) = 1 - G_0(t) - \int_0^t H(t-s) F(u_1, u_2, \ldots, u_m; t-s) dG_0(s) + \int_0^t F(u_1, u_2, \ldots, u_m; t-s) dG_0(s) , \tag{4.4}$$

where

$$H(t-s) = 1 - h_0(\Phi(u_1, \ldots, u_m; t-s), \ldots, \Phi(u_1, \ldots, u_m; t-s)) .$$

Let $x(t) = F(u_1, u_2, \ldots, u_m; t)$ and $y(t) = 1 - G_0(t) - I(t)$ where

$$I(t) = \int_0^t H(t-s)F(u_1, u_2, \ldots, u_m; t-s)dG_0(s) .$$

Then

$$x(t) = 1 - G_0(t) - I(t) + \int_0^t x(t-s)dG_0(s) . \tag{4.5}$$

In order to apply the renewal theorem, it follows from J a g e r s [3] that it is sufficient to show that $I(t)$ is directly Riemann integrable, which follows on showing that $I(t)$ is bounded by a monotone function which is integrable on $[0, \infty)$.

For $0 \leqslant u_i < 1$ $(i = 1, 2, \ldots, m)$

$$\begin{aligned} H(t)F(u_1, u_2, \ldots, u_m; t) &\leqslant H(t) \leqslant \\ &\leqslant \sum_{i=1}^{m} m_i(1 - \Phi_i(u_1, u_2, \ldots, u_m; t)) \leqslant \\ &\leqslant \sum_{i=1}^{m} m_i \sum_{j=1}^{m} \mu_{ij}(t)(1-u_j) \quad , \end{aligned} \tag{4.6}$$

where

$$m_i = \sum_{n_1=0}^{\infty} \ldots \sum_{n_m=0}^{\infty} n_i p_{n_1, n_2, \ldots, n_m} .$$

Thus

$$\begin{aligned} 0 \leqslant I(t) &\leqslant \sum_{i=1}^{m} m_i \sum_{j=1}^{m} (1-u_j) \int_0^t \mu_{ij}(t-s)dG_0(s) \leqslant \\ &\leqslant \sum_{i=1}^{m} m_i \sum_{j=1}^{m} (1-u_j)\left(1 - G_0(t) + \int_0^t \mu_{ij}(t-s)dG_0(s)\right) , \end{aligned}$$

which is monotone non-decreasing in t.

Thus $y(t)$ is directly Riemann integrable and applying the renewal theorem

$$x(t) \to \frac{1}{\lambda_0} \int_0^\infty y(t)dt ,$$

where $\lambda_0 = \int_0^\infty t dG_0(t)$, which establishes the existence of the limit $F(u_1, u_2, \ldots, u_m)$.

A simple extension of Jagers [3] shows that $M_i(t)$ is finite for all t. In order to prove that $F(u_1, u_2, \ldots, u_m)$ is a probability generating function it is sufficient to show that $M_i(t)$ tends to a finite limit. Let

$$(4.8) \qquad f(t) = \sum_{j=1}^{m} m_j \int_0^t \mu_{ij}(t-s)dG_0(s) .$$

The equation for $M_i(t)$ for the multitype process corresponding to equation (2.1) is

$$(4.9) \qquad M_i(t) = f(t) + \int_0^t M_i(t-s)dG_0(s) .$$

By the renewal theorem

$$M_i(t) \to \frac{1}{\lambda_0} \int_0^\infty f(t)dt$$

which reduces to $\frac{1}{\lambda_0} \sum_{j=1}^{m} a_{ij} m_j$ on applying Lemma 1. Thus in the subcritical case there is a limit distribution with means

$$(4.10) \qquad M_i = \lim_{t \to \infty} M_i(t) = \frac{1}{\lambda_0} \sum_{j=1}^{m} a_{ij} m_j \qquad (i = 1, 2, \ldots, m) .$$

REFERENCES

[1] W. Feller, *An introduction to probability theory and its applications,* Vol. II., John Wiley and Sons, New York, 1966.

[2] T.E. Harris, *The theory of branching processes,* Springer Verlag, Berlin, 1963.

[3] P. Jagers, Age-dependent branching processes allowing immigratior *Teorija Verojatn. i Primenen.*, 13 (1968), 230-242; Correction ibid 13 (1968), (Letters to Ed.).

[4] C.J. Mode, *Multitype branching processes: theory and application:* American Elsevier, New York, 1971.

[5] J. Radcliffe, The convergence of a super-critical age-dependent branching process allowing immigration at the epochs of a renewal process, *Math. Biosciences,* 14 (1972), 37-44.

[6] J. Radcliffe, The asymptotic frequencies of types in a multitype age-dependent branching process allowing immigration, *J. Appl. Prob.* (to appear).

COLLOQUIA MATHEMATICA SOCIETATIS JÁNOS BOLYAI
9. EUROPEAN MEETING OF STATISTICIANS, BUDAPEST (HUNGARY), 1972.

SOME INEQUALITIES IN QUEUING THEORY

T. ROLSKI

1. INTRODUCTION

The paper deals with queuing systems GI/M/N under incomplete information concerning inter-arrival times distributions. It is assumed that the distribution function (d.f.) $G(t)$ of the inter-arrival time belongs to some set of d.f.'s and the d.f. of the service time is the exponential with the fixed expected value $\frac{1}{\mu}$. The bounds for some characteristics, in various systems, will be given here. The author has encountered some publications dealing with similar problems, among the others, the papers of Kingman [4], Rogozin [7], Marshall [5].

As an introduction into our consideration let us quote the theorem proved by Rogozin [7].

Denote by

$$\mathscr{F}^{m_1} = \left\{ F\colon F \text{ is a d.f.},\ F(0) = 0,\ \int_0^\infty x dF(x) = m_1 \right\}$$

and let us consider the class of systems GI/G/1 in which the d.f. of the inter-arrival time $G(x)$ belongs to $\mathscr{F}^{m_1}$ and the d.f. of the service time $B(x)$ is fixed (it is assumed that $\int_0^\infty x^2 dB(x) < \infty$ and $\int_0^\infty x dG(x) > > \int_0^\infty x dB(x)$). Let $\mathsf{E}W$ denote the expected value of the waiting time for service in the stationary conditions. Then the theorem of Rogozin may be formulated as follows

Theorem 1. $\inf\limits_{G \in \mathscr{F}^{m_1}} \mathsf{E}W$ *is reached for the d.f.* G_{m_1} *having only one jump of height* 1 *in* m_1.

This theorem one can interpret in two ways:

a) among all possible distributions of inter-arrival times, the constant inter-arrival time is the optimal one (i.e. it minimizes $\mathsf{E}W$);

b) having incomplete information concerning the inter-arrival time distribution we would like to estimate the expected value of the waiting time.

The second interpretation will correspond to our considerations. The theorem of Rogozin gives us the lower bound only, as the information concerning exclusively the first moment is not full. In various problems two side inequalities are interesting but more assumption are needed.

The following classes of d.f.'s will be considered further

$$\mathscr{F}^{(m_1, \dots, m_n)} = \left\{ F\colon F(0) = 0, \ \int_0^\infty x^i dF(x) = m_i, \ \ i = 1, 2, \dots, n \right\},$$

and

$$\mathscr{F}_{IFR}(\mathscr{F}_{DFR}) = \left\{ F\colon \frac{F(x+t) - F(x)}{1 - F(x)} \ \text{is increasing (decreasing)} \ \text{in } x \text{ for } F(x) < 1 \right\}.$$

2. BOUNDS FOR GI/M/1 QUEUES

In Takacs' book [8] it is proved that size η of the queue at the moment immediately before the arrival of a customer to a GI/M/1 system in stationary conditions has geometrical distribution

$$\mathsf{P}(\eta = k) = (1 - \delta_G)\delta_G^k, \qquad k = 1, 2, \dots .$$

The parameter δ_G fulfills the equation

$$x = \Phi_G(\mu(1 - x)); \qquad 0 \leqslant \delta_G \leqslant 1 ,$$

where $\Phi_G(s)$ denotes the Laplace transform of the d.f. $G(t)$ and μ is the service rate. It is assumed here that $r = m_1\mu > 1$. The following theorem gives us bounds for δ_G.

Theorem 2.

(1) $$\inf_{G \in \mathscr{F}^{(m_1, m_2)}} \delta_G = l$$

(2) $$\max_{G \in \mathscr{F}^{(m_1, m_2)}} \delta_G = 1 + \frac{m_1^2}{m_2}(l - 1) = \delta_{\max}$$

where $l \neq 1$ *is the root of the equation*

(3) $$x = e^{-r + rx} .$$

The proof of the theorem 2 is given in my paper [6]. The equation (3) depends only on r and it may be easily solved numerically. We can draw a graph of the solution depending on r (see Fig. 1).

Using (1) and (2) it is easy to estimate the expected value of the waiting time $\mathsf{E}W$

$$\frac{l}{\mu(1 - l)} < \mathsf{E}W \leqslant \frac{\delta_{\max}}{\mu(1 - \delta_{\max})} ,$$

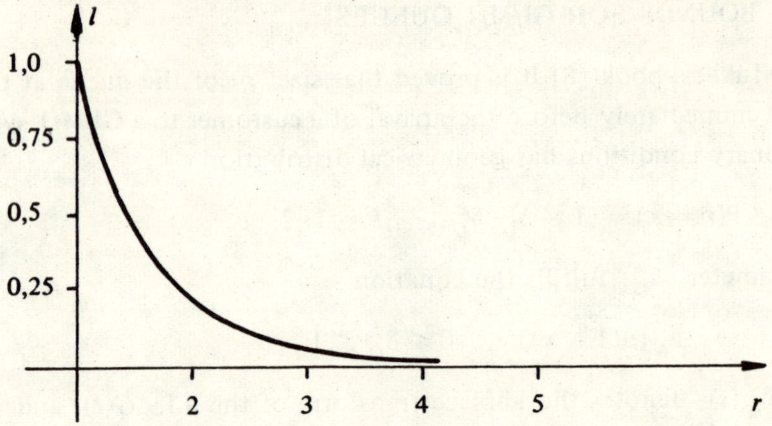

Figure 1.

Sometimes we known that the distribution of the inter-arrival time belongs to $\mathcal{F}^{m_1} \cap \mathcal{F}_{IFR}$. Then the following theorem may be useful.

Theorem 3.

$$\inf_{G \in \mathcal{F}^{m_1} \cap \mathcal{F}_{IFR}} \delta_G = l ,$$

$$\max_{G \in \mathcal{F}^{m_1} \cap \mathcal{F}_{IFR}} \delta_G = \frac{1}{r} .$$

where $r = m_1 \mu > 1$.

The proof of this theorem follows immediately from the estimations of Laplace transform given by Barlow and Proschan [2].

3. BOUNDS FOR MANY SERVER QUEUES

The following theorem can be proved by using the Tchebycheff systems method (see Karlin and Studden [3]).

Theorem 4.

$$(4) \qquad \sup_{G \in \mathcal{F}^{(m_1, \dots, m_n)}} \int_0^\infty e^{-sx} dG(x) = \int_0^\infty e^{-sx} dG_0(x) \qquad (s > 0)$$

where $G_0(x)$ *is a d.f. of a distribution concentrated in the points* x_i *with the masses* p_i *where* x_i *and* p_i *fulfil the following system of equations:*

$$
(5) \qquad
\begin{aligned}
&\sum_{i=1}^{r} p_i = 1\,,\\
&\sum_{i=1}^{r} p_i x_i = m_1\,,\\
&\quad\cdots\cdots\\
&\sum_{i=1}^{r} p_i x_i^k = m_k\,,\\
&\quad\cdots\cdots\\
&\sum_{i=1}^{r} p_i x_i^n = m_l
\end{aligned}
$$

$(p_i \geqslant 0,\ \ 0 = x_1 \leqslant x_2 \leqslant \ldots \leqslant x_r < \infty) \quad r = \left[\frac{n}{2} + 1\right], \quad l = 2\left[\frac{n}{2}\right].$

$$(6) \qquad \inf_{G \in \mathscr{F}^{(m_1, \ldots, m_n)}} \int_0^\infty e^{-sx}\, dG(x) = \int_0^\infty e^{-sx}\, dG_1(x)$$

where $G_1(x)$ *is a d.f. of a distribution concentrated in points* x_i *with masses* p_i *where* x_i, p_i *fulfil* (5) *with* $r = \left[\frac{n+1}{2}\right]$, $l = 2\left[\frac{n-1}{2}\right] + 1$ *and* $0 < x_i \leqslant \ldots \leqslant x_r < \infty$.

In the special case with $n = 2$ *we have*

$$(7) \qquad \sup_{G \in \mathscr{F}^{(m_1, m_2)}} \int_0^\infty e^{-sx}\, dG(x) = 1 - \frac{m_1^2}{m_2} + \frac{m_1^2}{m_2}\, e^{\left(-\frac{m_2}{m_1} s\right)},$$

$$(8) \qquad \inf_{G \in \mathscr{F}^{(m_1, m_2)}} \int_0^\infty e^{-sx}\, dG(x) = e^{(-m_1 s)}\,.$$

One can use (4), (6) or (7), (8) to get estimations of some characteristics of queuing systems because many of them may be expressed in terms of Laplace transform of the d.f. of inter-arrival time. However, it is rather difficult to obtain the explicite results. Thus, e.g. the rth binomial moment B_r of steady-state distribution of queue size in GI/M/∞ fulfills the inequalities

$$B_r^{\min} = \prod_{i=1}^{r} \frac{\Phi_{G_1}(i\mu)}{1 - \Phi_{G_1}(i\mu)} \leqslant B_r \leqslant \prod_{i=1}^{r} \frac{\Phi_{G_0}(i\mu)}{1 - \Phi_{G_0}(i\mu)} = B_r^{\max} \; ;$$

the expected value ES of the average distance between consecutive lost calls in a telephone traffic process fulfills

$$\frac{1}{\mu} \sum_{j=0}^{m} \binom{m}{j} \prod_{r=0}^{j} B_r^{\min} \leqslant \mathrm{E}S \leqslant \frac{1}{\mu} \sum_{j=0}^{m} \binom{m}{j} \prod_{r=0}^{j} B_r^{\max}$$

or

$$\delta_{G_1} \leqslant \delta_G \leqslant \delta_{G_0} ,$$

provided that the d.f. of the inter-arrival time $G(x)$ belongs to $\mathscr{F}^{(m_1, \ldots, m_n)}$.

One can also look for the estimation when the d.f.'s of the inter-arrival times belong to another set, for example to the class $\mathscr{F}^{(m_1, \ldots, m_n)} \cap \mathscr{F}_{IFR}$ or $\mathscr{F}^{(m_1, \ldots, m_n)} \cap \mathscr{F}_{DFR}$. Some results obtained by Barlow [1], and Proschan [2] turn out to be useful in these investiagations.

REFERENCES

[1] R. Barlow, Bounds on integrals with applications to realibility problems, *Ann. Math. Statist.*, 36 (1965), 565-574.

[2] R. Barlow – F. Proschan, *Mathematical Theory of Reliability.* John Wiley Inc., New York, 1965.

[3] S. Karlin – W. Studden, *Tchebycheff Systems with Applications in Analysis and Statistics,* New York, 1966.

[4] J.F.C. Kingman, Some inequalities for the queue GI/G/1, *Biometrica,* 49 (1962), 315-324.

[5] K.T. Marshall, Some inequalities in queuing, *Oper. Res.*, 16 (1968), 651-665.

[6] T. Rolski, On some inequalities for GI/M/n queue, *Zastosow. Mat.*,, 13 (1972), 43-47.

[7] B.A. Rogozin, Some extremal problems in the queuing theory, (in Russian) *Teorija Verojatn. i Primenen.*, 11 (1966), 161-169.

[8] L. Takacs, *Introduction to the Theory of Queue,* New York, 1962.

COLLOQUIA MATHEMATICA SOCIETATIS JÁNOS BOLYAI
9. EUROPEAN MEETING OF STATISTICIANS, BUDAPEST (HUNGARY), 1972.

TIME SERIES ANALYSIS ON CDC 3300

M. RUDA — L. SZEIDL — G. TUSNÁDY

At the beginning of this year a team has been organized by M. Arató to develop a comprehensive computer program package for time series analysis. The members of this team are M. Arató, A. Benczúr, I. H. Gaudi, T. N. Gyurácz and the authors. We are working on CDC 3300 our programs are written in Fortran language.

The main aim of our computer program package is the statistical analysis of the real valued multidimensional stochastic processes with discrete time parameter (usually they are called time series).

Our program-package has the following parts:

1. Transformations: seasonal and non-seasonal differences.

2. Elementary statistics: sample mean, sample variance, autocovariance function, autocorrelation function, partial autocorrelation coefficients, and roots of the characteristic equation.

3. Cross-covariances: the multidimensional extensions of the statistics

mentioned in the previous point.

4. Estimation of the spectral density function according to the following spectral windows: periodogram, Bartlett, Daniell, Tuckey – Hamming, and Parzen.

5. Estimation of the parameters of the autoregressive, moving average and mixed processes.

6. Estimation of the regression parameters in the following cases:

a) the error term is a white noise,

b) the error term is an arbitrary process with known spectral density,

c) the error term is an autoregressive process with unknown parameters.

7. Linear prediction of processes with known or unknown covariances.

We should like to present here some problems which arise in connection of the parameters of stationary stochastic processes. Let us denote the process by $\mathscr{X} = \{\xi_t,\ t = 0, \pm 1, \pm 2, \ldots\}$, the autocovariances of $\mathscr{X}$ by $B(t)$ $(t = 0, \pm 1, \pm 2, \ldots)$:

$$B(t) = \mathsf{E}(\xi_0 - \mu)(\xi_t - \mu)$$

(where $\mu = \mathsf{E}\xi_0$), and the spectral density of $\mathscr{X}$ by $f(\lambda)$:

$$f(\lambda) = \frac{1}{2\pi} \sum_{t=-\infty}^{\infty} B(t) e^{it\lambda} .$$

For estimating the autocovariances $B(t)$ and the spectral density $f(\lambda)$ we used the usual methods. In connection with the estimation of the spectral density we have the following

Problem. What is the asymptotic distribution of the relative difference

$$\alpha_n \sup_{0 \leqslant \lambda \leqslant \pi} \frac{|f_n(\lambda) - f(\lambda)|}{f(\lambda)} - \beta_n$$

for suitable chosen norming factors α_n, β_n? Some results are known in this direction ([7], [8]). It seems that the situation is similar to the estimation of the density function from samples with independent, identically distributed elements, namely the norming factors are the following:

$$\alpha_n = \sqrt{2 \log n - \log \log n}, \quad \beta_n = 2 \log n - \log \log n ;$$

and the asymptotic distribution is of the form $\exp\{-e^{-x}\}$. However our efforts to find a fairly general theorem in the literature have been unsuccessful.

As a second step, the process $\mathscr{X}$ is assumed to be the solution of the difference equation

$$\sum_{i=0}^{p} a_i \xi_{t-i} = \sum_{j=0}^{q} b_j \epsilon_{t-j} ,$$

where the series $\{\epsilon_t : t = 0, \pm 1, \pm 2, \ldots\}$ is a white noise (i.e. the ϵ_i's are independent, identically distributed random variables with expectation 0 and varaince σ^2). The numbers p, q are the autoregressive and the moving average dimension of the process, respectively. In case $q = 0$ the process is autoregressive, in case $p = 0$ it is moving average, if $p > 0$, and $q > 0$, then the process is called a mixed process. There is a great difference between autoregressive, and mixed processes. If μ and σ are known, $q = 0$, and the distribution is normal, then the system of statistics

$$\{\xi_i\}_{t=0}^{p-1}, \qquad \left\{ \sum_{\nu=0}^{n-p} \xi_{i+\nu} \xi_{j+\nu} \right\}_{i,j=0}^{p}$$

is sufficient, and therefore the maximum likelihood equation is relatively simple. In this case the conditional maximum likelihood equations coincide with the Youle – Walker equations, which are usually deduced by the method of moments. Hence the asymptotic distribution of the roots of the Youle – Walker equations is normal. However – as a previous investigation of M. Arató and A. Benczúr [2] have shown – the rate of the convergence is very slow, and that is why the asymptotic distribution is inefficient in the majority of practical cases. There is a permanent need to develop more and more tables and other representations of exact distributions of the parameter-estimators.

The problem of producing the distributions of the estimators is more cumbersome in case of mixed processes. We started with Monte-Carlo methods, the first results showed some kind of irregularity. The roots of the Youle – Walker equations were very sensitive, very inaccurate in cases when the hypothetical dimensions were over their theoretical values. This effect was completely clearified later by T. N. Gyurácz who proved that the Youle – Walker equations become singular if for the hypothetical autoregressive dimension P, hypothetical moving average dimension Q, theoretical autoregressive dimension p, and theoretical moving average dimension q the inequalities

$$P > p , \qquad Q > q$$

hold, and the empirical covaraინces in Youle – Walker equations are replaced by their expectations. This effect is rather surprising, and so is the fact it is not mentioned in the literature ([1], [3], [4], [5], [6]).

This effect shows, that there is a need for other methods to supply estimations in mixed models. The maximum likelihood method is practically inefficient since there is no suitable system of sufficient statisties. Our method is based on the following remark. In most cases of practical life the values of a mixed process are equal to the values in some discrete time points of a process with continuous time parameter, which is the solution of the stochastic differential equation of order n

$$d\xi^{(n-1)}(t) + [c_1 \xi^{(n-1)}(t) + \ldots + c_n \xi(t)]dt = dw(t) ,$$

where $w(t)$ is a Wiener process. That is why we investigate directly the problem of estimation of the parameters c_i of this equation.

An other possibility to develop a new estimation is to apply the methods of stochastic approximation. The difficulties in this case arise from the fact, that the classical theorems of stochastic approximation have been proved under the condition of the independence of the erros. We are working on the following

Problem. Assume that ξ_t is a stationary stochastic process, $F(x, y)$ is a real function with the property that the function

$$R(x) = \mathsf{E}F(x, \xi_0)$$

is monotone increasing. We find the root $\tilde{x}$ of the equation

$$R(x) = 0$$

using the Robbins – Monroe iteration

$$x_{n+1} = x_n - \lambda_n F(x_n, \xi_n).$$

The problem is to present conditions ensuring the convergence of x_n to $\tilde{x}$.

REFERENCES

[1] T.W. Anderson, *An introduction to multivariate statistical analysis,* Wiley, New York, 1970.

[2] M. Arató – A. Benczúr, On the distribution function of the estimation of the damping parameter of a Gauss – Markov process, *Studia Sci. Math. Hungar.,* 5 (1970), 445-456.

[3] G. Box – G. Jenkins, *Time series analysis, Forecasting and control,* Holden-Day, 1970.

[4] V. Grenander – J. Rosenblatt, *Statistical analysis of stationary time series,* Almquist and Wiksell. Stockholm, 1956.

[5] E.T. Hannan, *Multiple time series,* Wiley, New York, 1970.

[6] M.G. Kendall – A. Stuart, *The advanced theory of statistics,* Griffin, London, 1958.

[7] M. Rosenblatt, Curve estimates, *Ann. Math. Statist.,* 42 (1971), 1815-1842.

[8] M. Woodroofe – J. Van Ness, The maximum deviation of sample spectral densities, *Ann. Math. Statist.,* 38 (1967), 1558-1569.

COLLOQUIA MATHEMATICA SOCIETATIS JÁNOS BOLYAI
9. EUROPEAN MEETING OF STATISTICIANS, BUDAPEST (HUNGARY), 1972.

A GLOBALLY CONVERGENT ALGORITHM FOR NONLINEAR LEAST SQUARES

L.E. SCHWARTZ

1. INTRODUCTION

Available algorithms, such as those of Levenberg [9] and Marquardt [10] among a number of later ones, cannot guarantee global solutions to nonlinear regression problems. To remedy this situation for a broad class of nonlinear regression problems that can be cast in the form of fixed-point problems, a new non-gradient algorithm that obtains global solutions is given in this paper. For convenience in exposition, results reported are limited to the special case of inequality-constrained nonlinear least squares problems, having constant unitary weights. Such problems are important in econometric and certain other statistical applications. A broader class of constrained nonlinear regression problems will be the subject of a later paper.

The nonlinear least squares problem dealt with in Section 3 is of the form,

$$\text{(P)} \quad \begin{array}{ll} \text{Minimize:} & \| y - F(x) \| \\ \text{Subject to:} & G_i(x) \leqslant h \quad (i = 1, 2, \ldots, m;\ x \geqslant 0)\ , \end{array}$$

where the transformation $F(x)$ is a twice continuously differentiable, lower semi-continuous, nonconvex, single-valued mapping from E^p to E^1; x is a vector of parameters to be estimated in E^p; y is a fixed n-vector of observations, such that $p \leqslant n$; the Euclidean norm is in E^p; the right hand side of the a priori inequality restrictions is the m-vector h; and, the transformations $G_i(x)$, $i = 1, 2, \ldots, m$, are also twice continuously differentiable, lower semi-continuous, convex, single valued mappings from E^p to E^1. The last assumption is due to the fact that the possible nonconvexity of the $G_i(x)$ is not an intrinsic property of (P), but the nonconvexity of $F(x)$ can be demonstrated by examining the Hessian of the objective of (P). Furthermore, it is assumed that the set of global optima $\{x^*\}$ is nonempty. This assumption is for convenience in proofs, since (P) may in general have many, or possibly an infinite number, of local solutions, but no global solutions.

The algorithm of Section 3 applies iteratively another algorithm that uses continuous self-maps of p-dimensional simplices, for the case of the Brouwer fixed point theorem, due to Kuhn [6]. The combinatorial topological approach of the Kuhn algorithm was suggested by some earlier work of Scarf [14] on the solution of highly nonlinear systems of equations in mathematical economics for which gradient methods were ill-suited; and the Kuhn algorithm was later elaborated and proved by Hansen and Scarf [5]. Before each applications of the Kuhn algorithm, one of a sequence of convexified subproblems of (P) is formed. The problem (P) is convexified by means of a Tihonov regularization [15], using a generalized parametric Lagrangean penalty function of a form suggested by Gould [3]. The procedure can handle nonconvex constraints $G_i(x)$ $(i = 1, 2, \ldots$ $\ldots, m)$ if the feasible region defined by the intersection of the constraints of (P) can be divided into convex subregions. An alternative possibility not pursued in this paper, is to modify the Kuhn algorithm so that convexified constraints replace nonconvex $G_i(x)$ $(i = 1, 2, \ldots, m)$. The Tihonov regularization is changed in a systematic way by changing the values of a parameter t. The procedure may thus be viewed as the solution of (P) through the use of a parameter imbedding method that is a generalization to dynamic systems with unstable structures of a procedure due to Meyer [11]. The Kuhn algorithm is applied iteratively to conjugate dynamic systems related to the convexified subproblems derived from problem (P), rather than to those subproblems themselves. The use of dynamic systems preserving the solution of (P) is suggested by related work of Brannin and Hoo [1] on the solution of unconstrained problems by means of a variable-step, quasi-Newton method.

The plan of the paper is as follows: Section 2 contains most of the notation required throughout the paper; the algorithm is presented in summary form, along with a brief discussion, in Section 3; Section 4 gives a proof of convergence in outline-form; and finally, some lines of future research are the subject of Section 5.

2. NOTATION

Transform the vectors x and y, in terms of which the problem (P) is defined, to barycentric coordinates. If the new coordinates $x_i \geqslant 0$ $(i = 1, 2, \ldots, p)$, $\sum_{i=1}^{p} x_i = 1$ and $y_j \geqslant 0$ $(j = 1, 2, \ldots, n)$, $\sum_{j=1}^{n} y_j = 1$ then the vector x lies on the unit p-simplex

$$S = \left\{ x_i \geqslant 0 \,\middle|\, \sum_{i=1}^{p} x_i = 1 \right\}.$$

Let x^k denote the solution obtained from the kth iterative application of the algorithm to the convexified problem obtained from problem (P) defined below. Define the finite indexing set $I(x) = \{i \,|\, F_i(x) \geqslant 0,$ $G_i(x) \leqslant 0,\ i = 0, 1, 2, \ldots, m\}$, with h transformed to zero. The Jakobia matrix is then defined as

$$(1) \qquad J(x) = [\nabla\lambda_i(F_i(x),\ G_i(x),\ t]_{i \in I(x)}\ ,$$

where $\lambda_i(x, t)$ is an appropriate generalized parametric Lagrangean pena ty function, a special case of the Tihonov regularization used to convexi problem (P). The generalized inverse of $J(x)$ is denoted by $J^-(x)$ fo lowing Rao and Mitra [13], where $J^{-}(x) = (J^T J)^{-1} J^T$ (assuming that $J^T J$ is nonsingular) and J^T denotes the transpose of J. After each iteration, the value of parameter t of $\lambda_i(x, t)$ is changed so that the error involved in solving the convexified problem instead of (P) is reduced in a monotonic fashion. Normally, x^{k+1} would be calculated as the optimum point of

$$\text{(2)} \qquad \begin{aligned} &\min_x \Big\{ F^c(x, t) = F(x^k) + \sum_{\mu} \mu_i^k \lambda_i(F_i(x), t) \mid F_i(x) \geqslant 0 \,, \\ &\qquad G_i(x) \leqslant 0, \; i = 0, 1, 2, \ldots, m; \; \mu_i^k > 0; \; x \in T \Big\} \end{aligned}$$

where by a theorem of Falk [16], $F^c(x, t)$ is the convex envelope [2] of (P) in the case where the $G_i(x)$ $(i = 1, 2, \ldots, m)$ are convex mappings, and T is a compact subset of E^p. At each iteration k, each violated constraint $G_i(x) \leqslant 0$, $i = 1, 2, \ldots, m$ is added to the objective $F(x)$ with the penalty $\mu\lambda_i(x, t)$. Given x^k,

$$\text{(3)} \qquad \mu^k = J^{-}(x^k) \nabla F(x^k) \,,$$

where $\nabla F(x^k)$ denotes the gradient, a column vector.

However, the conjugate dynamic systems with unstable structure,

$$\text{(4)} \qquad \frac{dF^c(x, t)}{dt} = \mp H^{+}(x) J(x) \,,$$

are solved instead, where the scalar differential equation with the p-vector x as argument is shown in (4), H denotes the Hessian matrix of $F^c(x, t)$, and the superscript $+$ on H indicates the generalized inverse computed by means of the singular value decomposition. The matrix H^{+} is defined, following Rao and Mitra [13],

$$\text{(5)} \qquad H^{+} = U \Sigma V^T \,,$$

with U and V orthogonal matrices consisting of the orthonormalized eigenvectors of HH^T in the case of U and the orthonormalized eigenvectors of $H^T H$ in the case of V, and Σ is a partitioned matrix, consisting of diagonal elements that are non-negative square roots of the eigenvalues of $H^T H$ form a diagonal matrix above the partition, and there is a null matrix below.

3. ALGORITHM

A simplified summary of the algorithm $A: T \to 2^T$, where T is a compact subset of E^p.

Step 0. Compute an $x^0 \in T$; set $k = 0$.

Step 1. Choose $\lambda_i^{(1)}(x, t)$ $(i = 1, 2, \ldots, m)$.

Step 2. Compute $x^j \in A(x^k)$ for one of the conjugate dynamic systems related to $F^c(x, t)$ and then x^{j+1} for the other conjugate dynamic system related to $F^c(x, t)$, for j odd integers, using the Kuhn algorithm.

Step 3. Choose an improved Lagrangean $\lambda_i^{(k)}(x, t)$ $(i = 1, 2, \ldots, m)$ by changing t, based on information obtained in Step 2 on the error between (P) and $F^c(x, t)$ used instead of (P); if $k \leqslant K$, go to step 2 and set $k = k + 1$; otherwise go to step 4 and set $k = k + 1$.

Step 4. Extrapolate then solutions $\{x^k\}$, using an appropriate extrapolation or spline curve fitting technique.

Step 5. If $\{x^k\}$ is optimal in (P), stop; otherwise set $k = k + 1$ and go to Step 3.

A brief discussion of each step follows:

Step 0. Use a feasible point if one is known; otherwise any arbitrary point will do as a starting point for the algorithm.

Step 1. Choose any arbitrary functions $\lambda_i^{(1)}(x, t)$ such that the function $\lambda: E^1 \to E^p$ is coordinatewise monotonic decreasing, and for some $x \in T$, $\lambda_i^{(1)}(x, t) < \infty$ $(i = 1, 2, \ldots, m)$.

Step 2. Each of the conjugate dynamic systems shown in (4) is on opposite sides of a zero of the Newton vector, the right hand side of (4). Finding a zero of (4) is clearly equivalent to finding the minimum of $F^c(x, t)$. Since finding the fixed point of the scalar differential equations (4) is a special case of finding their zeros, the Kuhn algorithm [6] for approximating fixed points in the case of Brouwer's theorem is well-suited to this calculation since its computations are algebraic. Kuhn's algorithm uses a particular simplicial subdivision [7], and an analogue to the change of bases in linear programming in which a vertex is removed from a subsimplex and replaced by a unique adjacent subsimplex containing the remaining $p - 1$ vertices. The Lemke – Howson technique [8] is used to select the vertex to be removed. Both conjugate dynamic systems are solved by means of the Kuhn algorithm, since Brannin and Hoo [1] found distinct zeros of (4) on each of these branches.

Step 3. The choice of improved Lagrangeans $\lambda_i^{(k)}(x, t)$ $(i = 1, 2, \ldots, m)$ is in general nontrivial as suggested by Tihonov [15] in a closely related context. However, a reasonable approximation can be obtained by using a spline curve fitting technique to minimize the error incurred by replacing (P) by the convexified problem (2), i.e. (P) $-F^c(x, t)$ such that these errors are monotonically decreasing as $k \to \infty$. This process is interpreted as a structural change in the Lagrangean, dependent on values of the parameter t.

Step 4. After some number, K, of iterations (a prespecified number) have been reached, the algorithm can be appreciably accelerated by using an appropriate extrapolation or spline curve fitting technique. Such a procedure is essential, since the Kuhn algorithm is computation-limited.

Step 5. If the $\{x^*\}$ of the Problem (P) are the same as some subproblem of the convexified problem (2), except for a numerical tolerance, the procedure can be terminated since the optimal solutions of (P) have been found provided that they exist. If an optimal solution of (P) has not been found, additional iterations of Steps 3 and 4 can be initiated in an attempt to refine further the single trajectory on which all global optima lie.

A sharp bound on the error incurred by approximating the fixed points of the continuous mappings (2) with Kuhn's algorithm at each iteration k, depending on a measure of the approximate continuity of (2), is presented in Halpern [4].

4. CONVERGENCE PROOF

A convergence theorem and a brief outline of a proof are developed in this section for the case of convex constraints. A slightly modified theorem of Polak is needed to accomplish this.

Theorem (Polak [12]). *If* (i) *a mapping* $F^c(x, t)$: $T \to E^p$ *exists such that* $F^c(x, t)$ *is bounded from below on* T; *and if* (ii) *for every* $x \in T$ *which is not a global minimum, there exists an* $\epsilon(x) > 0$ *and a* $\delta(x) < 0$ *such that*

$$(6) \qquad F^c(x^{k+1}, t) - F^c(x^k, t) \leqslant \delta(x) < 0$$

for all $\{x^k \in T \mid \|x^k - x\| \leqslant \epsilon(x)\}$, *and for all* $x^{k+1} \in A(x^k)$. *Then, the sequence* $\{x^k\}$ *constructed by the algorithm of Section* 3 *is infinite, and every accumulation point of* $\{x^k\}$ *is a global optimum of* (P).

Outline of the proof. To verify hypothesis (i) of the theorem above it would be necessary to show that the assumption of compactness for T, and the continuity of $F(\cdot)$, $G(\cdot)$ imply the boundedness of $F^c(x, t)$ for any value of the parameter t taken on in the algorithm of Section 3.

The verification of hypothesis (ii) of the theorem would involve a demonstration that a monotonically nonincreasing sequence $\{x^k\}$ of solutions are produced by the algorithm of Section 3. This could be shown by noting that teach of the mappings $F^c(x, t)$ is convex and the parameter t is changed in such a way that no increasing x^k occur.

The rest of the proof consists of verifying that $F^c(x, t)$ is well-defined and deriving the conclusion of the theorem above from a number of facts. First, is the fact that finding the zeros of the dynamic systems (4) preserves the solution of $F^c(x, t)$, which is in turn one-to-one with the solution of (P) under the stated assumptions, according to Falk (personal communication). It would next be necessary to verify that the primitive sets [6] of the dynamic systems (4) exist, satisfying the hypothesis of the theorem underlying the Kuhn algorithm, use the Kuhn technique to approximate their fixed points. The procedure to be used is that sketched by Hansen and Scarf [5] for the case of Brouwer's theorem, using integral equation images of the dynamic systems of (4), with the system $Ax = b$ of the Hansen and Scarf Theorem 1 interpreted as the related linear algebraic system, e.g., a bi-diagonal or tri-diagonal algebraic system, the solution of which simulates the solution of the dynamic systems (4). Finally, the solution set obtained could be shown to be global, since $\{x^*\}$ of solutions was initially assumed to be nonempty.

5. RESEARCH PERSPECTIVES

There is a great deal of additional work that must be done before research reported here has been completed. A computer code, implementing the conceptual algorithm presented in Section 3 is the primary line of research that is being pursued now. As soon as the routine is developed, a number of test problems, and later real econometric and other statistical problems, will be run. Another task for the future is to develop versions of the combinatorial topological approach to the computation of fixed points of a continuous mapping that are better suited to the computation of (P) and other nonlinear regression problems, e.g., maximum likelihood formulations. The maximum likelihood formulation would be useful for those applications in which the variance is not constant, so that the assumption of fixed unitary weights developed here would not be applicable. Extensions of proofs of convergence to the case of nonconvex constraints should also be pursued.

REFERENCES

[1] F.H. Branin – S.K. Hoo, A method for finding multiple extrema of a function of n variables in F. A. Lootsma (ed.), *Numerical methods for non-linear optimization,* Academic Press, London, 1972, 231-237.

[2] J.E. Falk, Lagrange multipliers and nonconvex programs, *SIAM Journal on Control,* 7 (1969), 534-545.

[3] F.J. Gould, Extensions of Lagrange multipliers in nonlinear programming, *SIAM Journal on Applied Mathematics,* 17 (1969), 1280-1297.

[4] B. Halpern, Almost fixed points for subsets of Z^n, *Journal of Combinatorial Theory,* 11 (1971), 251-257.

[5] T. Hansen – H. Scarf, *On the Applications of a Recent Combinatorial Algorithm,* Cowels Foundation Discussion, No. 272, Yale University, New Haven, Connecticut, 1969.

[6] H.W. Kuhn, Simplicial approximation of fixed points, *Proceedings of the National Academy of Sciences,* 61 (1968), 1238-1242.

[7] H.W. Kuhn, Some combinatorial lemmas in topology, *IBM Journal of Research and Development,* 4 (1960), 518-524.

[8] C.E. Lemke – J.T. Howson, Equilibrium points of bimatrix games, *SIAM Journal on Applied Mathematics,* 12 (1964), 413-423.

[9] K. Levenberg, A method for the solution of certain nonlinear problems in least squares, *Quarterly of Applied Mathematics,* 2 (1944), 164-168.

[10] D.W. Marquardt, An algorithm for leas-squares estimation of non-linear parameters, *SIAM Journal on Applied Mathematics,* 11 (1963), 431-441.

[11] G.H. Meyer, On solving nonlinear equations with a one-parameter imbedding, *SIAM Journal on Numerical Analysis,* 5 (1968), 739-752.

[12] E. Polak, On the implementation of conceptual algorithms, *Nonlinear Programming: Proceedings of a Symposium Conducted by the Mathematics Research Center,* Academic Press, New York, 1970.

[13] C.R. Rao – S.K. Mitra, *Generalized inverses of matrices and its applications,* Wiley, New York, 1971.

[14] H. Scarf, The approximation of fixed points of a continuous mapping, *SIAM Journal on Applied Mathematics,* 15 (1967), 1328-1343.

[15] A.N. Tihonov, Methods for the regularization of optimal control problems, *Doklady Akad. Nauk.,* 162 (1965), 761-763.

COLLOQUIA MATHEMATICA SOCIETATIS JÁNOS BOLYAI
9. EUROPEAN MEETING OF STATISTICIANS, BUDAPEST (HUNGARY), 1972.

D_A-OPTIMALITY AND DUALITY

R. SIBSON

1. INTRODUCTION. THE OPTIMAL DESIGN PROBLEM.

The fundamentals of optimal experimental design theory have been set out in a number of papers from about 1950 onwards. In particular, clear accounts of the background and motivation are given by Whittle [8] and Wynn [9], both of whom give numerous references to earlier work in this area, and there is much material in the book by Fedorov [3]. We shall take advantage of this by declining to repeat the full background of the subject, and we shall pose the mathematical problem directly. It is as follows. $\mathscr{X}$ is a set, and f is a function on $\mathscr{X}$ which takes values in k-dimensional Euclidean space. We assume that $\mathscr{X}$ is a measurable space and that f is a Borel measurable function, and moreover that $f(\mathscr{X})$ is a compact set $\mathscr{V}$. If ξ is a probability measure on $\mathscr{X}$, then the induced probability measure on $\mathscr{V}$ is ξf^{-1}. If λ is a probability measure on $\mathscr{V}$, then we define

$$\mu(\lambda) = \int_{\mathscr{V}} vv^T \lambda(dv)$$

where superscript T denotes the transpose of a vector or matrix. Thus $\mu(\lambda)$ is a $k \times k$ positive-semidefinite (nonnegative-definite) symmetric matrix. In particular, $\mu(\xi f^{-1})$ is such a matrix; in experimental design theory it represents the inverse-covariance of the MVULE of the system parameters given by the design ξ, up to a multiplicative constant. Thus, informally speaking, we would like to make $\mu(\xi f^{-1})$ as large as possible in some appropriate sense, so as to make the estimator covariance as small as possible. The problem of optimal experimental design can thus be expressed as the problem of maximizing some criterion of the largeness of $\mu(\xi f^{-1})$ by appropriate choice of the probability measure ξ.

Let $\mathscr{P}_r$ denote the set of $r \times r$ positive-definite symmetric matrices. We may consider $\mathscr{P}_r$ to be a subset of $\frac{1}{2}r(r+1)$-dimensional space, in which space it is an open convex cone. Its closure $\mathscr{Q}_r$ is the closed convex cone of $r \times r$ positive-semidefinite symmetric matrices. We define

$$\mathscr{M}(\mathscr{V}) = \{\nu\nu^T : \nu \in \mathscr{V}\}.$$

This is a subset of $\mathscr{Q}_k$, and since $\mathscr{V}$ is compact, so is $\mathscr{M}(\mathscr{V})$. It follows that the convex hull of $\mathscr{M}(\mathscr{V})$, which we write as $\mathscr{H}(\mathscr{V})$, is also compact (Eggleston [1], p. 22 and p. 38) and hence closed and bounded. Every matrix $\mu(\lambda)$ can be approximated arbitrarily closely (in the usual norm) by a matrix $\mu(\lambda')$ for which λ' has finite support; thus $\mu(\lambda)$ lies in the closure of $\mathscr{H}(\mathscr{V})$ and hence, by the above, in $\mathscr{H}(\mathscr{V})$ itself. But $\mathscr{H}(\mathscr{V})$ is precisely the set of all finite nonnegative linear combinations of points in $\mathscr{M}(\mathscr{V})$ with total weight unity (*convex* combinations) and so for any given probability measure λ on $\mathscr{V}$ there is a finitely supported probability measure λ' on $\mathscr{V}$ such that $\mu(\lambda') = \mu(\lambda)$. We shall call such a λ' a *finite design;* this phrase will carry no connotation of equal or rational weighting on the support points of λ', and we note that it applies to probability measures on $\mathscr{V}$ rather than on $\mathscr{X}$. If λ' is a finite design, then it may be represented in the form $\xi' f^{-1}$ where ξ' is a probability measure on $\mathscr{X}$; ξ' may, if so desired, be chosen to have no more support points than λ'. Sometimes we are required to optimise subject to certain conditions, for example symmetry conditions on

ξ. These conditions may be strong enough to prevent ξf^{-1} from ranging through all the probability measures on $\mathscr{V}$. We shall not consider this case; we shall rather assume that ξf^{-1} ranges through all probability measures on $\mathscr{V}$, and we have seen from the above arguments about finite reducibility that if no constraints are placed on ξ, then this assumption holds good, and that moreover we need only consider finite designs on $\mathscr{V}$. We note in passing that if we were concerned with the actual number of support points needed for λ, then we could bound this as $\frac{1}{2}k(k+1)+1$ by using Carathéodory's Theorem (Eggleston [1], p. 35). Having obtained a reduction of the problem of optimal experimental design to the problem of choosing a design λ on $\mathscr{V}$ to make $\mu(\lambda)$ in some sense as large as possible, we shall henceforward consider only this reduced problem, and we shall make use of the fact that it is always possible to replace an arbitrary design by a finite design yielding the same matrix.

2. CRITERIA FOR OPTIMALITY. WHITTLE'S QUASILINEAR REPRESENTATIONS

Numerous criteria for the largeness of the matrix $\mu(\lambda)$ have been proposed. Of these the best-known are c-optimality (minimization of $c^T[\mu(\lambda)]^{-1}c$ for a fixed nonzero vector c), D-optimality (minimization of $\det\{[\mu(\lambda)]^{-1}\}$), and D_s-optimality (minimization of $\det\{\boldsymbol{P}_s^T[\mu(\lambda)]^{-1}\boldsymbol{P}_s\}$, where $\boldsymbol{P}_s$ is the $k \times s$ matrix which has the $s \times s$ unit matrix as its principal submatrix and zeroes elsewhere; that is, $\boldsymbol{P}_s$ is the matrix of the projection operator onto the first s coordinates). We may unify these three criteria in the proposed new criterion of D_A-optimality, defined by minimization of $\det\{A^T[\mu(\lambda)]^{-1}A\}$, where A is a fixed $k \times s$ matrix of full rank s. c-, D-, and D_s-optimality are obtained by taking A to be $c, \boldsymbol{I}$, and $\boldsymbol{P}_s$ respectively, and in effect D_A-optimality is simply a coordinate-free version of D_s-optimality.

Unless we wish to deal only with D-optimality we must next consider how to interpret the inverse of $\mu(\lambda)$ when $\mu(\lambda)$ is singular. The "generalised inverse" construction for a positive-semidefinite symmetric matrix

M is well-known, and is essentially what we shall be performing, but in order to retain a coordinate-free approach we shall have to attack the problem somewhat obliquely. The matrix A is a $k \times s$ matrix of full rank s. So it is possible to choose a $k \times (k-s)$ matrix a with the property that the partitioned $k \times k$ matrix $[A\ a]$ is nonsingular; a can of course be chosen in many ways. The inverse of this matrix we partition correspondingly as $[B^T\ b^T]^T$, where B is an $s \times k$ matrix and b is a $(k-s) \times k$ matrix. We construct an interpretation for $A^T M^{-1} A$ in the case where the kernel of M is contained in that of A^T. Now the condition $A^T u = 0$ is equivalent to the condition that u can be expressed in the form $b^T t$ for some vector t, and so our condition is that the kernel of M should be spanned by the columns of b^T. Let the matrix z be chosen so that the columns of $b^T z^T$ from a basis for the kernel of M. Then z is of full rank and can be filled out by choice of Z to a nonsingular $s \times s$ matrix $[Z^T\ z^T]^T$. Let $[Y\ y]$ be the correspondingly partitioned inverse of this. Then

$$[A\ aY][B^T\ b^T Z^T]^T M[B^T\ b^T Z^T][A\ aY]^T =$$

$$= (AB + aYZb)M(AB + aYZb)^T = M\ ,$$

since $zbM = 0$ and $Mb^T z^T = 0$. Let L denote the matrix $[B^T\ b^T Z^T]^T M[B^T\ b^T Z^T]$, that is,

$$\begin{bmatrix} BMB^T & BMb^T Z^T \\ ZbMB^T & ZbMb^T Z^T \end{bmatrix},$$

so that $M = [a\ aY]L[A\ aY]^T$. L is clearly a positive-semidefinite symmetric matrix, and we now show that it is nonsingular, that is, positive-definite. Suppose that $q^T L q = 0$, where q is a vector. Partition q into $[q_1^T\ q_2^T]^T$. Then

$$(B^T q_1 + b^T Z^T q_2)^T M(B^T q_1 + b^T Z^T q_2) = 0$$

so

$$M(B^T q_1 + b^T Z^T q_2) = 0$$

and so

$$A^T(B^T q_1 + b^T Z^T q_2) = 0 ,$$

by the assumption that the kernel of M is contained in that of A^T. But $A^T B^T = I_s$, $A^T b^T = 0$, so $q_1 = 0$. By choice of z, $Mb^T Z^T q_2 = 0$ implies $q_2 = 0$, so $q = 0$ and L is nonsingular. It follows that the submatrices BMB^T and $ZbMb^T Z^T$ are both nonsingular, and that

$$\gamma_A(M) = BMB^T - BMb^T Z^T [ZbMb^T Z^T]^{-1} ZbMB^T$$

is also nonsingular, and thus positive-definite symmetric, since it is the inverse of the $s \times s$ principal submatrix of L^{-1}. We shall interpret $[A^T M^{-1} A]^{-1}$ as $\gamma_A(M)$ even when M is singular, since it will appear that we are only concerned with precisely the case where the kernel of M is contained in that of A^T. It is easy to check that this interpretation agrees with that which we would have arrived at by dealing with the generalized inverse in the usual way. It may be noted that the notation $\gamma_A(M)$ contains no reference to the arbitrary choices of a, z, Z which we have made; we show below that $\gamma_A(M)$ is in fact independent of such choices.

The set $\mathcal{Q}_r$ of $r \times r$ positive-semidefinite matrices has a natural partial order. If $M_1, M_2 \in \mathcal{Q}_r$, we write $M_1 \geqslant M_2$ to mean that $M_1 - M_2$ lies in $\mathcal{Q}_r$; since $\mathcal{Q}_r$ is a closed convex cone we may expect this partial ordering to relate well to convexity properties. We shall certainly expect that any criterion function extends this natural partial ordering. We note in passing that it is easy to prove that the inversion operation on $\mathcal{P}_r$ is order-reversing and $\mathcal{Q}_r$-convex in the strict sense.

Whittle [18] has pointed out that

$$c^T M^{-1} c = \max_h (2c^T h - h^T M h)$$

and that this quasilinear representation holds good without modification even in the case of singular M. In our notation the result is

$$[\gamma_c(M)]^{-1} = \max_h (2c^T h - h^T M h) .$$

This representation generalizes to a form suitable for use with $[\gamma_A(M)]^{-1}$, and this is the content of the following lemma.

Lemma (Generalized Whittle Representation). *If the kernel of* M *is contained in the kernel of* A^T *then*

$$[\gamma_A(M)]^{-1} \geqslant A^T H + H^T A - H^T M H$$

with equality for some H.

Corollary. $\gamma_A(M)$ *depends only on* A *and* M.

Proof. With the above notation, put $K = [B^T\ b^T Z^T]L^{-1}[B^T\ b^T Z^T]^T$. Then

$$MKA = M[B^T\ b^T Z^T]L^{-1}[B^T\ b^T Z^T]^T A =$$
$$= [A\ aY]LL^{-1}[B^T\ b^T Z^T]^T A = [A\ aY][B^T\ b^T Z^T]^T A = A\,.$$

Thus

$$A^T KA - (A^T K - H^T)M(KA - H) =$$
$$= A^T KA - A^T KMKA + A^T KMH + H^T MKA - H^T MH =$$
$$= A^T H + H^T A - H^T MH\,.$$

Now $(A^T K - H^T)M(KA - H) \in \mathcal{Q}_s$, so we have

$$A^T KA \geqslant A^T H + H^T A - H^T MH$$

with equality if $H = KA$.

But

$$A^T KA = A^T[B^T\ b^T Z^T]L^{-1}[B^T\ b^T Z^T]^T A =$$
$$= [I_s\ 0]L^{-1}[I_s\ 0]^T$$

which is the $s \times s$ principal submatrix of L^{-1} and this is, of course, $[\gamma_A(M)]^{-1}$, so the proof is complete.

Whittle (pers. comm.) has proved a result which becomes, in our present notation, the theorem that

$$\varphi_A(M) = \log\det\gamma_A(M)$$

is a concave function of M. The following two lemmas and theorem are his results in the notation used above.

Lemma (Whittle). *Let $M \in \mathscr{P}_r$. Then if S is an $r \times r$ matrix and $\det S = 1$, we have*

$$\operatorname{trace} SMS^T \geqslant r(\det M)^{\frac{1}{r}}$$

with equality for some S.

Proof.

$$\inf\{\operatorname{trace} SMS^T:\ \det S = 1\} \geqslant$$
$$\geqslant \inf\{\operatorname{trace} R:\ \det R = \det M, R \in \mathscr{P}_r\} \geqslant r(\det M)^{\frac{1}{r}}$$

by a diagonalization argument. This gives the desired inequality; equality is achieved with $S = (\det M)^{\frac{1}{2r}} M^{-\frac{1}{2}}$.

Lemma (Whittle). *Let S be an $s \times s$ matrix. Then if R is a matrix with s rows and the same number of columns as Z has rows, we have*

$$[SBMB^TS^T - SBMb^TZ^TR^T - RZbMB^TS^T + RZbMb^TZ^TR^T] \geqslant$$
$$\geqslant S\gamma_A(M)S^T$$

with equality for some R.

Proof. By completing the square in the variable R, the expression on the left hand side of the inequality can be put into the form $S\gamma_A(M)S^T + QQ^T$, where Q depends on R. The inequality follows immediately, and for the last part just note that there is a choice of R for which Q is zero.

Theorem (Whittle).

$\varphi_A(M) = \log\det\gamma_A(M)$ *is a concave function of M.*

Proof. Combining the last two lemmas yields a representation for the function $s(\det \gamma_A (M))^{\frac{1}{s}}$ as the lower envelope of a family of linear functions; this function is thus concave. The log function is concave and increasing, and it follows that φ_A is concave.

We shall make use of the concavity of $\log \det M$ in the proof of the duality theorem. The more general result is of considerable interest, in that it is always reassuring to know that a function which one is trying to maximize is concave, but we shall make no direct use of this result here. I am particularly indebted to Professor Whittle for drawing my attention to the quasilinear representations employed above.

3. SILVEY'S MINIMAL ELLIPSOID PROBLEM. THE DUALITY THEOREM.

The problem originally posed by Silvey [6] was that of finding the minimal-content ellipsoid centred at the origin and containing the compact set $\mathscr{V}$; he conjectured that this problem was dual in some suitable sense to the D-optimal design problem, and further that this relationship extended to a duality between the D_s-optimal design problem and a version of the minimal-content problem for ellipsoidal cylinders. Sibson [5] established that the D-optimal design problem is the strong Lagrangian dual of the minimal ellipsoid problem, and Silvey and Titterington (unpublished, pers. comm.) have since extended this result. Here we prove that the D_A-optimal design problem is the strong Lagrangian dual of the A-minimal ellipsoid problem, which is an appropriate generalization of the minimal-content problem for ellipsoidal cylinders. As above, A will be a fixed $k \times s$ matrix of full rank s. Then the A-minimal ellipsoid problem is the problem of finding $N \in \mathscr{Q}_k$ to maximize

$$\psi_A (N) = \log \det A^T N A$$

subject to the conditions

$$v^T N v \leqslant s \qquad \text{for all} \qquad v \in \mathscr{V}.$$

The conditions are clearly precisely that the quadric

$$u^T N u = s\,,$$

which since $N \in \mathcal{Q}_k$ is an ellipsoid degenerating off $\mathcal{P}_k$ to an ellipsoidal cylinder, should contain $\mathcal{V}$; maximization of $\psi_A(N)$ is clearly a kind of generalized minimal-content condition.

To obtain a clean statement of the duality theorem we shall reformulate the D_A-optimal design problem slightly. The basic version of the problem is the maximization of $\varphi_A(\mu(\lambda))$ over all probability measures λ on $\mathcal{V}$. This is easily seen to be equivalent to the minimization of

$$-\varphi_A(\mu(\lambda)) + s\lambda(\mathcal{V}) - s$$

over all positive measures λ on $\mathcal{V}$, and we shall refer to this minimization problem as the D_A-optimal design problem throughout this section.

We now state and prove our main theorem.

Duality theorem for D_A-optimality. *Let $\mathcal{V}$ be a compact set in k-dimensional Euclidean space and let A be a fixed $k \times s$ matrix of full rank s. Then the A-minimal ellipsoid problem, of maximizing*

$$\psi_A(N) = \log\det A^T N A$$

for $N \in \mathcal{Q}_k$, subject to the conditions

$$v^T N v \leqslant s \qquad \text{for all} \qquad v \in \mathcal{V}\,,$$

and the D_A-optimal design problem, of minimizing

$$-\varphi_A(\mu(\lambda)) + s\lambda(\mathcal{V}) - s$$

over all positive measures λ on $\mathcal{V}$ share a common extremal value.

Moreover, if $\bar{N}$ is any solution to the A-minimal ellipsoid problem and $\bar{\lambda}$ is any solution to the D_A-optimal design problem, then $\bar{\lambda}$ is supported by the set $\{u\colon u^T \bar{N} u = s\}$.

Proof. We begin by considering a finite subset of the compact set $\mathcal{V}$, say

$$\mathcal{W} = \{w_1, \ldots, w_j, \ldots, w_J\}\,.$$

The A-minimal ellipsoid problem for $\mathscr{W}$ is the problem of maximizing $\psi_A(N)$ with $N \in \mathscr{Q}_k$, subject to the finitely many constraints

$$w_j^T N w_j \leqslant s \qquad (j = 1, \ldots, J).$$

We can rephrase this as the problem of maximizing $\psi_A(N)$ with $N \in \mathscr{Q}_k$, $\zeta \geqslant 0$ subject to the constraints

$$w_j^T N w_j + \zeta_j = s \qquad (j = 1, \ldots, J).$$

Now the region defined by $N \in \mathscr{Q}_k$, $\zeta \geqslant 0$ is a convex cone, and ψ_A is a concave function on this region, and the finitely many constraints are linear equality constraints, it follows (Whittle [7], p. 61) that the strong Lagrangian principle applies, in the sense that for suitably chosen constants $\lambda_1, \ldots, \lambda_J$ every solution to the A-minimal ellipsoid problem for $\mathscr{W}$ also maximizes

$$(1) \qquad \log\det A^T N A + \sum \lambda_j (s - w_j^T N w_j - \zeta_j)$$

to the same maximal value. There may be other solutions to the problem of maximizing (1) which do not satisfy the constraints for the original problem. The finiteness of the maximum in the original problem in the case in which we are interested immediately allows us to place constraints on the possible values of $\lambda_1, \ldots, \lambda_J$. First, we must have $\lambda_1, \ldots, \lambda_J \geqslant 0$, otherwise by suitable choice of ζ we can make (1) arbitrarily large. For maximality, for each j at least one of ζ_j, λ_j must be zero, that is, whenever $\lambda_j > 0$ we must have $w_j^T N w_j = s$. In other words λ_j cannot be positive unless w_j lies on the ellipsoids $u^T N u = s$ where N ranges through all solutions to the A-minimal ellipsoid problem for $\mathscr{W}$. The partially maximized value of (1) is then

$$(2) \qquad \log\det A^T N A + \sum \lambda_j (s - w_j^T N w_j).$$

We can rearrange this expression as

$$(3) \qquad \log\det A^T N A - \operatorname{trace} N\mu(\lambda) + s \sum \lambda_j$$

where we have written $\mu(\lambda)$ for $\sum \lambda_j w_j w_j^T$. For the moment we shall abbreviate this still further to

(4) $$\log\det A^T NA - \operatorname{trace} NM + s\sum \lambda_j$$

with $M = \mu(\lambda)$. M is a $k \times k$ positive-semidefinite symmetric matrix. There is no guarantee that it will be nonsingular, and our next task is to establish what degree of singularity is possible. Let u be an arbitrary k-vector. Then

$$\log\det A^T(N + \alpha uu^T)A = \log\det A^T NA +$$
$$+ \log(1 + \alpha u^T A[A^T NA]^{-1} A^T u)$$

where we assume that $\log\det A^T NA$ is finite, that is, that $A^T NA$ is nonsingular. Similarly

$$\operatorname{trace}(N + \alpha uu^T)M = \operatorname{trace} NM + \alpha u^T Mu$$

and so the increment in (4) on replacing N by $N + \alpha uu^T$ is

(5) $$\log(1 + \alpha u^T A[A^T NA]^{-1} A^T u) - \alpha u^T Mu\,.$$

If $N \in \mathcal{Q}_k$ then $N + \alpha uu^T \in \mathcal{Q}_k$ for all $\alpha \geqslant 0$. Now (5) is bounded as a function of $\alpha \geqslant 0$ unless $u^T Mu = 0$, and then to guarantee boundedness we must also have $u^T A[A^T NA]^{-1} A^T u = 0$. This reduces to the condition that the kernel of M be contained in that of A^T, and this is familiar as the condition under which we worked in the preceding section. Thus $\lambda_1, \ldots, \lambda_J$ must be nonnegative and must have the property that the kernel of $\sum \lambda_j w_j w_j^T$ is contained in that of A^T. This is the desired singularity constraint. Under these circumstances $\gamma_A(M)$ is well-defined and is an $s \times s$ positive-definite symmetric matrix. We have

(6) $$\log\det A^T NA = \log\det A^T NA\gamma_A(M) - \log\det \gamma_A(M)\,.$$

In the notation of the preceding section,

$$\operatorname{trace} NM = \operatorname{trace} NM(I_k - b^T Z^T [ZbMb^T Z^T]^{-1} ZbM) +$$
$$+ \operatorname{trace} NMb^T Z^T [ZbMb^T Z^T]^{-1} ZbM =$$
$$= \operatorname{trace}(B^T A^T + b^T a^T)NM(I_k - b^T Z^T [ZbMb^T Z^T]^{-1} ZbM) +$$
$$+ \operatorname{trace}[ZbMb^T Z^T]^{-1} ZbMNMb^T Z^T =$$

$$= \text{trace}\,(A^T NMB^T - A^T NMb^T Z^T [ZbMb^T Z^T]^{-1} ZbMB^T) +$$

$$+ \text{trace}\,[ZbMb^T Z^T]^{-1} ZbMNMb^T Z^T +$$

$$+ \text{trace}\, a^T N(I_k - Mb^T Z^T [ZbMb^T Z^T]^{-1} Zb) Mb^T$$

Now $Z^T Y^T + z^T y^T = I_s$ and $Mb^T z^T = 0$, so the third term is

$$\text{trace}\, Y^T a^T N(I_k - Mb^T Z^T [ZbMb^T Z^T]^{-1} Zb) Mb^T Z^T = \text{trace}\, 0 = 0$$

and trace NM is consequently equal to the sum of the first two terms. We have also $A^T NMB^T = A^T NABMB^T + A^T NabMB^T$ and

$$A^T NMb^T Z^T [ZbMb^T Z^T]^{-1} ZbMB^T =$$

$$= A^T NABMb^T Z^T [ZbMb^T Z^T]^{-1} ZbMB^T +$$

$$+ A^T NabMb^T Z^T [ZbMb^T Z^T]^{-1} ZbMB^T .$$

Insert $I_s = YZ + yz$ between a and b in the second term and observe that $zbM = 0$. Cancel $ZbMb^T Z^T$ with its inverse, and then contract again using $I_s = YZ + yz$, to obtain

$$A^T NMb^T Z^T [ZbMb^T Z^T]^{-1} ZbMB^T =$$

$$= A^T NABMb^T Z^T [ZbMb^T Z^T]^{-1} ZbMB^T + A^T NabMB^T .$$

This gives an expression for trace NM in the form

$$\text{trace}\, NM = \text{trace}\, A^T NA\gamma_A(M) +$$

$$+ \text{trace}\,[ZbMb^T Z^T]^{-1} ZbMNMb^T Z^T .$$

Now the expression

$$\log\det A^T NA\gamma_A(M) - \text{trace}\, A^T NA\gamma_A(M)$$

is maximized to $-s$ uniquely when $A^T NA\gamma_A(M) = I_s$, and the expression

$$- \text{trace}\,[ZbMb^T Z^T]^{-1} ZbMNMb^T Z^T$$

is maximized to 0 uniquely when $ZbMNMb^T Z^T = 0$. These conditions are satisfied simultaneously by the $k \times k$ positive-semidefinite matrix

$$N_0 = (I_k - b^T Z^T [ZbMb^T Z^T]^{-1} ZbM) B^T [\gamma_A(M)]^{-1} \times$$
$$\times B(I_k - b^T Z^T [ZbMb^T Z^T]^{-1} ZbM)^T$$

and this gives the maximized value of (4) as

$$-\log\det \gamma_A(M) + s \sum \lambda_j - s \,. \tag{6}$$

Thus the necessary and sufficient conditions on N that it should maximize (4) are

$$\begin{aligned} &A^T N A \gamma_A(M) = I_s \\ &ZbMNMb^T Z^T = 0 \end{aligned} \tag{7}$$

although, as we have pointed out, there is no guarantee that all such matrices are solutions of the A-minimal ellipsoid problem, although at least one must be. In particular, we offer no guarantee that N_0 as defined above is a solution to the minimal ellipsoid problem.

The dual problem to the A-minimal ellipsoid problem is the problem of minimizing (6) subject to the constraints that $\lambda \geqslant 0$ and that $\gamma_A(M)$ has its kernel contained in that of A^T, and this is simply the D_A-optimal design problem for $\mathscr{W}$. Since the primal A-minimal ellipsoid problem is amenable to strong Lagrangian methods, it follows that the extremal values attainable in the two problems are equal (Whittle [7], p. 65).

So far we have considered only a finite set $\mathscr{W}$. We next observe that, by reducibility to finite designs, there is a subset

$$\bar{\mathscr{W}} = \{\bar{w}_1, \ldots, \bar{w}_J\}$$

of $\mathscr{V}$ for which the D_A-optimal design problem has the same extremum as the D_A-optimal design problem for the whole of $\mathscr{V}$. Now suppose that the A-minimal ellipsoid problem for the whole of $\mathscr{V}$ (that is, with more constraints than for $\bar{\mathscr{W}}$) has a smaller maximum than for $\bar{\mathscr{W}}$. Then by adding a suitable point to $\bar{\mathscr{W}}$ we could construct an A-minimal ellipsoid problem for a finite subset of $\mathscr{V}$ on which the maximum was smaller than for $\bar{\mathscr{W}}$; by the duality for finite subsets established above, the D_A-optimal design problem for this set would have a smaller minimum than for $\bar{\mathscr{W}}$, and this is impossible because $\bar{\mathscr{W}}$ was chosen to support an overall minimizing measure for the D_A-optimal design problem on $\mathscr{V}$. This is the first and main conclusion of the theorem.

For the last part of the theorem we again deal first with the finite case and then with the general case. Again, let $\mathscr{W}$ be a finite subset of $\mathscr{V}$. Clearly $\lambda_1, \ldots, \lambda_J$ are such that *either every* solution N, ζ to the minimal ellipsoid problem for $\mathscr{W}$ is a solution to the problem of maximizing (1), or *no* solution maximizes (1), and we saw that λ_j could not be positive unless the corresponding ζ_j was always zero. Thus we can certainly say that every finitely supported λ solving the D_A-optimal design problem for $\mathscr{W}$ is supported by each of the ellipsoids $u^T\bar{N}u = s$ where $\bar{N}$ ranges through the solutions to the A-minimal ellipsoid problem for $\mathscr{V}$. Now suppose that λ is an arbitrary probability measure on $\mathscr{V}$, and that $\bar{N}$ is some solution to the A-minimal ellipsoid problem for $\mathscr{V}$. Now

$$\int_{\mathscr{V}} v^T\bar{N}v\lambda(dv) = \int_{\mathscr{V}} \operatorname{trace} \bar{N}vv^T\lambda(dv) = \operatorname{trace} \bar{N}\mu(\lambda)$$

so $\lambda(\{u\colon u^T\bar{N}u < s\}) > 0$ if and only if $\operatorname{trace} N\mu(\lambda) < s$. If λ solves the D_A-optimal design problem for $\mathscr{V}$, then choose a finitely supported λ' such that $\mu(\lambda) = \mu(\lambda')$; this is possible by the arguments given in Section 1 above. By the above arguments for finitely supported solutions to the D_A-optimal design problem, we have $\lambda'(\{u\colon u^T Nu < s\})$ equal to zero, hence the same is true for λ. This completes the proof ot the last part of the theorem.

4. TWO COROLLARIES OF THE DUALITY THEOREM

The general equivalence theorem for *D*-optimality (Kiefer and Wolfowitz [4]) states that the following conditions on the probability measure λ on $\mathscr{V}$ are equivalent.

(i) λ maximizes $\log\det\mu(\lambda)$,

(ii) λ minimizes $\max\{v^T[\mu(\lambda)]^{-1}v: v\in\mathscr{V}\}$,

(iii) λ is such that $\max\{v^T[\mu(\lambda)]^{-1}v: v\in\mathscr{V}\}=k$.

This result follows immediately from the duality theorem and the conditions on N derived in the course of its proof. Put $A=I_k$. Then the first of the conditions (7) on N becomes $NM=I_k$ so N is just the inverse of M. The function $\log\det N$ is strictly concave, so the solution of the minimal ellipsoid problem is unique, and consequently $\mu(\lambda)$ is unique, although of course λ need not be. It then follows immediately from the duality theorem that each of (i), (ii), (iii) above is simply a different way of characterizing the optimizing $\mu(\lambda)$, and so all the conditions are equivalent in the sense that they lead to the same matrix $\mu(\lambda)$. It is perhaps worth pointing out that even in the D-optimality case, with $A=I_k$, the duality theorem is a strictly stronger result than the general equivalence theorem; in effect the general equivalence theorem states that λ solves the D-optimal design problem if and only if $[\mu(\lambda)]^{-1}$ is feasible for the minimal ellipsoid problem, whereas the duality theorem guarantees that in this case $[\mu(\lambda)]^{-1}$ actually solves the minimal ellipsoid problem.

Elfving [2] characterized c-optimal designs as follows. Let $\mathscr{R}$ be the convex hull of $\mathscr{V}\cup-\mathscr{V}$. Let $c^*=\frac{c}{\kappa}$ be the point at which the vector in the c-direction cuts the boundary of $\mathscr{R}$, and suppose that $c^*=$ $=\sum\lambda_j\chi_j v_j$ where $\chi_j=\pm 1$, $\lambda_j\geqslant 0$, $\sum\lambda_j=1$, $v_j\in\mathscr{V}$. This is an expression for c^* as a convex combination of points in $\mathscr{V}\cup-\mathscr{V}$. Elfving's result is that the extremal value of $-\log\det\gamma_c(M)$ where $M=\mu(\lambda)$ and λ is a probability measure on $\mathscr{V}$ is $\log\kappa^2$ and is attained by taking $M=\sum\lambda_j v_j v_j^T$. The duality theorem with $A=c$ provides a simple proof of this result, as follows. c^* is a boundary point of the convex set $\mathscr{R}$, so by the supporting hyperplane theorem we can choose a hyperplane, say $d^Tu=1$, which supports $\mathscr{R}$ at c^*, that is, $d^Tu\leqslant 1$ for $u\in\mathscr{R}$. Then clearly $N=dd^T$ satisfies the constraints for the c-minimal ellipsoid problem for $\mathscr{V}$, and the value attained is $\log\nu^2$. Since $c^*=\sum\lambda_j\chi_j v_j$ and $d^Tc^*=1$ and for all $u\in\mathscr{R}$ we have $d^Tu=1$, it follows that $d^T\chi_j v_j=1$

for all j. Put $M = \sum \lambda_j v_j v_j^T$. Then

$$\begin{aligned} -\log \det \gamma_c(M) &= \log \max_h (2c^T h - h^T M h) = \\ &= \log \max_h \left(2\kappa \sum \lambda_j \chi_j v_j^T h - \sum \lambda_j h^T v_j v_j^T h\right) = \\ &= \log \left[\kappa^2 + \max\{-\sum \lambda_j (\kappa \chi_j - v_j^T h)^2\}\right]. \end{aligned}$$

Taking $h = \kappa d$ maximizes the sum to zero and the whole expression to $\log \kappa^2$. Since $\log \kappa^2$ is a value attainable in both the c-minimal ellipsoid problem for $\mathcal{V}$ and in the c-optimal design problem for $\mathcal{V}$, it is extremal for each and the proof is complete.

REFERENCES

[1] H.G. Eggleston, *Convexity,* Cambridge Tract in Mathematics and Mathematical Physics, N. 47, Cambridge University Press, 1958.

[2] G. Elfving, Optimum allocation in linear regression theory, *Ann. Math. Statist.,* 23 (1952), 255-262.

[3] V.V. Fedorov, *The theory of optimal experiments,* (translated from the Russian), Academic Press, New York, 1972.

[4] J. Kiefer – J. Wolfowitz, The equivalence of two extremum problems, *Can. J. Math.,* 12 (1960), 363-366.

[5] R. Sibson, In discussion after Wynn [9].

[6] S.D. Silvey, In discussion after Wynn [9].

[7] P. Whittle, *Optimization under constraints,* John Wiley, London and New York, 1971.

[8] P. Whittle, Some general points in the theory of optimal experimental design, *J. Roy. Statist. Soc. Ser. B,* 35 (1973), 123-130.

[9] H.P. Wynn, Results in the theory and construction of D-optimum experimental designs, *J. Roy. Statist. Soc. Ser. B,* 34 (1972), 133-147.

COLLOQUIA MATHEMATICA SOCIETATIS JÁNOS BOLYAI
9. EUROPEAN MEETING OF STATISTICIANS, BUDAPEST (HUNGARY), 1972.

A CHAIN OF INEQUALITIES FOR SOME TYPES OF MULTIVARIATE DISTRIBUTIONS

Z. ŠIDAK

1. BASIC INEQUALITIES.

In this communication we try to extract the core of the argument used by Y. L. Tong [15] for probabilities in multivariate equicorrelated normal distributions, to generalize it as far as possible, and to apply it for many more cases.

Y. L. Tong [15], Theorem 1, proved the following assertion: If a random variable X is non-negative with probability 1, then

$$\mathsf{E}X^k \geqslant \left(\mathsf{E}^{\frac{k}{s}}\right)^s \geqslant (\mathsf{E}X)^k + \left[\mathsf{E}X^{\frac{k}{s}} - (\mathsf{E}X)^{\frac{k}{s}}\right]^s \qquad (k \geqslant s \geqslant 1). \tag{1}$$

An immediate consequence is the following chain of inequalities.

Lemma. *If a random variable* X *is non-negative with probability* 1, *then*

$$\mathsf{E}X^k \geqslant [\mathsf{E}X^{k-1}]^{\frac{k}{k-1}} \geqslant [\mathsf{E}X^{k-2}]^{\frac{k}{k-2}} \geqslant \ldots \geqslant [\mathsf{E}X^m]^{\frac{k}{m}} \geqslant \tag{2}$$

$$\geqslant [\mathsf{E}X^2]^{\frac{k}{2}} \geqslant [\mathsf{E}X]^k + [\mathsf{E}X^2 - (\mathsf{E}X)^2]^{\frac{k}{2}} \geqslant [\mathsf{E}X]^k$$

$$\textit{for} \qquad k \geqslant m \geqslant 2 .$$

Now, the main result of this communication is as follows.

Theorem. *Let* $X_i = (X_{i1}, \ldots, X_{ip})$ $(i = 1, 2, \ldots, k)$ *be a sample of* k *independent* p*-dimensional random vectors with the same distribution (but where the* p *components of each vector may be dependent), and let* $U = (U_1, \ldots, U_q)$ *be another* q*-dimensional random vector (with possibly dependent components) which is independent of all* X_{ij}*'s. Further, let* $f = f(x_1, \ldots, x_p; u_1, \ldots, u_q)$ *be any measurable* r*-dimensional vector function, and* A *any measurable* r*-dimensional set. Denoting*

$$\beta(k) = \mathsf{P}\{f(X_{i1}, \ldots, X_{ip}; U_1, \ldots, U_q) \in A; \quad i = 1, 2, \ldots, k\}, \tag{3}$$

we have the following chain of inequalities

$$\beta(k) \geqslant [\beta(k-1)]^{\frac{k}{k-1}} \geqslant [\beta(k-2)]^{\frac{k}{k-2}} \geqslant \ldots \geqslant [\beta(m)]^{\frac{k}{m}} \geqslant \tag{4}$$

$$\geqslant [\beta(2)]^{\frac{k}{2}} \geqslant \beta^k(1) + [\beta(2) - \beta^2(1)]^{\frac{k}{2}} \geqslant \beta^k(1)$$

$$\textit{for} \qquad k \geqslant m \geqslant 2 .$$

The proof follows very easily by putting

$$X = \mathsf{P}\{f(X_{i1}, \ldots, X_{ip}; U_1, \ldots, U_q) \in A \mid U_1, \ldots, U_q\}$$

in the preceding Lemma, and its idea is extracted from the proof of Theorem 2 in Tong [15].

The inequalities (4) embrace a large number of interesting special cases.

2. SOME SPECIAL CASES

We will mention here briefly some special cases of the inequalities (4), and discuss in more detail two of them whose comparison with known results is perhaps most interesting.

If the variables $Z_1, \ldots, Z_k$ have a k-variate equicorrelated normal distribution with mean values 0, variances 1, and all correlations $\rho \geqslant \geqslant 0$, it is well known that they may be represented as $Z_i = (1-\rho)^{\frac{1}{2}} X_i - \rho^{\frac{1}{2}} U$ $(i = 1, 2, \ldots, k)$, where $U, X_1, \ldots, X_k$ are independent $N(0, 1)$ variables. Then our Theorem may be applied with $p = q = 1$. This was the case studied by Tong [15], which served as a stimulus for the present paper. The inequality $\beta(k) \geqslant \beta^k(1)$ for A being an interval and for arbitrary correlations was proved by Khatri [7], Šidák [10], Slepian [12].

The matter proceeds quite similarly also for multivariate equicorrelated Student distributions. This case was also discussed by Tong [15], and $\beta(k) \geqslant \beta^k(1)$ proved for more general correlations by Khatri [7], Šidák [11].

The case of equicorrelated χ^2 *distributions.* Let $Z_j = (Z_{1j}, \ldots \ldots, Z_{kj})$ be a random vector having a k-variate normal distribution with mean values 0, variances 1, and all correlations equal to $\rho_j \geqslant 0$; let j run through $1, 2, \ldots, \nu$, and let the vectors $Z_1, Z_2, \ldots, Z_\nu$ be independent. Then we can use a representation $Z_{ij} = (1-\rho_j)^{\frac{1}{2}} X_{ij} - \rho_j^{\frac{1}{2}} U_j$ $(i = 1, 2, \ldots, k;\ j = 1, 2, \ldots, \nu)$, where all X_{ij}, U_j are independent $N(0, 1)$ variables. Putting now in our Theorem $p = q = \nu$, $r = 1$, $A = (d_1, d_2)$,

$$f(X_{i1}, \ldots, X_{i\nu};\ U_1, \ldots, U_\nu) = \sum_{j=1}^{\nu} \left[(1-\rho_j)^{\frac{1}{2}} X_{ij} - \rho_j^{\frac{1}{2}} U_j\right]^2,$$

we have the inequalities (4) for the probability

$$\beta(k) = \mathrm{P}\{d_1 < \sum_{j=1}^{\nu} Z_{ij}^2 < d_2;\ i = 1, \ldots, k\}.$$

This result may be compared with the following results obtained under different assumptions than ours: Jensen [5] proved $\beta(k) \geqslant \beta^k(1)$ when the assumption of independence of the vectors Z_j was relaxed, but only for $k = 2$; Khatri [7] proved $\beta(k) \geqslant \beta^k(1)$ for more general correlations, but only for one-sided intervals (d_1, d_2), i.e. where either $d_1 = 0$ or $d_2 = \infty$. All these inequalities can be used for finding conservative confidence intervals for k variances simultaneously, cf. Jensen – Jones [6], p. 328, or Khatri [7], p. 1855.

Continuing our survey of special cases, let us note that our Theorem can be used for special multivariate Poisson variables given by $Z_1 = X_1 + U, \ldots, Z_k = X_k + U$, where $X_1, \ldots, X_k, U$ are independent Poisson variables; this includes, in particular, the bivariate Poisson distribution, cf. Haight [3], Section 3.12, or Holgate [4].

A similar special case are multivariate exponential variables $Z_1 = \min(X_1, U), \ldots, Z_k = \min(X_k, U)$, where $X_1, \ldots, X_k, U$ are independent exponential variables; this gives for $k = 2$ the bivariate exponential distribution, cf. Marshall – Olkin [8], Theorem 3.2.

Further, we may mention the test statistics used for comparing k experimental groups of observations (receiving some treatments) with one control group. Testing in such a situation was discussed for normal variables by Dunnett [2] (or, cf. also Miller [9], Section 2.5), and in nonparametric set-up by Steel [13], [14] (cf. also Miller [9], Sections 4.1 and 4.3). Roughly speaking, the observations in experimental groups are taken for the vectors X_i $(i = 1, 2, \ldots, k)$, and the observations in the control group for U, and our Theorem may be applied.

Estimation of quantiles in equicorrelated normal distributions. Let $Y_j = (Y_{1j}, \ldots, Y_{kj})$ $(j = 1, 2, \ldots, n)$ be a sample of n independent vectors, each of which has a k-variate normal distribution with an unknown mean vector $(\mu_1, \ldots, \mu_k)$, an unknown vector of standard deviations $(\sigma_1, \ldots, \sigma_k)$, and with all correlations equal to $\rho \geqslant 0$. Let $y_{i\alpha}$ denote the α-quantile of the normal distribution $N(\mu_i, \sigma_i^2)$, and let $Y_i^{(1)} \leqslant Y_i^{(2)} \leqslant \ldots \leqslant Y_i^{(n)}$ be the ordered observations $Y_{i1}, Y_{i2}, \ldots, Y_{in}$. We wish to find a bound for the confidence coefficient of simultaneous statements

$$Y_i^{(s)} < y_{i\alpha} < Y_i^{(t)} \qquad (i = 1, 2, \ldots, k)$$

for some fixed s, t $(1 \leqslant s < t \leqslant n)$. Define the variables $Z_{ij} = \dfrac{Y_{ij} - \mu_i}{\sigma_i}$ $(i = 1, \ldots, k;\ j = 1, \ldots, n)$ and let $Z_i^{(s)}$ has an analogous meaning as $Y_i^{(s)}$ above. The variables Z_{ij} may be represented as $Z_{ij} = (1-\rho)^{\frac{1}{2}} X_{ij} - \rho^{\frac{1}{2}} U_j$, where all X_{ij}, U_j are independent $N(0, 1)$ variables. Further, if z_α is the α-quantile of the distribution $N(0, 1)$, then $y_{i\alpha} = \sigma_i z_\alpha + \mu_i$. Now, in the Theorem, let $p = q = n$, $r = 2$, let the coordinates of the two-dimensional function f be

$$f_1(X_{i1}, \ldots, X_{in};\ U_1, \ldots, U_n) = Z_i^{(s)}$$

$$f_2(X_{i1}, \ldots, X_{in};\ U_1, \ldots, U_n) = Z_i^{(t)} ,$$

and let $A = (-\infty, z_\alpha) \times (z_\alpha, \infty)$. Then the probability $\beta(k)$ becomes

$$\begin{aligned} \beta(k) &= \mathsf{P}\{Z_i^{(s)} < z_\alpha < Z_i^{(t)};\ i = 1, 2, \ldots, k\} = \\ &= \mathsf{P}\left\{\frac{Y_i^{(t)} - \mu_i}{\sigma_i} < z_\alpha < \frac{Y_i^{(t)} - \mu_i}{\sigma_i};\ i = 1, 2, \ldots, k\right\} = \\ &= \mathsf{P}\{Y_i^{(s)} < y_{i\alpha} < Y_i^{(t)};\ i = 1, 2, \ldots, k\} , \end{aligned}$$

and our Theorem gives the desired bounds for $\beta(k)$. In particular, if $y_{i\alpha}$ are the medians, our result $\beta(k) \geqslant \beta^k(1)$ may be compared with those presented by O. J. Dunn [1]: she showed that $\beta(2) \geqslant \beta^2(1)$ for any bivariate population with continuous marginal distributions, but that $\beta(k) \geqslant \beta^k(1)$ need not hold for $k \geqslant 3$. Our result shows that, for a very special type of distributions, $\beta(k) \geqslant \beta^k(1)$ holds for any k.

Remark. This communication is an abridged version of a paper, the detailed version of which will appear in *Aplikace matematiky*, 18 (1973), 110-118.

REFERENCES

[1] O.J. Dunn, Estimation of the medians for dependent variables, *Ann. Math. Statist.*, 30 (1959), 192-197.

[2] C.W. Dunnett, A multiple comparison procedure for comparing several treatments with a control, *J. Amer. Statist. Assoc.*, 50 (1955), 1096-1121.

[3] F.A. Haight, *Handbook of the Poisson distribution*, J. Wiley, 1967.

[4] P. Holgate, Estimation for the bivariate Poisson distribution, *Biometrika*, 51 (1964), 241-245.

[5] D.R. Jensen, An inequality for a class of bivariate chi-square distributions, *J. Amer. Statist. Assoc.*, 64 (1969), 333-336.

[6] D.R. Jensen – M.Q. Jones, Simultaneous confidence intervals for variances, *J. Amer. Statist. Assoc.*, 64 (1969), 324-332.

[7] C.G. Khatri, On certain inequalities for normal distributions and their applications to simultaneous confidence bounds, *Ann. Math. Statist.*, 38 (1967), 1853-1867.

[8] A.W. Marshall – I. Olkin, A multivariate exponential distribution, *J. Amer. Statist. Assoc.*, 62 (1967), 30-44.

[9] R.G. Miller, *Simultaneous statistical inference*, McGraw-Hill Book Company, 1966.

[10] Z. Šidák, Rectangular confidence regions for the means of multivariate normal distributions, *J. Amer. Statist. Assoc.*, 62 (1967), 626-633.

[11] Z. Šidák, On probabilities of rectangles in multivariate Student distributions: their dependence on correlations, *Ann. Math. Statist.*, 42 (1971), 169-175.

[12] D. Slepian, The one-sided barrier problem for Gaussian noise, *Bell System Tech. J.*, 41 (1962), 463-501.

[13] R.G.D. Steel, A multiple comparison rank sum test: treatments versus control, *Biometrics*, 15 (1959), 560-572.

[14] R.G.D. Steel, A multiple comparison sign test: treatments versus control, *J. Amer. Statist. Assoc.*, 54 (1959), 767-775.

[15] Y.L. Tong, Some probability inequalities of multivariate normal and multivariate *t*, *J. Amer. Statist. Assoc.*, 65 (1970), 1243-1247.

COLLOQUIA MATHEMATICA SOCIETATIS JÁNOS BOLYAI
9. EUROPEAN MEETING OF STATISTICIANS, BUDAPEST (HUNGARY), 1972.

OPTIMAL TESTS FOR THE NON-EXISTENCE OF IMMORTALS

R.S. SILVERMAN – A. NÁDAS

1. INTRODUCTION AND SUMMARY

Let X be a non-negative random variable having a (possibly defective) distribution function G such that $0 \leqslant 1 - p = \mathsf{P}(X = +\infty) \leqslant 1$. We are interested in devising sampling schemes and corresponding decision rules (test functions) for testing the hypothesis that $p = 1$. It is clear that simple random sampling of X is possible only if the outcome $X = +\infty$ is observable, otherwise one is forced to consider other schemes.

In this paper we consider this problem in the context of "life testing", where X denotes the life (age-at-death) of objects some of which may be immortal. We treat only the case of simple time-censored testing; i.e. a simple random sample of n objects are tested from birth (time zero) to a fixed time $T > 0$ and the number of deaths r during the time interval $(0, T)$ is observed together with the lifetimes $x_1, \ldots, x_r$.

Areas where this model is of interest include biomedicine, where X is the time to death in the presence of an infection and $1 - p$ is the probability of immunity; physics, where X is the distance traveled by certain particles born of a subatomic collision in a cloud chamber of length T and $1-p$ is the probability of those that must travel a long distance ($> T$). Also included are various problems in reliability, e.g. if transistors are used in parallel so that a device will continue to function so long as at least one transistor functions. In this case it is of interest to decide whether $1-p > > 0$, that is whether there exist immortal devices.

Under our assumption of time-censored testing it is clear that the problem is hopeless unless something is known about G prior to sampling. On the other hand, if the distribution F of mortals, $G = pF$, is known, then the problem is easy in that it reduces to the problem of testing the hypothesis that $q = F(T)$ versus the alternative $q < F(T)$ based on an observation from a binomial distribution having parameters n and q. Thus the problem is of interest in intermediate situations, particularly when $F = F_\theta$ for some θ in a parameter space which is contained in a k-dimensional Cartesian space.

Even in this case insurmountable difficulties arise if one cannot distinguish between the various distributions pF_θ. For example if F_θ is the uniform distribution on the interval $(0, \theta)$ then observations up to time T from F_{2T} are indistinguishable from those that come from $\frac{1}{2}F_T$. Thus it is necessary in testing for the non-existence of immortals, that is $p = 1$, that the distributions F_θ and $pF_{\theta'}$ $(p < 1)$ are different on the interval $(0, T)$.

Our model can be reformulated as follows. The family of distributions (F_θ) is such that F_θ and pF_θ are different on the interval $(0, T)$ for all p, θ, θ' with $p < 1$. H is a distribution whose support is contained in $[T, \infty)$ and G is a mixture of the form $pF_\theta + (1-p)H$, a proper distribution. We wish to test the hypothesis that $p = 1$ based on observing n samples from G for no more than T units of time per sample.

In the present paper we consider only the family of negative exponential distributions. We assume that the distribution of mortals F is in $\mathscr{F} = \{F_\theta : F_\theta(x) = 1 - e^{-\theta x}$ for $x \geqslant 0, \theta > 0\}$. This family is of special interest in the life-testing problem and the cloud chamber problem. Danziger [3] has shown that under these assumptions the generalized likelihood ratio test is biased. He also continues the work of Anscombe [1] on the maximum likelihood estimation of p and θ. The general problem of maximum likelihood estimation of p and θ for an arbitrary parametric family $\mathscr{F}$ has been examined by Bandes and Nádas [2]. One of our students, K. Owen, has some results for testing the hypothesis $p = 1$ when $\mathscr{F}$ is the general exponential (Darmois – Koopman) family. We also have results for testing $p = 1$ with other sampling schemes. We again assume a fixed amount of time T to collect data but in this case allow the removal of objects being tested at any time $t < T$ and their replacement by new samples. We investigate the relative merits of various sampling schemes and, under reasonable assumptions, find best schemes. These results will appear elsewhere [6].

In Section 2 we introduce notation, define criteria for "good" tests and give some lemmas needed in the sequel. In Section 3 we obtain a best test when the sample size $n = 1$, and a best unbiased test for $n = 2$. In Section 4 we prove the non-existence of best unbiased tests for $n = 3$ by constructing different admissible unbiased tests. In particular we construct a locally (near $p = 1$) best unbiased test. This Section is intended to reveal the nature of the maximization problem for all $n \geqslant 3$. In Section 5 we treat the problem for all $n \geqslant 3$. We describe the construction of a locally best unbiased test and indicate that similarly to the $n = 3$ situation there exist no best unbiased tests.

Section 6 is devoted to some peripheral observations in connection with the main problem. First we illustrate the nature of the problem for "large" sample sizes by rephrasing it in terms of a hypothesis testing problem about the parameters of the asymptotic bivariate normal distribution of the sufficient statistics. Secondly we digress to indicate the existence of best unbiased tests for testing one and two-sided hypotheses on θ. Lastly we show that the best or locally best tests constructed in the previous sections cannot be Bayes solutions for any prior probability distribution on the parameter space.

2. PRELIMINARIES

The problem considered in this paper can be phrased as follows: Let X_1 have the distribution function

$$(1) \qquad {}_pF_\theta(x) = \begin{cases} 0 & x < 0 \\ p(1 - e^{-\theta x}) & 0 \leq x < 1 \\ 1 & 1 \leq x \end{cases} \qquad \begin{array}{l} 0 \leq p \leq 1 \\ 0 < \theta < \infty \end{array}$$

having the probability element (Lebesgue measure on $[0, 1)$ plus the point mass at 1)

$$(2) \qquad d_pF_\theta(x) = \begin{cases} 0 & x < 0 \quad \text{or} \quad 1 < x \\ p\theta e^{-\theta x} & 0 \leq x < 1 \\ 1 - p + pe^{-\theta} & x = 1 \end{cases}$$

With no loss of generality we have taken the cutoff time $T = 1$. Let $X_1, \ldots, X_n$ be independent and distributed as X_1. We seek "good" tests (e.g. admissible or best, etc.) of the composite hypothesis $p = 1,\ \theta > 0$ against the composite alternative $p < 1,\ \theta > 0$, based on $X_1, \ldots, X_n$. In applications where $T \neq 1$, X_i is unity if the ith item does not fail before time T; otherwise X_i is the observed lifetime of the ith object divided by T.

Let I be the indicator function of the half open unit interval $[0, 1)$. Put $R = I(X_1) + \ldots + I(X_n)$ and $X = X_1 I(X_1) + \ldots + X_n I(X_n)$; $R = r$ if exactly r objects die prior to time 1 and $X = x$ is then the sum of the lifetimes of these r objects ($X = 0$ if $r = 0$). The joint distribution of the sample then has the probability element

$$(3) \qquad \binom{n}{r} (p\theta)^r (1 - p + pe^{-\theta})^{n-r} e^{-\theta x}$$

for $r = 0, 1, \ldots, n$ and $0 \leq x \leq r$. The corresponding distribution has its support on the unit n-cube. By the Halmos – Savage factorization theorem the pair (R, X) is a sufficient statistic for all (p, θ) and without loss of generality we consider only those tests which are based on the sufficient statistic (see. e.g. [5]). A simple but tedious computation shows that the distribution of (R, X) has the probability element (i.e. density with respect to Lebesgue measure crossed with counting measure)

$$(4) \qquad d_p \mathsf{P}_\theta(r, x) = \binom{n}{r} (p\theta)^r (1 - p + pe^{-\theta})^{n-r} f_r(x) e^{-\theta x}$$

for $r = 0, 1, \ldots, n$ and $0 \leqslant x \leqslant r$ where $f_0(0) = 1$ and

$$(5) \qquad f_r(x) = r \sum_{i=0}^{j-1} (-1) \frac{(x - i)^{r-1}}{i!(r - i)!} \qquad 0 \leqslant j - 1 < x \leqslant j \leqslant r\,.$$

Perhaps the simplest way to obtain (4) is to first note that, conditional on $R = r > 0$, the distribution of X is the r-fold convolution of the distribution $\frac{1 - e^{-\theta x}}{1 - e^{-\theta}} I(x) + 1 - I(x)$ $(x \geqslant 0)$ with itself, and then use Laplace transforms.

Let Φ denote the class of all (randomized) test functions (tests, rules). This means that φ in Φ is a measurable map from the space of the sufficient statistic to the closed unit interval and $\varphi(r, x)$ is interpreted as the conditional probability of rejecting the hypothesis having observed $(R, X) = (r, x)$. Our criteria of goodness for φ are defined in terms of the power function β_φ of φ:

$$(6) \qquad \beta_\varphi(p, \theta) = {}_p\mathsf{E}_\theta \varphi(R, X) = \int \varphi d_p \mathsf{P}_\theta = {}_p\mathsf{P}_\theta(\varphi \text{ rejects})\,.$$

For any α in $(0, 1)$ φ is called level α if $\beta_\varphi(1, \theta) \leqslant \alpha$ and φ is size α if $\sup(\beta_\varphi(1, \theta)\colon \theta > 0) = \alpha$. If Φ_α is the class of level α tests, then for φ_1 and φ_2 in Φ_α we deem φ_1 as good as φ_2 if $\beta_{\varphi_1}(p, \theta) \geqslant \beta_{\varphi_2}(p, \theta)$ for all $p < 1$ and $\theta > 0$. If this inequality holds and is in fact strict for some (p, θ) with $p < 1$, then φ_1 is better than φ_2. We call φ in Φ_α admissible if there exists no φ^* in Φ_α which is better; φ in Φ_α is best if it is as good as any φ^* in Φ_α. If φ in Φ_α is such that $\beta_\varphi(p, \theta) \geqslant \alpha$ for all $p < 1$ and $\theta > 0$, then φ is called unbiased. For details on this notion of goodness see e.g. [4] or [5].

Since the family of distributions ${}_p\mathsf{P}_\theta$ is continuous in (p, θ) it is clear that the power $\beta_\varphi(p, \theta)$ of any φ in Φ is a continuous function of (p, θ). Hence for any unbiased φ in Φ_α we must have $\beta_\varphi(1, \theta) = \alpha$ for all $\theta > 0$, i.e. φ is a so-called similar test of size α.

It is convenient to use the notation $D_r(x) = \varphi(r, x) - \alpha$ and

$$(7) \qquad {}_jQ_r(x) = \begin{cases} D_r(x + j - 1)f_r(x + j - 1) & 0 \leqslant x \leqslant 1 \\ 0 & \text{otherwise} \end{cases}$$

for $1 \leqslant j \leqslant r \leqslant n$. For functions g integrable on the interval $(0, \alpha)$ we write ${}^1g(x) = \int_0^x g(u)du$ and ${}^{k+1}g(x) = \int_0^x {}^k g(u)du$ where $0 \leqslant x \leqslant a$ and $k = 1, 2, \ldots$.

With these notations our problem is to find good *unbiased* tests φ in Φ_α, i.e., find the functions ${}_jQ_r$ on $[0, 1]$ and $D_0(0)$ that in some sense maximize the shifted power

$$(8) \qquad \begin{aligned} {}_p\mathsf{E}_\theta(D_R(X)) &= (1 - p + pe^{-\theta})^n D_0(0) + \\ &+ \sum_{r=1}^{n} \binom{n}{r} (p\theta)^r (1 - p + pe^{-\theta})^{n-r} \int_0^r D_r(x) f_r(x) e^{-\theta x} dx = \\ &= (1 - p + pe^{-\theta})^n D_0(0) + \\ &+ \sum_{r=1}^{n} \sum_{j=1}^{r} \binom{n}{r} (p\theta)^r (1 - p + pe^{-\theta})^{n-r} e^{\theta(1-j)} \int_0^1 {}_jQ_r(x) e^{-\theta x} dx = \\ &= (1 - p + pe^{-\theta})^n D_0(0) + \\ &+ \sum_{k=0}^{n} \sum_{r=1}^{k} \sum_{j=1}^{r} \binom{n}{k} p^k (1 - p)^{n-k} \times \\ &\times \binom{k}{r} e^{\theta(r+1-k-j)} \theta^r \int_0^1 {}_jQ_r(x) e^{-\theta x} dx \end{aligned}$$

subject to $-\alpha \leqslant D_r \leqslant 1 - \alpha$ and the unbiasedness condition ${}_1\mathsf{E}_\theta(D_R(X)) \geqslant 0$ and the resulting similarity condition

$$
(9) \qquad \begin{aligned} 0 &= {}_1\mathrm{E}_\theta(D_R(X)) = e^{-n\theta} D_0(0) + \\ &+ \sum_{r=1}^{n} \sum_{j=1}^{r} \binom{n}{r} \theta^r e^{\theta(r+1-n-j)} \int_0^1 {}_jQ_r(x) e^{-\theta x}\, dx\,. \end{aligned}
$$

These constraints must be statisfied for all $\theta > 0$. The index k in (8) may be regarded as the number (unobservable) of mortals among the n objects in the sample.

We shall refer to the following theorems (see e.g. [7]).

A. There is a bi-unique relation between functions and their Laplace transforms.

B. Liouville's theorem: Bounded analytic functions are constant.

C. The Phragmen – Lindelöf theorem: Let $\varphi(z)$, $z = re^{i\theta}$, be analytic on the closed wedge W between two straight lines making an angle $\frac{\pi}{a}$ at the origin. If φ is bounded on the lines and if $\varphi(z) = O(e^{rb})$ as r tends to infinity, uniformly in W for $b < a$, then φ is bounded on W.

D. The zeros of a nonzero analytic function are isolated.

In addition we shall use the following lemmas.

Lemma 1. *If f is integrable and $e^\theta \int_0^1 f(x)e^{-\theta x}\, ds$ is bounded for $\theta \geqslant 0$ then $f = 0$ a.e.*

Proof. By the dominated convergence theorem the function

$$
\varphi(z) = e^z \int_0^1 f(x) e^{-zx}\, dx
$$

is entire and it is bounded on the real and imaginary axes. Thus we may let $a = 2$ in Theorem C and them $|\varphi(z)| \leqslant e^r \|f\|_1$ implies that φ is bounded. Thus by Theorem B φ is constant. But $\varphi(\theta)$ tends to zero as θ tends to $-\infty$ hence $\varphi = 0$. Thus $\int_0^1 f(x)e^{-\theta x}\, dx = 0$ for all $\theta > 0$

so that Theorem A implies $f = 0$ a.e.

Lemma 2. *If* f_i, g_i $(i = 1, 2)$ *are bounded and if*

$$\varphi(\theta) = e^{\theta} \int_0^1 (\theta f_1(x) + f_2(x)) e^{-\theta x} dx = \int_0^1 (\theta g_1(x) + g_2(x)) e^{-\theta x} dx$$

then φ *is constant.*

Proof. $\varphi(z)$ is analytic and for $z = u + iu$ the right side is bounded for $u > 0$ while the left side is bounded for $u < 0$. Similarly if $z = u - iu$. The angle between these two lines is $\frac{\pi}{2}$ and $|\varphi(z)| \leqslant Mre^{r} \leqslant Me^{\frac{3r}{2}}$. Hence φ is bounded and thus a constant.

3. THE CASES $n = 1$ AND $n = 2$

The case $n = 1$. Here φ in Φ_α means

$$(1) \qquad {}_1\mathrm{E}_\theta(\varphi(R, X)) = e^{-\theta}\varphi(0, 0) + \int_0^1 \varphi(1, x)\theta e^{-\theta x} dx \leqslant \alpha$$

and the power function can be written as

$$(2) \qquad {}_p\mathrm{E}_\theta(\varphi(R, X)) = (1 - p)\varphi(0, 0) + p_1\,\mathrm{E}_\theta(\varphi(R, X)) .$$

Letting θ tend to zero in (1) shows $\varphi(0, 0) \leqslant \alpha$ so that the power function of any level α test is bounded above by α. Hence the trivial test $\varphi(r, x) = \alpha$ for all r and x is a best test.

The case $n = 2$. In this case, as in Sections 4 and 5, we first examine the similarity constraint (1.9). We have for all $\theta > 0$

$$(3) \qquad \begin{aligned} -e^{-2\theta} D_0(0) &= 2\theta e^{-\theta} \int_0^1 {}_1Q_1(x) e^{-\theta x} dx + \theta^2 \int_0^1 {}_1Q_2(x) e^{-\theta x} dx + \\ &+ \theta^2 e^{-\theta} \int_0^1 {}_2Q_2(x) e^{-\theta x} dx . \end{aligned}$$

Letting θ tend to zero we see that $D_0(0) = 0$. Thus for all $\theta > 0$

(4) $$e^{\theta}\int_0^1 {}_1Q_2(x)e^{-\theta x}\,dx = -\int_0^1 ({}_2Q_q(x) + 2\theta^{-1}{}_1Q_1(x))e^{-\theta x}\,dx\,.$$

Since the right side of (4) is bounded as θ tends to infinity, we can apply Lemma 1 to the left side to conclude that

(5) $${}_1Q_2(x) = 0 \qquad \text{for} \qquad \text{a.e. } x\,.$$

Thus (4) becomes

(6) $$\int_0^1 {}_2Q_2(x)e^{-\theta x}\,dx = \int_0^1 2{}_1Q_1(x)\theta^{-1}e^{-\theta x}\,dx\,.$$

Since this function is bounded as θ tends to zero we conclude that ${}_1^1Q_1(1) = 0$. Hence integrating the right side of (6) by parts we obtain

(7) $$\int_0^1 {}_2Q_2(x)e^{-\theta x}\,dx = -\int_0^1 2{}_1^1Q_1(x)e^{-\theta x}\,dx$$

and hence, using the unicity of Laplace transforms, we have

(8) $${}_2Q_2(x) = -2{}_1^1Q_1(x) \qquad \text{for} \qquad \text{a.e. } x\,.$$

Using these relations the shifted power function can be written as

(9) $${}_p\mathsf{E}_\theta(D_R(X)) = p^2{}_1\mathsf{E}_\theta(D_R(X)) + 2p(1-p)\int_0^1 D_1(x)\theta e^{-\theta x}\,dx =$$
$$= 2p(1-p)\theta^2\int_0^1 {}_1^1Q_1(x)e^{-\theta x}\,dx\,.$$

Evidently we can maximize the power by simply maximizing the function ${}_1^1Q_1$ subject to the remaining constraints. Now since $-\alpha \leqslant D_2(x) \leqslant 1-\alpha$, we have, using the polynomials (1.5),

(10) $${}_1^1Q_1(x) \leqslant \frac{\alpha(1-x)}{2}\,.$$

Then, since ${}_1Q_1(x) \leqslant 1-\alpha$ we get

(11) $${}_1^1Q_1(x) \leqslant (1-\alpha)x\,.$$

Thus ${}_1^1Q_1$ is maximized if it coincides with the lower envelope of

$\dfrac{\alpha(1-x)}{2}$ and $(1-\alpha)x$; i.e., by

$$
(12) \qquad {}_1^1Q_1^*(x) = \begin{cases} (1-\alpha)x & 0 \leqslant x \leqslant \dfrac{\alpha}{2-\alpha} \\[2ex] \dfrac{\alpha(1-x)}{2} & \dfrac{\alpha}{2-\alpha} \leqslant x \leqslant 1\,. \end{cases}
$$

Hence ${}_1Q_1^*$ will uniformly maximize the power if

$$
(13) \qquad {}_1Q_1^*(x) = \begin{cases} 1-\alpha & 0 \leqslant x < \dfrac{\alpha}{2-\alpha} \\[2ex] -\dfrac{\alpha}{2} & \dfrac{\alpha}{2-\alpha} < x < 1\,. \end{cases}
$$

It follows that the corresponding test function φ^* is given by

$$
(14) \qquad \varphi^*(0,0) = \alpha
$$

$$
\varphi^*(1,x) = \begin{cases} 1 & 0 < x < \dfrac{\alpha}{2-\alpha} \\[2ex] \dfrac{\alpha}{2} & \dfrac{\alpha}{2-\alpha} \end{cases}
$$

$$
\varphi^*(2,x) = \begin{cases} \alpha & 0 < x < 1 \\[1ex] \dfrac{2(1-\alpha)(x-1)}{2-x} & 1 < x < \dfrac{2}{2-\alpha} \\[2ex] 0 & \dfrac{2}{2-\alpha} < x < 2\,. \end{cases}
$$

By construction φ^* is best among size α similar tests. But the power of φ^* is

$$
(15) \qquad \begin{aligned} {}_p\mathsf{E}_\theta\varphi^*(R,X) &= \alpha + 2p(1-p)\theta^2 \int_0^1 {}_1^1Q_1^*(x)e^{-\theta x}\,dx = \\ &= \alpha + p(1-p) + 2p(1-p)\left(1-\alpha+\frac{\alpha}{2}e^{-\theta} - (1-\alpha)e^{-\frac{\alpha\theta}{2-\alpha}}\right) \end{aligned}
$$

for all (p,θ), so that φ^* is in fact a best unbiased size α test as well. As such, it is automatically and admissible test. Note that for each fixed

θ the power is a unimodal function of p having its maximum value at $p = \frac{1}{2}$. Note also that

(16) $$\sup ({}_p \mathrm{E}_\theta \varphi^*(R, X) \colon 0 \leqslant p \leqslant 1,\ \theta > 0) = \frac{1 + \alpha}{2}\,.$$

We now show that φ^* is not a best test. Let $D^* = \varphi^* - \alpha$ and define another test $\varphi = \alpha + D$ by

$$D_0(0) = 0$$

(17) $$D_1(x) = \begin{cases} 1 - \alpha & 0 \leqslant x \leqslant \alpha \\ -\alpha & 0 < x \leqslant 1 \end{cases}$$

$$D_2(x) = -\alpha \qquad 0 \leqslant x \leqslant 2\,.$$

The difference in power can be written as

(18) $$\begin{aligned} &{}_p\mathrm{E}_\theta (D_R(X) - D_R^*(X)) = \\ &= 2p(1-p) \int_0^1 (D_1(x) - D_1^*(x))\theta e^{-\theta x}\, dx + p^2\, {}_1\mathrm{E}_\theta (D_R(X))\,. \end{aligned}$$

Now φ is a level α test because

(19) $${}_1\mathrm{E}_\theta (D_R(X)) = -\alpha(1 - e^{-2\theta}) + 2e^{-\theta}(1 - e^{-\alpha\theta}) \leqslant 0$$

as can be checked using the convexity of e^θ. We have ${}^1D_1(1) = {}^1D_1^*(1) = 0$ and

(20) $$\int_0^1 (D_1(x) - D_1^*(x))\theta e^{-\theta x}\, dx > 0$$

for all $\theta > 0$. Hence for all $\theta > 0$ and p sufficiently small relative to θ

(21) $$2(1-p)\int_0^1 (D_1(x) - D_1^*(x))\theta e^{-\theta x}\, dx + p\, {}_1\mathrm{E}_\theta (D_R(x)) > 0\,,$$

showing that the (biased) rule φ has greater power than φ^* in a neighbourhood of the line $(p = 0,\ \theta > 0)$.

We mention that the best unbiased size α test φ^* is not a Bayes rule (see Section 6).

4. THE CASE $n = 3$

In this case the similarity condition is

$$(1)\qquad \begin{aligned} &e^{-3\theta} D_0(0) + \int_0^1 {}_1Q_3(x)\theta^3 e^{-\theta x}dx + e^{-\theta}\int_0^1 {}_2Q_3(x)\theta^3 e^{-\theta x}dx + \\ &+ e^{-2\theta}\int_0^1 {}_3Q_3(x)\theta^3 e^{-\theta x}dx + e^{-\theta}\int_0^1 3_1Q_2(x)\theta^2 e^{-\theta x}dx + \\ &+ e^{-2\theta}\int_0^1 3_2Q_2(x)\theta^2 e^{-\theta x}dx + e^{-2\theta}\int_0^1 3_1Q_1(x)\theta e^{-\theta x}dx = 0 \end{aligned}$$

for all $\theta > 0$. Letting θ tend to zero shows $D_0(0) = 0$. Multiplying (1) by $\dfrac{e^\theta}{\theta^3}$ and then using Lemma 1 we get ${}_1Q_3(x) = 0$ a.e. Next, dividing (1) by θ and then letting θ tend to zero shows that ${}_1^1Q_1(1) = 0$, thus

$$(2)\qquad 0 = {}_1^1Q_1(1) \qquad 0 = {}_1Q_3(x) \qquad \text{for} \quad \text{a.e. } x\,.$$

Now substituting (2) into (1) and integrating by parts we obtain

$$(3)\qquad \begin{aligned} &e^\theta \int_0^1 (\theta_2Q_3(x) + 3_1Q_2(x))e^{-\theta x}dx = \\ &= -\int_0^1 (\theta_3Q_3(x) + 3_2Q_2(x) + 3_1^1Q_1(x))e^{-\theta x}dx \end{aligned}$$

for all $\theta > 0$. Hence, by Lemma 2, both sides of (3) are equal to a constant which we denote by c. Letting θ tend to zero in the left side of (3) we get $3_1^1Q_2(1) = c$. Using this and integrating the left side of (3) by parts gives $0 = \int_0^1 ({}_2Q_3(x) + 3_1^1Q_1(x))e^{-\theta x}dx$. Hence by the unicity of the Laplace transform

$$(4)\qquad {}_2Q_3(x) = -3_1^1Q_2(x) \qquad \text{a.e.}$$

Since ${}_2Q_3(x) = D_3(x+1)\left(x - x^2 + \frac{1}{2}\right)$ we have the bounds

$$-\frac{1-\alpha}{3}\left(x - x^2 + \frac{1}{2}\right) \leqslant {}_1^1Q_2(x) \leqslant \frac{\alpha}{3}\left(x - x^2 + \frac{1}{2}\right),$$

which gives, at $x = 1$,

$$-\frac{1-\alpha}{2} \leqslant c \leqslant \frac{\alpha}{2}. \tag{5}$$

Letting θ tend to zero in the right side of (3) gives the relation $3{}_2^1Q_2(1) + 3{}_1^2Q_2(1) = -c$. This, together with another integration by parts, yields

$$\int_0^1 ({}_3Q_3(x) + 3{}_2^1Q_2(x) + 3{}_1^2Q_1(x))e^{-\theta x}\,dx = -c\,\frac{1-e^{-\theta}}{\theta}$$

for all $\theta > 0$. Hence, by inversion,

$${}_3Q_3(x) + 3{}_2^1Q_2(x) + 3{}_1^2Q_1(x) = -c \qquad \text{for} \qquad \text{a.e. } x. \tag{6}$$

Subject to the above constraints we wish to maximize the shifted power ${}_p\mathsf{E}_\theta(D_R(X))$ or, equivalently, to maximize

$$\begin{aligned}\frac{{}_p\mathsf{E}_\theta(D_R(X))}{3\theta^2 p(1-p)} &= p\int_0^1 {}_1Q_2(x)e^{-\theta x}\,dx + pe^{-\theta}\int_0^1 {}_2Q_2(x)e^{-\theta x}\,dx + \\ &+ 2pe^{-\theta}\int_0^1 {}_1^1Q_1(x)e^{-\theta x}\,dx + (1-p)\int_0^1 {}_1^1Q_1(x)e^{-\theta x}\,dx\end{aligned} \tag{7}$$

where we have used $0 = p^3\,{}_1\mathsf{E}_\theta(D_R(X))$. Integrating by parts we arrive at a more tractable form for (7); to wit,

$$\begin{aligned}&p\left(\int_0^1 {}_1^1Q_2(x)\theta e^{-\theta x}\,dx + e^{-\theta}\int_0^1 ({}_2Q_2(x) + 2{}_1^1Q_1(x))e^{-\theta x}\,dx\right) + \\ &+ (1-p)\int_0^1 {}_1^1Q_1(x)e^{-\theta x}\,dx + p\,\frac{ce^{-\theta}}{3}.\end{aligned} \tag{8}$$

In order to maximize (8) we first examine the form of a maximal ${}_1^1Q_2(x)$ for different values of c. Since ${}_1Q_2(x) = xD_2(x)$ we must have $-\alpha x \leqslant {}_1Q_2(x) \leqslant x(1-\alpha)$. From (3) and (4) we also have ${}_1^1Q_2(x) \leqslant \frac{\alpha}{3}\left(x - x^2 + \frac{1}{2}\right)$. Hence

(9) $${}_1^1Q_2(x) \leqslant \min\left[(1-\alpha)x^2, \left(\frac{\alpha}{3}\,x - x^2 + \frac{1}{2}\right)\right]$$

and ${}_1^1Q_2(1) = \frac{c}{3}$. It is clear from (9) that the maximal ${}_1^1Q_2(x)$ increases as c increases.

Since the ${}_jQ_r$ may depend on c, we rewrite (8) as follows

(10) $$p\left(\frac{ce^{-\theta}}{3} + \theta\int_0^1 f(x,c)e^{-\theta x}\,dx + e^{-\theta}\int_0^1 g(x,c)e^{-\theta x}\,dx\right) + \\ + (1-p)\int_0^1 h(x,c)e^{-\theta x}\,dx$$

where $f(x,c)$ increases with c and where $|g(x,c)| \leqslant 3$ and $|h(x,c)| \leqslant 1$. We observe that the coefficient of p in (10) increases with c for all sufficiently large θ. To see this note that $\left|e^{-\theta}\int_0^1 g(x,c)e^{-\theta x}\,dx\right| \leqslant \frac{3e^{-\theta}}{\theta}$, so that for $c_1 > c_2$ and θ sufficiently large we have $e^{-\theta}\frac{c_1 - c_2}{3} > \frac{6e^{-\theta}}{\theta}$. Since $\theta\int_0^1 f(x,c)e^{-\theta x}\,dx$ also increases with c we conclude that for all θ sufficiently large the coefficient of p in (10) is greatest when $c = \frac{\alpha}{2}$. In this case maximizing ${}_1^1Q_2(x)$ is fairly simple since

$${}_1^1Q_2(x) \leqslant \left[\left(x - x^2 + \frac{1}{2}\right)\right]_{x=1} = \frac{\alpha}{6} = \frac{c}{3} = {}_1^1Q_2(1)\,.$$

Hence ${}_1^1Q_2$ is maximized by ${}_1^1Q_2^*$ where

(11) $${}_1^1Q_2^*(x) = \begin{cases} \dfrac{(1-\alpha)x^2}{2} & 0 \leqslant x \leqslant b \\[2ex] \dfrac{\alpha}{3}\left(x - x^2 + \dfrac{1}{2}\right) & b < x \leqslant 1 \end{cases}$$

with $b = \dfrac{\alpha + 3\sqrt{\dfrac{2\alpha}{3} - \dfrac{\alpha^2}{4}}}{3-\alpha}$. This choice of ${}_1^1Q_2$ in no way affects our choice of ${}_2Q_2$ and ${}_1^1Q_1$, but of course it determines ${}_2Q_3$.

Next we proceed to maximize $\int_0^1 {}_1^1Q_1(x)e^{-\theta x}\,dx$ with $c = \frac{\alpha}{2}$. Using

(6) and noting that ${}_3Q_3(x) = D_3 \dfrac{(x+2)(1-x)^2}{2}$ we have

$$-\frac{(1-\alpha)(1-x)^2}{2} - \frac{\alpha}{2} \leqslant 3{}_2^1Q_2(x) + 3{}_1^2Q_1(x) \leqslant$$
$$\leqslant \frac{\alpha(1-x)^2}{2} - \frac{\alpha}{2}$$

with ${}_2Q_2(x) = (x-1)D_2(x+1)$. By letting $D_2(x+1) = -\alpha$ we arrive at

$$(12)\qquad \begin{aligned} &-\frac{(1-\alpha)(1-x)^2}{6} + \alpha\left(x - \frac{x^2}{2}\right) - \frac{\alpha}{6} \leqslant \\ &\leqslant {}_1^2Q_1(x) \leqslant \frac{2\alpha}{3}\left(x - \frac{x^2}{2}\right). \end{aligned}$$

We also have

$$(13)\qquad -\alpha \leqslant {}_1Q_1(x) \leqslant 1 - \alpha \quad \text{and} \quad {}_1^1Q_1(1) = 0\,.$$

A tedious calculation yields a maximal ${}_1^2Q_1$, statisfying (12) and (13) as follows

$$(14)\qquad {}_1^2Q_1^*(x) = \begin{cases} \dfrac{(1-\alpha)x^2}{2} & 0 \leqslant x \leqslant b_1 \\ -\dfrac{\alpha}{2}x^2 + b_1 x - \dfrac{x^2}{2} & b_1 < x \leqslant b_2 \\ \dfrac{2\alpha}{2}\left(x - \dfrac{x^2}{2}\right) & b_2 < x \leqslant 1 \end{cases}$$

with

$$b_1 = \frac{2\alpha}{3}\left(1 + \sqrt{\frac{\alpha}{1+\alpha}}\right) \quad \text{and} \quad b_2 = 2\sqrt{\frac{\alpha}{1+\alpha}}\,,$$

provided $\alpha \leqslant \frac{1}{3}$. For $\alpha > \frac{1}{3}$ we get

$$(14a)\qquad {}_1^2Q_1^*(x) = \begin{cases} \dfrac{(1-\alpha)x^2}{2} & 0 \leqslant x \leqslant \alpha \\ \alpha\left(x - \dfrac{x^2}{2}\right) - \dfrac{\alpha^2}{2} & \alpha \leqslant x \leqslant 1\,. \end{cases}$$

Now since $\int_0^1 {}_1^1Q_1(x)e^{-\theta x}\,dx = \int_0^1 {}_1^2Q_1(x)\theta e^{-\theta x}\,dx + {}_1^2Q_1(1)e^{-\theta}$ and since we have maximized ${}_1^2Q_1$, we have thus maximized $\int_0^1 {}_1^1Q_1(x)e^{-\theta x}\,dx$ subject ot $c = \frac{\alpha}{2}$. And for $\alpha \leqslant \frac{1}{3}$

$$(15) \qquad {}_1Q_1^*(x) = \begin{cases} 1-\alpha & 0 \leqslant x < b_1 \\ -\alpha & b_1 \leqslant x < b_2 \\ -\dfrac{2\alpha}{3} & b_2 \leqslant x \leqslant 1\,. \end{cases}$$

Subject to this choice of ${}_1Q_1$ (with $\alpha \leqslant \frac{1}{3}$ we find the maximal ${}_2Q_2$ is (since ${}_2Q_2(x) = D_2(x+1)(1-x)$)

$$(16) \qquad {}_2Q_2^*(x) = \begin{cases} \dfrac{2\alpha}{.3} + x\left(\dfrac{\alpha}{3} - 1\right) & 0 \leqslant x < b_1 \\ -\dfrac{\alpha(1-x)}{3} + \alpha x - b_1 & b_1 \leqslant x < b_2 \\ -\alpha(1-x) & b_2 \leqslant x \leqslant 1\,. \end{cases}$$

Hence we have

$$(17) \qquad {}_3Q_3^*(x) = D_3^* \frac{(x+2)(1-x)^2}{2} = -\frac{\alpha(1-x)^2}{2}$$

for $0 \leqslant x \leqslant 1$. Summarizing: In attempting to maximize (8) we first showed that by choosing $c = \frac{\alpha}{2}$ we get better power for p near 1 than for any other choice of c. Next we maximized ${}_1^1Q_2$. Then for this choice of c we maximized $\int_0^1 {}_1^1Q_1(x)e^{-\theta x}\,dx$ and, finally, subject to this choice of ${}_1^1Q_1$ we maximized ${}_2Q_2$. This explicitely determines a size α similar test φ^*.

We now argue that φ^* is admissible in the class of size a similar tests. For any other size α similar test φ_1, if the corresponding constant c_1 is less than $\frac{\alpha}{2}$, then the power of φ_1, for p near 1 and large θ, is less than the power of φ^*. Thus if φ_1 is as good as φ^* then $c_1 = = \frac{\alpha}{2}$. But then φ_1 determines the same Q functions so $\varphi_1 = \varphi^*$ (up to equivalence).

By construction φ^* is a locally (near $p = 1$) best size similar test. An examination of the power (7) determined by (1), (15), (16) and (17) shows that φ^* is in fact unbiased and hence it is admissible in the class of level α unbiased tests. But then φ^* is automatically admissible in the class of all level α tests.

Nevertheless, φ^* is not a best test in Φ_α, in fact it is not even best in the class of size α similar tests. This is due to the fact that if we choose c less than $\frac{\alpha}{2}$ then we are free to choose a larger function ${}_1Q_1$ on account of (6). For example if $c = 0$ then the similar test $\varphi^\#$ such that $\varphi^\#(0, 0) = \alpha < (\sqrt{3} - 1)^2$ and which is determined by

$$ {}_1Q_1^\#(x) = \begin{cases} 1 - \alpha & 0 \leqslant x < \alpha \\ -\alpha & \alpha \leqslant x < 1 \end{cases} $$

(18)
$$ {}_1Q_2^\#(x) = \begin{cases} (1 - \alpha)x & 0 \leqslant x < b \\ \dfrac{\alpha(1 - 2x)}{3} & b \leqslant x < \sqrt{3} - 1 \qquad \text{(see (11))} \\ -\alpha x & \sqrt{3} - 1 \leqslant x \leqslant 1 \end{cases} $$

$$ {}_2Q_2^\#(x) = \begin{cases} (1 - \alpha)(1 - x) & 0 \leqslant x < b_3 \\ -\alpha(1 - x) & b_3 \leqslant x \leqslant 1 \end{cases} \qquad b_3 = 1 - \sqrt{1 - \alpha^2} $$

is more powerful than φ^* for p near zero. It can be shown that $\varphi^\#$ is also unbiased, so that there exists no best unbaised test. Incidentally, the above argument shows that each choice of the constant c yields a different admissible similar test. For any such test φ we get

(19)
$$ \sup({}_p\mathrm{E}_\theta(\varphi(R, X)): 0 \leqslant p \leqslant 1,\ \theta > 0) = \frac{3 + \alpha}{4}. $$

Lastly we mention that φ^* is not a Bayes solution for any prior distribution on the parameter space (see Section 6).

5. THE CASE $n \geqslant 3$.

Letting θ tend to zero in (1.9) shows $D_0(0) = 0$ so

$$\varphi(0, 0) = \alpha . \tag{1}$$

Let

$$H_k(x) = \binom{n}{k} D_{n-k}(x) f_{n-k}(x) I_{[0, n-k]}(x) \tag{2}$$

where $I_{[a,b]}$ is the indicator function of the interval $[a, b]$. From (1.9) we have upon division by θ^n,

$$0 = \sum_{k=0}^{n-1} e^{-\theta k} \int_0^{n-k} H_k(x) \theta^{-k} e^{-\theta x} dx . \tag{3}$$

Iterated integration by parts yields

$$\begin{aligned} 0 &= \sum_{k=0}^{n-1} e^{-\theta k} \left\{ \int_0^{n-k} {}^k H_k(x) e^{-\theta x} dx + \right. \\ &\quad \left. + e^{-\theta(n-k)} \sum_{i=1}^{k} {}^{k+1-i} H_k(n-k) \theta^{-i} \right\} = \\ &= \sum_{k=0}^{n-1} e^{-\theta k} \int_0^{n-k} {}^k H_k(x) e^{-\theta x} dx + \\ &= e^{-n\theta} \sum_{i=1}^{n-1} \theta^{-i} \sum_{k=i}^{n-1} {}^{k+1-i} H_k(n-k) . \end{aligned} \tag{4}$$

Letting θ tend to zero in (4), we see that the coefficient if θ^{-i} must vanish, thus for $i = 1, 2, \ldots, n-1$,

$$\sum_{k=i}^{n-1} {}^{k+1-i} H_k(n-k) = 0 . \tag{5}$$

Hence for all $\theta > 0$

$$0 = \sum_{k=0}^{n-1} e^{-\theta k} \int_0^{n-k} {}^kH_k(x)e^{-\theta x}\,dx =$$

$$(6) \qquad = \sum_{k=0}^{n-1} \int_0^{\infty} I_{[0,n-k]}(x)\,{}^kH_k(x)e^{-\theta x}\,dx =$$

$$= \int_0^{\infty} I_{[0,n]}(x) \sum_{k=0}^{n-1} {}^kH_k(x-k)e^{-\theta x}\,dx .$$

Since (6) is a Laplace transform with argument θ, we have by inversion

$$(7) \qquad -H_0(x) = \sum_{k=1}^{n-1} {}^kH_k(x-k) \qquad \text{for} \quad \text{a.e. } 0 \leqslant x \leqslant n .$$

The functions H_k are related to the functions of interest ${}_jQ_r$ by

$$(8) \qquad H_{n-r}(x) = \binom{n}{r} \sum_{j=1}^{r} {}_jQ_r(x-j+1) = \binom{n}{r}{}_{[x+1]}Q_r(x-[x]) ;$$

in particular

$$(9) \qquad H_0(x) = {}_{[x+1]}Q_n(x-[x])$$

where $[a]$ denotes the greatest integer less than or equal to a. A further computation shows that

$$(10) \qquad \begin{aligned} {}^kH_k(x) &= \binom{n}{k} \sum_{j=1}^{n-k} I_{[j-1,j]}(x)\,{}^k_jQ_{n-k}(x-j+1) + \\ &+ \binom{n}{k} \sum_{j=1}^{[x]} \sum_{u=1}^{k} {}^{k-u+1}_{\;\;j}Q_{n-k}(1)\,\frac{x^{u-1}}{(u-1)!} . \end{aligned}$$

Now, using (7) and (9), we have for $0 \leqslant x \leqslant n$

$$(11) \qquad -{}_{[x+1]}Q_n(x-[x]) = \sum_{k=1}^{[x]} {}^kH_k(x-k)$$

so that for $j = 1, 2, \ldots, n$ (using (10))

$$(12) \qquad -{}_jQ_n(x) = \sum_{r=n-j+1}^{n-1} \binom{n}{r}{}^{n-r}_{j-n+r}Q_r(x) + h_j(x)$$

where h_j is a polynomial of degree at most $j-2$. We omit this calculation. The polynomial h_j is given by

$$h_j(x) = n \sum_{r,u,v} \frac{(x-j+1-n+r)^u}{r!(n-r)!} \,{}^{n-r-u+1}_{\quad v}Q_r(1) \tag{13}$$

where the summation extends over the integers

$$n-j+1 \leqslant r \leqslant n-1\,, \quad 1 \leqslant v \leqslant r+j-n-1\,, \quad 1 \leqslant u \leqslant n-r\,, \tag{14}$$

with the convention that empty sums are zero. Equations (5) and (12) express the similarity constraint (1.9) in terms of the functions ${}_jQ_r$. Evidently ${}_1Q_n(x) = 0$ a.e. so that $\varphi(n, x) = \alpha$ for $0 \leqslant x \leqslant 1$. Now, using the above together with (1.9), the power function can be written as

$$\begin{aligned} &{}_pE_\theta(\varphi(R, X)) = \\ &= \alpha + \sum_{k=1}^{n-1} \binom{n}{k} p^k(1-p)^{n-k} \sum_{r=1}^{k} \sum_{j=1}^{r} \binom{k}{r} \theta^r e^{\theta(r+1-k-j)} \times \\ &\times \int_0^1 {}_jQ_r(x) e^{-\theta x}\,dx = \alpha + \sum_{k=1}^{n-1} p^k(1-p)^{n-k} G_k(\theta) \end{aligned} \tag{15}$$

say. The functions G_k can be put in a more tractable form via repeated integrations by parts, as in (4.8). A locally (near $p = 1$) best size α similar test φ^* can then be obtained by first maximizing G_{n-1} subject to $0 \leqslant \varphi(r, x) \leqslant 1$ and to (5) and (12) to obtain G^*_{n-1}; next maximizing G_{n-2} subject to the additional constraint of $G_{n-1} = G^*_{n-1}$ to obtain G^*_{n-2} and so on; finally maximizing G_1 subject to having fixed $G_k = G^*_k$ for $k = 2, \ldots, n-1$ to obtain $G^*_1(x) = n\theta \int_0^1 D^*_1(x) e^{-\theta x}\,dx$. It can be shown that the G^*_k are non-negative functions and hence that φ^* is locally best in the class of size α unbiased tests. As in the case of $n = 3$, φ^* is neither best nor best unbiased but it is admissible. The power function of φ^* (and in fact of any admissible similar test) satisfies

$$\sup({}_p\mathrm{E}_\theta \varphi^*(R, X)\colon 0 \leqslant p \leqslant 1,\ \theta > 0) = 1 - (1-\alpha)2^{1-n}\,. \tag{16}$$

We mention that φ^* is not Bayes for any prior distribution (see Sec. 6).

6. FURTHER OBSERVATIONS

The form of the problem in the large-sample case.

For n sufficiently large the distribution of the sufficient statistic $\frac{1}{n}(R, X)$ is approximately bivariate normal with parameters

$$\mu_1 = \mathsf{E}\frac{1}{n}R = p(1 - e^{-\theta}) \qquad \mu_2 = \mathsf{E}\frac{1}{n}X = \mu_1 f(\theta) \tag{1}$$

where

$$f(\theta) = \frac{1}{\theta} - \frac{1}{e^{\theta} - 1} \qquad (\theta > 0) \tag{2}$$

and

$$\begin{aligned} \sigma_{11} &= \operatorname{var}\left(\frac{1}{n}R\right) = \frac{1}{n}\mu_1(1 - \mu_1) \\ \sigma_{22} &= \operatorname{var}\left(\frac{1}{n}X\right) = -\mu_1 f'(\theta)\frac{1}{n} \\ \sigma_{12} &= \operatorname{cov}\left(\frac{1}{n}R, \frac{1}{n}X\right) = \frac{1}{n}\mu_1(1 - \mu_1)f(\theta) . \end{aligned} \tag{3}$$

In this formulation the parameter space is a closed and bounded region in the plane

$$\{(\mu_1, \mu_2)\colon 0 \leqslant \mu_1 < 1, \ 0 \leqslant \mu_2 \leqslant \mu_1 f(-\log(1 - \mu_1))\} . \tag{4}$$

The hypothesis that $p = 1$ transforms to the hypothesis

$$\{(\mu_1, \mu_2)\colon 0 \leqslant \mu_1 < 1, \ \mu_2 = \mu_1 f(-\log(1 - \mu_1))\} . \tag{5}$$

K. Owen is now examining this and more general problems in this setting.

Existence of best unbiased tests for testing θ.

The probability element (2.4) of the sufficient statistic (R, X) can be written in the Darmois – Koopman form

$$c(\pi_1, \pi_2)h(r, x)e^{\pi_1 x + \pi_2 r}$$

where $\pi_1 = -\theta$ and $\pi_2 = \log\theta - \log\left(\frac{1}{p} - 1 + e^{-\theta}\right)$ are the so-called natural parameters of the distribution. Consequently (see e.g. Lehmann [5]) there exist best unbiased tests for testing one and two-sided hypotheses on θ. These are obtained as best tests based on the conditional distribution of (R, X) given R.

We observe the obvious contrast in difficulty (due to being a natural parameter versus not being a natural parameter) between the problem of testing the usual sorts of hypotheses about θ on the one hand and about p on the other.

Non-Bayes nature of similar tests.

In the Bayesian formulation, the family of distributions $({}_p\mathsf{P}_\theta : 0 \leqslant 0 \leqslant p \leqslant 1,\ \theta > 0)$ of the sufficient statistic represent a family of conditional distributions given that the random parameter takes the value (p, θ). Let M denote the "prior" (i.e. marginal) probability measure on the parameter space so that the joint probability element of the sufficient statistic and the random parameter is given by

$$d\mathsf{P}(r, x, p, \theta) = d_p\mathsf{P}_\theta(r, x)\, d\mathsf{M}(p, \theta)\ .$$

The "posterior" (i.e. conditional) distribution $\mathsf{P}(\cdot \mid r, x)$ of the random parameter (given that $(R, X) = (r, x)$) the determines all Bayes solutions φ_{M} to the problem of testing $p = 1$ versus $p < 1$. In the case where the losses are zero or one accordingly as the correct decision has or has not been made, the Bayes solutions have the form

$$\text{(1)} \qquad \varphi_{\mathsf{M}}(r, x) = \begin{cases} 1 & \text{if} \quad \mathsf{P}(p = 1 \mid r, x) < \mathsf{P}(p < 1 \mid r, x) \\ 0 & \text{if} \quad \mathsf{P}(p = 1 \mid r, x) > \mathsf{P}(p < 1 \mid r, x) \\ \text{arbitrary} & \text{if} \quad \mathsf{P}(p = 1 \mid r, x) = \mathsf{P}(p < 1 \mid r, x)\ . \end{cases}$$

We now show that if φ is any size α similar test then there exists no probability measure M such that φ is a Bayes solution for M. We know that for any similar φ, $\varphi(0, 0) = \alpha$ is strictly between zero and one. Thus if $\varphi = \varphi_{\mathsf{M}}$ then we must have

$$\text{(2)} \qquad \mathsf{P}(p = 1 \mid 0, 0) = \mathsf{P}(p < 1 \mid 0, 0)$$

or equivalently

(3) $$\int_{\{p=1\}} e^{-\theta n} d\mathsf{M} = \int_{\{p<1\}} (1-p+pe^{-\theta})^n d\mathsf{M}\,.$$

We also know that for any similar φ, $\varphi(n,x)=\alpha$ a.e. for $0 \leqslant x \leqslant 1$ (see Section 5). Thus if $\varphi = \varphi_{\mathsf{M}}$, then we must also have

(4) $$\mathsf{P}(p=1\,|\,n,x) = \mathsf{P}(p<1\,|\,n,x)$$

for a.e. x, $0 \leqslant x \leqslant 1$. Equivalently we have for such x

(5) $$\int_{\{p=1\}} \theta^n e^{-\theta x} d\mathsf{M} = \int_{\{p<1\}} p^n \theta^n e^{-\theta x} d\mathsf{M}\,.$$

Now integrating (5) from zero to x, $0 \leqslant x \leqslant 1$, we get

(6) $$\int_{\{p=1\}} \theta^{n-1}(1-e^{-\theta x}) d\mathsf{M} = \int_{\{p<1\}} p^n \theta^{n-1}(1-e^{-\theta x}) d\mathsf{M}\,.$$

Consider now the left side minus the right side of (6) as a function of x. It is clearly analytic, yet it vanishes for real x in the unit interval. Hence by theorem D of Section 2 such a function is identically zero whence (6) must hold for all $x \geqslant 0$. It follows that

(7) $$\left\{\begin{aligned} &\int_{\{p=1\}} \theta^k d\mathsf{M} = \int_{\{p<1\}} \theta^k p^n d\mathsf{M} \\ &\int_{\{p=1\}} \theta^k e^{-\theta x} d\mathsf{M} = \int_{\{p<1\}} p^n \theta^k e^{-\theta x} d\mathsf{M} \end{aligned}\right. \qquad \begin{aligned}&(k=0,1,\ldots,n;\\ &0 \leqslant x < \infty)\,.\end{aligned}$$

In particular,

(8) $$\int_{\{p=1\}} e^{-\theta n} d\mathsf{M} = \int_{\{p<1\}} p^n e^{-\theta n} d\mathsf{M}\,.$$

Thus, using (3), we find

(9) $$0 = \int_{\{p<1\}} \sum_{k=0}^{n-1} \binom{n}{k} (1-p)^{n-k} p^k e^{-k\theta} d\mathsf{M}$$

so that

(10) $\mathsf{M}(\{(p, \theta): p < 1\}) = 0$.

But then $\varphi_{\mathsf{M}}(r, x) = 0$ for all $r = 0, 1, \ldots, n$ and $0 \leqslant x \leqslant r$ (a.e.) so that $\varphi_{\mathsf{M}} \neq \varphi$, a contradiction.

We are not aware of other examples of best unbiased tests that fail to be Bayes.

REFERENCES

[1] F.J. Anscombe, Estimating a mixed exponential law, *J. Amer. Statist. Ass.*, 56 (1961), 493-505.

[2] S.H. Bandes – A. Nádas, Estimating the life distribution of a population containing immortals, *IBM Technical Report*, 22.1345 (1971).

[3] L. Danziger, *A Life Distribution Containing Immortals*, Ph. D. Thesis, New York University, 1971.

[4] T.S. Ferguson, *Mathematical Statistics*, Academic Press, 1967.

[5] E.L. Lehmann, *Testing Statistical Hypotheses*, J. Wiley, 1959.

[6] R.S. Silverman – A. Nádas, *Optimal Time-Constrained Tests.* (to appear).

[7] E.C. Titchmarsh, *Theory of Functions*, Oxford Univ. Press, 1939.

COLLOQUIA MATHEMATICA SOCIETATIS JÁNOS BOLYAI
9. EUROPEAN MEETING OF STATISTICIANS, BUDAPEST (HUNGARY), 1972.

ON DISCRETIONARY QUEUING DISCIPLINE

A. SIMONOVITS

SUMMARY

The *discretionary* (or semipreemptive) priority queuing discipline is the common generalization of the nonpreemptive discipline and the preemptive discipline. An arriving unit of the ith priority class $(L_i \in \mathscr{C}_i)$ preempts the service of L_j if and only if the latter has received less service than the (i, j)-th *discretionary value*.

Generalizing the results of Avi-Itzhak, Brosh and Naor [2], Etschmaier [7], Jaiswal [8], Bronstein and Veklerov [3] we determine the expected queuing time of each priority class if they arrive in independent and homogeneous Poisson streams. Hence we determine the *optimal system* of the discretionary values minimizing the total expected queuing loss. Finally, we present some related theorems discussed in an earlier paper [10] of the author.

1. INTRODUCTION

The discretionary (or semipreemptive) priority queuing discipline is the common generalization of the nonpreemptive discipline and the preemptive one. There are r priority classes, where the ith priority class $\mathscr{C}_i$ is the ith most important class, $1 \leqslant i \leqslant r$. An arbitrary element of $\mathscr{C}_i$ is denoted by L_i. An arriving unit L_i interrupts (or preempts) the service of L_j if and only if the total elapsed service time of L_j is smaller than the (i, j)-th discretionary value; notation: $u_j < u_{ij}^0$.

We assume no arriving unit may preempt a unit of non-lower priority: $u_{ij}^0 = 0$ for $i \geqslant j$, further if an arriving unit preempted a unit a given moment, then an arriving unit of higher priority would preempt the same unit at the same moment: $u_{i+1,j}^0 \leqslant u_{ij}^0$, $1 \leqslant i, j \leqslant r$.

Of course, if all the non prefixed discretionary values are zero (or infinite), then we get the nonpreemptive (or the preemptive) discipline. The main advantage of the discretionary discipline over the nonpreemptive and the preemptive discipline is the following: Using its discretion, the discretionary discipline can avoid the preemption of a unit with lower priority the service of which is almost completed; and it can allow a unit with higher priority to preempt the service of a unit just entering service.

Except a single paragraph, we assume that the ith priority units *arrive in a Poisson stream* with intensity a_i, $a_i > 0$ $(i = 1, 2, \ldots, r)$; and these streams are *independent*, i.e. the united stream is also a Poisson one with intensity $a = \sum_{i=1}^{r} a_i$.

The service time distribution of L_i is denoted by $B_i(t)$. We assume that it has finite first and second moments β_i and γ_i, respectively; its standard deviation is $\sigma_i = \sqrt{\gamma_i - \beta_i^2}$, $1 \leqslant i \leqslant r$. We assume any preempted unit resumes the service from the point where it has been interrupted (preemptive resume discipline). Consequently the total service time, the total busy period and the utilization factor of the system are independent of the priority discipline. We assume the queue is not saturated, i.e.

$$R_r = \sum_{i=1}^{r} a_i\beta_i < 1.$$

Every moment the *state* of the system is totally descibed by the elapsed service time of each unit. But at most one unit of each priority class has positive elapsed service time at a given moment, thus we shall denote only these times in the order of the priority indices: $u = (u_i)_{i=1}^r$. To complete the description of the state we denote the actual number of ith priority units in the queue and in service: $m = (m_i)_{i=1}^r$; the state-vector is $x = (u, m)$. It is evident $x(t)$ is a homogeneous Markov process, but the proof of the existence and ergodicity of a unique stationary (steady state) distribution seems to be hard. For the two-class process Jaiswal [8] has given an integro-partial differential equation system of infinite order for the density function of the state-vector. Using generating functions and Laplace – Stieltjes transformations explicit formulae can be derived. The generalization of this seems to be too complicated, thus we follow a usual method: we assume the existence of the steady state.

The *waiting time* of a unit is the period beginning from the arrival and finishing at the first start of the service of our unit. The *interruption time* is the union of those periods when the interrupted unit stands in the queue.

The time loss due to the interruptions and the interruption time are totally different notions. The first quantity is negligible in this paper, the second one is the difference of the well-known *completion* (or residence) time and the service time. We investigate only the expected values of these quantities at the ergodic distribution for each class; their notations are w_i, z_i and $v_i = w_i + z_i$, respectively, $1 \leqslant i \leqslant r$.

2. THE DETERMINATION OF THE QUEUING TIMES

Cobham [5] determined the expected values v_i at the nonpreemptive case by a rather simple method for arbitrary service-time distributions. Bronstein and Veklerov [3] used this method at the discretionary discipline which is more general than the nonpreemptive one; but for the special case of *non-random service time.* The generalization of the latter for the general case of *random* service times is the aim of this part and it is a rather easy consequence of the mentioned papers.

To determine the queuing times we determine the waiting times and the interruption times separately. As the second task is easier than the first, let us begin with the interruption times.

The Determination of the Expected Interruption Times

The service of $L_j(u_j)$ may be interrupted in $u_j \in [u^0_{h+1,j}, u^0_{hj})$ exactly by the units of higher priority then $h+1$: $\mathscr{C}^{[h]} = \bigcup_{i=1}^{h} \mathscr{C}_i$, $(h<j)$. The united arrival process of $\mathscr{C}^{[h]}$ is a Poisson stream with intensity $a^{[h]} = \sum_{i=1}^{h} a_i$. If the actual value of the random service time of L_j is t, where $t \geqslant u_{h+1,j}$; then the stream of the arriving $L^{[h]}$'s preempts its service expectedly $a^{[h]}[\min(t, u^0_{h,j}) - u^0_{h+1,j}]$ times, which is the parameter of the corresponding Poisson distribution.

It is well-known that the expected length of a busy period is $\dfrac{R_h}{(1-R_h)a^{[h]}}$, where $\rho_i = a_i\beta_i$ and $R_h = \sum_{i=1}^{h} \rho_i$. According to the Wald-identity the expected total length is equal to the expected busy period multiplied by the expected number of the preemptions: $\dfrac{R_h}{1-R_h} \times$ $\times [\min(t, u^0_{h,j}) - u^0_{h+1,j}]$. The actual service time of L_j equals to t with probability $dB(t)$, thus the expected value of $\min(t, u^0_{h,j}) - u^0_{h+1,j}$ is $B^*_j(u^0_{h+1,j}) - B^*_j(u^0_{h,j}+0)$, where $B^*(u) = \int_u^\infty (t-u)dB(t) = \int_u^\infty \bar{B}(t)dt$, $\bar{B}(t) = 1 - B(t)$. Consequently, the expected (j,h)-th interruption time is $z_{jh} = \dfrac{R_h}{1-R_h}\{B^*_j(u^0_{h+1,j}) - B^*_j(u^0_{h,j}+0)\}$, and the expected jth interruption time is $z_j = \sum_{h=1}^{j-1} z_{jh}$. According to the Abel's sum-identity the interruption time of L_j is given by

$$(1)\qquad z_j = \frac{R_{j-1}\beta_j}{1-R_{j-1}} - \sum_{h=1}^{j-1} \frac{\rho_h}{(1-R_{h-1})(1-R_h)} B_j^*(u_{h,j}^0 + 0)$$

$$(1 \leqslant j \leqslant r)\,.$$

The determination of the Waiting Times

As we have remarked, we assume the existance of the stationary distribution. First we shall prove the existence of the expected waiting times. Let ω_j be the actual value of the jth waiting time at a moment $(1 \leqslant j \leqslant r)$. Evidently, $\omega_j \leqslant \omega_r$. But ω_r is identical with the rth waiting time of the nonpreemptive discipline denoted by $\widetilde{\omega}_r$. The finiteness of $\widetilde{w}_r = \mathrm{E}\widetilde{\omega}_r$ was proved by Cobham [5], which implies the finiteness of $w_j = \mathrm{E}\omega_j$, $(i \leqslant j \leqslant r)$.

We look for an equation system where w_j's are the unknowns, following Cobham and Bronstein – Veklerov. From the arrival of a fixed L_j^* to the first start of its service the following events take place:

1. The completion of the service of the unit $L_i(u_i)$ which is serviced at the arrival of L_j^* if $L_i(u_i)$ is prior to L_j^*, $1 \leqslant i \leqslant r$.

2. The completion of the service of the preempted units $L_k(u_k)$'s standing in the queue at the arrival of L_j^* if they are prior to L_j^*.

3. The service of units waiting in the queue at the arrival of L_j^* if they are prior to L_j^* before their start of the service: $\mathscr{C}^{[j]} = \bigcup_{h=1}^{j} \mathscr{C}_h$.

4. The service of units arriving during the waiting period of L_j^* if they are serviced during this period: $\mathscr{C}^{[j-1]}$.

We shall determine the duration time of these events in the order of the complicatedness.

Event 4. The expected number of hth priority units arriving during the waiting interval of L_j^* is the arrival intensity multiplied by the expected length of this interval: $a_h w_j$. The expected duration of these services equals to $\beta_h a_h w_j$ by the Wald-identity. A newly arriving unit

L_h is serviced before L_j^* if and only if $h<j$, hence the total expected duration of Event 4 is

$$(2) \qquad w_j(4) = \sum_{h=1}^{j-1} \rho_h w_j = R_{j-1} w_j \qquad (1 \leqslant j \leqslant r).$$

Event 3. The expected number of hth priority units waiting in the queue for the first start of their service at the arrival of L_j^* is $a_h w_h$, hence the total expected duration of Event 3 is

$$(3) \qquad w_j(3) = \sum_{h=1}^{j} \rho_h w_h \qquad (1 \leqslant j \leqslant r).$$

Event 1. At the arrival of L_j^* a unit of $\mathscr{C}_i$ is serviced with probability ρ_i; $L_i(u_i)$ is prior to L_j^* if and only if $u_{ji}^0 \leqslant u_i$. If the actual service time of L_i is t, then *in average* $t - u_i$ divides the interval $[u_{ji}^0, t]$ on two equal parts, i.e. $t - u_i = \dfrac{t - u_{ji}^0}{2}$. The probability of the arrival of L_j^* is $a_i(t - u_{ji}^0)$, hence the *expected remaining service time* of $L_i(u_i)$ equals to $\dfrac{a_i}{2} \int_{u_{ij}^0}^{\infty} (t - u_{ji}^0)(t - u_{ij}^0) dB_i(t)$. Let us have $\Gamma(u) = \dfrac{1}{2} \int_u^{\infty} (t-u)^2 dB(t)$, then the total expected duration of Event 1 is

$$(4) \qquad w_j(1) = \sum_{i=1}^{r} a_i \Gamma_i(u_{ji}^0) \qquad (1 \leqslant j \leqslant r).$$

Event 2. It is easy to see there is at most one preempted unit $L_k(u_k)$ in the queue which is prior to L_j^* and $k \geqslant j$. (If $L_k(u_k)$ was preempted by L_i, then $u_{jk}^0 \leqslant u_k < u_{ik}^0$, hence $i<j$.)

We shall determine $w_{j+1}(2) - w_j(2)$ instead of $w_j(2)$. Evidently, each preempted unit of $\mathscr{C}^{[j]}$ is prior to L_j^* as well as to L_{j+1}^*, therefore we concentrate on the unique preempted unit $L_k(u_k)$, if it is prior to L_{j+1}^*, while L_j^* is prior to $L_k(u_k)$: $u_{j+1,k}^0 \leqslant u_k < u_{j,k}^0$, $(j \leqslant k \leqslant r)$.

The probability of this event is $\rho_k \sum_{l=1}^{\infty} R_j^l = \dfrac{\rho_k R_j}{1 - R_j}$. The conditional expected remaining service time of $L_k(u_k)$ is $\frac{1}{\beta_k}\{\Gamma_k(u^0_{j+1,k}) - - \Gamma_k(u^0_{j,k}+0)\}$. Consequently, (see also (4))

$$w_{j+1}(2) - w_j(2) = \frac{R_j}{1-R_j}[w_{j+1}(1) - w_j(1)]\,. \tag{5}$$

Now we shall solve the linear system $w_j = w_j(1) + w_j(2) + w_j(3) + + w_j(4)$ $(1 \leqslant j \leqslant r)$, hence (see also (2) and (3))

$$(1 - R_{j-1})w_j = \sum_{h=1}^{j} w_h \rho_h + w_j(1) + w_j(2) \qquad (1 \leqslant j \leqslant r)\,. \tag{6}$$

Substituting (5) into (6), we get

$$\begin{aligned}(1-R_{j+1})(1-R_j)w_{j+1} &= (1-R_j)(1-R_{j-1})w_j + \\ &+ w_{j+1}(1) - w_j(1) \qquad (1 \leqslant j \leqslant r-1)\,.\end{aligned} \tag{7}$$

If $j = 1$, then $w_1(2) = w_1(4) = 0$, thus (6) gives $(1 - R_j)w_1 = w_1(1)$. These two equations imply together $(1 - R_j)(1 - R_{j-1})w_j = w_j(1)$, which implies with (4):

$$w_j = \frac{\sum_{i=1}^{r} a_i \Gamma_i(u^0_{ji})}{(1-R_j)(1-R_{j-1})} \qquad (1 \leqslant j \leqslant r)\,. \tag{8}$$

We have

Theorem 1. *The expected waiting times and the expected interruption times are given by* (8) *and* (1).

Remark. The expected waiting time, the expected interruption time and the expected queuing time of the jth priority class do not depend on the discretionary value of any other pair of classes: w_j depends only on $(u^0_{jk})_{k>j}$ and z_j on $(u^0_{hj})_{h<j}$. This fact is evident for such pairs of classes where both classes are prior to L^*_j, because the order of the service of $L_h(u_h)$ and $L_{h'}(u_{h'})$ does not influence the order of L_h, $L_{h'}$ and L_j. However, this independence is interesting for the remaining discretionary values.

The method of the solution of the linear system (6) used here differs from Cobham's and Bronstein – Veklerov's method. We need not guess w_j's (i.e. (8)) in advance and we need not apply the mathematical induction to prove (8).

3. THE OPTIMAL DISCRETIONARY DISCIPLINE

To define the optimality of the disciplines we must define the objective function of the system. Following others, the objective function will be the *expected queuing loss.* Let c_i be the queuing loss of L_i for one time unit, then the objective function is $F = \sum_{i=1}^{r} c_i a_i v_i$.

We have $\frac{r(r-1)}{2}$ control variables: $u^0_{h,j}$ $(1 \leqslant h < j \leqslant r)$. (The remaining discretionary values are *prefixed:* $u^0_{kj} = 0$ $(1 \leqslant j \leqslant k \leqslant r)$.)

The system of the discretionary values is *locally (or globally) optimal* if F has local (or global) minimum at this vector in the domain of the *feasible control variables:* $(u^0_{hj})_{h<j}$, $u^0_{i+1,j} \leqslant u^0_{i,j}$, $1 \leqslant i, j \leqslant r$. As we shall use the gradient condition, we assume each service time distribution is *absolutely continuous*, the corresponding density function is $b_j(t) = B'_j(t)$ $(1 \leqslant j \leqslant r)$. Though the distribution function of a non-random service time is not continuous, the whole proof is correct in this case, too. As F is differentiable by all its variable u^0_{hj}, if F has a local minimum at $(\hat{u}^0_{hj})_{1 \leqslant h < j \leqslant r}$, then $\left.\frac{\partial F}{\partial u^0_{hj}}\right|_{u^0_{hj}=\hat{u}^0_{hj}} = 0$ $(1 \leqslant h < j \leqslant r)$. We have already remarked that only v_h and v_j depends on u^0_{hj}:

$$\frac{\partial v_h}{\partial u^0_{hj}} = \frac{a_j \Gamma'_j(u^0_{hj})}{(1 - R_{h-1})(1 - R_h)},$$

$$\frac{\partial v_j}{\partial u^0_{hj}} = - \frac{\rho_h B^{*\prime}_j(u^0_{hj})}{(1 - R_{h-1})(1 - R_h)}.$$

As $B^{*\prime}(u) = -\bar{B}(u)$ and $\Gamma'(u) = -\beta(u)\bar{B}(u)$, where $\beta(u)$ *is the conditional expected remaining service time, if the service has not yet finished* *during time* u, i.e. $\bar{B}(u) > 0$. Consequently, $\dfrac{\partial F}{\partial u_{hj}^0} = 0$ implies

$$(9) \qquad \frac{\beta_j(\hat{u}_{hj}^0)}{c_j} = \frac{\beta_h}{c_h}, \quad \hat{u}_{h+1,j}^0 \leqslant \hat{u}_{h,j}^0, \quad u_{kj}^0 = 0 \quad (1 \leqslant h < j < k \leqslant r).$$

If

$$\frac{\beta_j(u_{hj}^0)}{c_j} > \frac{\beta_h}{c_h}$$

holds for each value of $u_{hj}^0 \in [0, \infty)$, then only $\hat{u}_{hj}^0 = \infty$ (L_h always preempts $L_j(u_j)$) may be optimal. If

$$\frac{\beta_j(u_{hj}^0)}{c_j} < \frac{\beta_h}{c_h}$$

holds for each value of $u_{hj}^0 \in [0, \infty)$, then only $\hat{u}_{hj}^0 = 0$ may be optimal (L_h never preempts $L_j(u_j)$).

As the mixed second partial derivatives $\dfrac{\partial^2 F}{\partial u_{hj} \partial u_{\bar{h}\bar{j}}} \equiv 0$ $((h, j) \neq (\bar{h}, \bar{j}))$, $(\hat{u}_{hj})_{h<j}$ is locally strictly optimal if and only if (9) holds and $\beta_j(u_j)$ *decreases* at $\hat{u}_{hj}^0$ $(1 \leqslant h < j \leqslant r)$.

If c_j varies from zero to infinity, then $\hat{u}_{hj}^0$ varies from infinity to zero. Therefore, if we want a sufficient condition for the local optimum which is qualitatively independent of c_j's, then *we must assume* that $\beta_j(u_j)$ decreases (or at least, does not increases) in $u_j \in [0, \infty)$ $(1 \leqslant j \leqslant r)$.

Now, this assumption implies the existence of a unique locally and globally optimal system of the discretionary values, possibly infinite: Let

$$\hat{u}_{hj}^0 = \inf \left\{ u: \frac{\beta_j(u)}{c_j} = \frac{\beta_h}{c_h} \right\},$$

then

$$\frac{\beta_j(u_j)}{c_j} > \frac{\beta_h}{c_h} \qquad \text{if} \qquad u_j < \hat{u}_{hj}^0$$

and

$$\frac{\beta_j(u_j)}{c_j} < \frac{\beta_h}{c_h} \quad \text{if} \quad u_j \geqslant \hat{u}^0_{hj} \quad (1 \leqslant h < j \leqslant r)\,.$$

For $u_j = 0$ we get $\frac{\beta_j}{c_j} > \frac{\beta_h}{c_h}$ if $h < j$, which is the optimal order of the nonpreemptive discipline (see Aczél [1] and Cox – Smith [6]). Let us remark, that the condition of feasibility in (9) is also satisfied: $\hat{u}^0_{i+1,j} \leqslant \leqslant \hat{u}^0_{i,j}$.

We have

Theorem 2. (i) *If there exists a locally optimal system of the discretionary values, then it satisfies* (9).

(ii) *If the expected remaining service time* $\beta_j(u_j)$ *is decreasing, then there exists a unique optimal priority order and a system of discretionary values which is determined by* (9).

Remarks. The notion of the discretionary discipline was introduced by Avi-Itzhak, Brosh and Naor [2] *for two-class process with non-random service times* in 1964. This work was generalized by Etschmaier [7] and Jaiswal [8] if the *service times are random,* in 1966 and 1968, respectively. The *several-class discretionary discipline* was introduced by Bronstein and Veklerov [3] in 1967, in the special case of non-random service times. The mentioned papers proved Theorems 1 and 2 in special cases, with similar argumentation. Jaiswal failed to assume that $\beta_2(u_2)$ is nonincreasing and none of the authors proved that the optimal order of the priority indices is the optimal order at the nonpreemptive priority discipline, they simply assumed it implicitely.

Let us emphasize again that only Jaiswal proved the existence of the stationary ergodic distribution of the state of the system.

The present author [10] has proved that *the optimality condition*

$$\frac{\beta_2(\hat{u}^0_{12})}{c_2} = \frac{\beta_1}{c_1}$$

holds for arbitrary arrival process, if there is only two classes. I proved this directly, without any explicit formula like (1) and (8). The proof was very short and elementary, pointing out the close connection between the common sense and this mathematical proposition.

Finally, let us mention the following *conjecture* of Bronstein and Veklerov [4], published in 1969. At every moment the optimal general priority discipline (where the priority unit to be serviced at a given moment is arbitrarily determined by the state of the system at this moment) is to service a unit with minimal expected remaining service time/unit queuing loss. I have proved [10], this conjecture is generally false if $\beta_j(u_j)$ increases in some interval. If $\beta_i(u_i)$ is nonincreasing, then the optimum principle of Bronstein and Veklerov is identical with the optimality condition of the discretionary discipline (9). I have proved [10], *this general optimality principle is correct in two-class process with arbitrary arrival process if the completion rate of the service of* L_j, *i.e.* $\dfrac{b_j(u_j)}{\bar{B}_j(u_j)}$ $(1 \leqslant j \leqslant r)$ *is nondecreasing,* $\bar{B}_j(u_j) > 0$.

Let us remark nondecreasing completion rate implies nonincreasing expected remaining service time. According to Prékopa [9], if the density function $b(u)$ is log concave (for example: exponential, normal, gamma, beta, rectangular distributions) then $B(u)$ and $\bar{B}(u)$ are also log concave, i.e. the completion rate is nondecreasing, hence the remaining expected service time is nonincreasing.

We complete this paper with the following

Conjecture. *If units arrive in* r *independent Poisson streams, and if each service time distribution has nondecreasing completion rate, then there exists a unique optimal general priority discipline, which is determined by the optimum principle of* Bronstein *and* Veklerov: *the optimal discretionary discipline of Theorem* 2.

Addendum. This conjecture was proved by the author in revised version of [10] (Theorem 5iii). That proof is independent of the results of the present paper, so it is another, direct proof of Theorem 2 but under a slightly stronger condition.

REFERENCES

[1] M. Aczél, The effect of introducing priorities, *Oper. Res.*, 8 (1960), 730-733.

[2] B. Avi-Itzak – I. Brosh – P. Naor, On discretionary priority queuing, *Zeitschrift fur Angewandte Mathematik und Mechanik,* 44 (1964), 235-242.

[3] I.O. Bronstein – E.B. Veklerov, On a class of queuing discipline, (in Russian), *Tehn. Kibernetik,* 5 (1967), 101-107.

[4] I.O. Bronstein – E.B. Veklerov, On a class of discipline of queuing, (in Russian), *Queuing theory in the information transfer systems,* Nauka, Moskva, 1969, 54-59.

[5] A. Cobham, Priority assignement in waiting line problems, *Operations Research,* 2 (1954), 70-76.

[6] D.R. Cox – W.L. Smith, *Queues,* Methuen, London, 1961.

[7] M. Etschmaier, *Discretionary Priority Processes,* M. S. Thesis, Case Institute of Technology, Cleveland, Ohio, 1966.

[8] N.K. Jaiswal, *Priority Queues,* Academic Press, New York and London, 1968.

[9] A. Prékopa, *Optimization Problems of Stochastic Systems,* (Hungarian manuscript), Dissertation, 1971.

[10] A. Simonovits, Direct comparison of different priority queuing disciplines, *Studia Sci. Math. Hungar.*, (to appear).

COLLOQUIA MATHEMATICA SOCIETATIS JÁNOS BOLYAI
9. EUROPEAN MEETING OF STATISTICIANS, BUDAPEST (HUNGARY), 1972.

STATISTICS OF DIFFUSION PROCESSES

A.N. ŠIRJAEV

The estimation of the unknown components by the observed process of diffusion character is considered in Bayesian and non-parametric statistical problems. In the first paragraph some results are proved about an effective estimation in problems of optimal non-linear filtration. Second paragraph contains results about a sequential estimation of the unknown parameter of diffusion processes.

§1. EFFECTIVE CASES OF FILTRATION

1. Let $(\Omega, \mathscr{F}, \mathsf{P})$ be a complete probability field, $(\mathscr{F}_t)$ $t \geqslant 0$ an increasing family, continuous from the right, of sub σ-algebras of $\mathscr{F}$ and $w = (w_t, \mathscr{F}_t)$ a standard Wiener process. We assume that the diffusion process satisfying the stochastic differential equation

$$(1) \qquad d\xi_t = A(t, \omega)dt + B(t, \xi)dw_t, \qquad \xi_0 = 0$$

can be observed – where the stochastic process $A = (A(t, \omega), \mathscr{F}_t)$ contains the unobserved part (e.g. $A(t, \omega)$ is the unknown signal) which is to be estimated by the observed process ξ, and that $B(t, \xi)$ is $\mathscr{F}_t^\xi =$

$= \sigma(\omega\colon \xi_s\ s \leqslant t)$ measurable for every $t \geqslant 0$. Recently considerable progress has been achieved in solving the equations of optimal (in quadratic mean) nonlinear filtration (see [1]-[8]).

The main result can be formulated in the following way:

Let $\pi_t(g) = \mathsf{E}(g(t, \omega) \mid \mathcal{F}_t^{\xi})$ be the optimal (in quadratic mean) estimation of the function $g(t, \omega)$, $\mathsf{E}g^2(t, \omega) < \infty$, by the observation of ξ_s, $s \leqslant t$. If the estimated function $f = f(t, \omega)$ has the property that there exists a function $F(t, \omega)$ such that the process $x_t = f(t, \omega) - f(0, \omega) - \int_0^{\infty} F(s, \omega) ds$ is a square integrable martingale, then under natural assumptions, (detailed in [7], [8]) one has

$$\begin{aligned} \pi_t(f) = \pi_0(f) &+ \int_0^t \pi_s(F) ds + \int_0^t \{\pi_s(D) + \\ &+ [\pi_s(fA) - \pi_s(f)\pi_s(A)]\} B^{-1}(s, \xi) d\bar{w}_s \end{aligned} \tag{2}$$

where $\bar{w} = (\bar{w}_t, \mathcal{F}_t^{\xi})$, $t \geqslant 0$ is a Wiener process with

$$\bar{w}_t = \int_0^t \frac{d\xi_s - \pi_s(A)}{B(s, \xi)} ds$$

and $\langle X, W\rangle_t = \int_0^t D_s ds$.

Equation 2 is called the equation of optimal nonlinear filtration.

As an illustration for the application of equation (2), consider the following special case of process (1) when

$$d\xi_t = [A_0(t, \xi) + \Theta A_1(t, \xi)]dt + B(t, \xi) dw_t \,. \tag{3}$$

The random variable $\Theta = \Theta(\omega)$ is independent of the Wiener process w_t and has finite moments. Substituting into (2) $f(t, \omega) = \Theta$ we get that $\pi_t(\omega) = \mathsf{E}(\Theta \mid \mathcal{F}_t^{\xi})$ satisfies the stochastic differential equation

$$\begin{aligned} d\pi_t(\Theta) &= [A_0(t, \xi)\pi_t(\Theta) + A_1(t, \xi)\pi_t(\Theta^2)] B^{-1}(t, \xi) dt - \\ &- [A_0(t, \xi)\pi_t(\Theta) + A_1(t, \xi)\pi_t^2(\Theta)] B^{-1}(t, \xi)[d\xi_t - \pi_t(\Theta) dt], \end{aligned} \tag{4}$$

$$(F(t, \omega) \equiv 0, \ D_t \equiv 0) .$$

From (4) it can be seen that the first conditional moment $\pi_t(\Theta) = = \mathsf{E}(\Theta \mid \mathscr{F}_t^\xi)$ can not be determined without knowing the second conditional moment $\pi_t(\Theta^2) = \mathsf{E}(\Theta^2 \mid \mathscr{F}_t^\xi)$ and so on. Because of this "incompleteness" (the smaller moments depend on the order ones) it is important to find such cases when the equations of filtration can be solved.

Before formulating our main result (Theorem 2), let us consider a process with the differential (3). This example will show the idea used in the general case.

Theorem 1. *Let the distribution* $\mathsf{P}(\Theta \leqslant x \mid \xi_0)$ *be normal* $N(m, \gamma)$ *and*

(i) $$\mathsf{P}\left(\int_0^t [A_0^2(t, \xi) + B^2(t, \xi)]dt < \infty\right) = 1, \qquad t \geqslant 0 ,$$

(ii) $$|A_1(t, \xi)| \leqslant C < \infty \qquad \mathsf{P} \text{ a.e.},$$

(iii) $$B^2(t, \xi) \geqslant C > 0 .$$

Then for arbitrary $t > 0$ *the a posteriori distribution* $\mathsf{P}(\Theta \leqslant x \mid \mathscr{F}_t^\xi)$ *will also be normal with parameters* $m_t = \mathsf{E}(\Theta \mid \mathscr{F}_t^\xi)$, $\gamma_t = \mathsf{E}[(\Theta - m_t)^2 \mid \mathscr{F}_t^\xi]$ *satisfying the equations*

(5) $$dm_t = \gamma_t \frac{A_1(t, \xi)}{B^2(t, \xi)} [d\xi_t - (A_0(t, \xi) + A_1(t, \xi)m_t)dt] , \qquad m_0 = m ,$$

(6) $$\dot{\gamma}_t = - \gamma_t^2 \frac{A_1(t, \xi)}{B^2(t, \xi)} , \qquad \gamma_0 = \gamma ,$$

whose solutions can be given by the formulae

(7) $$m_t = m \frac{\gamma_t}{\gamma} + \gamma_t \int_0^t \frac{A_1(s, \xi)}{B^2(s, \xi)} (d\xi_s - A_0(s, \xi)ds) ,$$

(8) $$\gamma_t = \frac{\gamma}{1 + \gamma \int_0^t \frac{A_1^2(s, \xi)}{B^2(s, \xi)} ds} .$$

Remark. It is important to emphasize that whereas the pair (Θ, ξ) need not be normal, yet the conditional distribution $P(\Theta \leqslant x \mid \mathscr{F}_t^\xi)$ will be normal. It follows from the fact that the initial distribution $P(\Theta \leqslant x \mid \xi_0)$ is normal, and Θ appears in (3) in linear form in the coefficient of the drift.

The proof of this theorem is very simple. (In the case of Markov processes (Θ, ξ) see also [2]). Let μ_w and μ_ξ be the Wiener measure and the measure respectively corresponding to the process ξ. Write μ_{ξ^α} for the measure corresponding to the process with differential

$$d\xi_t^\alpha = (A_0(t_1 \xi^\alpha) + \alpha A_1(t, \xi^\alpha))dt + B(t, \xi^\alpha)dw_t$$

where $-\infty < \alpha < \infty$. Then (see [2], [8], [9]) the Random – Nikodym derivative $\frac{d\mu_{\xi^\alpha}}{d\mu_\xi}(t, \xi)$ of the measure μ_{ξ^α} by the measure μ_ξ on the interval $[0, t]$ is

$$\frac{d\mu_{\xi^\alpha}}{d\mu_\xi} = \exp\left[\int_0^t (\alpha - m_s)\frac{A_1(s, \xi)}{B^2(s, \xi)} \times \right.$$

$$\times [d\xi_s - (A_0(s, \xi) + m_s A_1(s, \xi))ds] -$$

$$\left. - \frac{1}{2}\int_0^t (\alpha - m_s)^2 \frac{A_1^2(s, \xi)}{B^2(s, \xi)} ds\right]$$

and consequently the density is

$$\rho_\alpha(t) = \frac{\partial P(\Theta \leqslant \alpha \mid \mathscr{F}_t^\xi)}{\partial \alpha} = \rho_\alpha(0) = \frac{d\mu_{\xi^\alpha}}{d\mu_\xi} =$$

$$= \frac{1}{\sqrt{2\pi\gamma}} e^{-\frac{(\alpha - m)^2}{2\gamma}} \exp\left[\int_0^t (\alpha - m_s)\frac{A_1(s, \xi)}{B^2(s, \xi)} \times \right.$$

$$\times [d\xi_s - (A_0(s, \xi) + m_s A_1(s, \xi))ds] -$$

$$\left. - \frac{1}{2}\int_0^t (\alpha - m_s)^2 \frac{A_1^2(s, \xi)}{B^2(s, \xi)} ds\right].$$

This shows that the conditional density $\rho_\alpha(t)$ is normal with parameters depending on ξ_s, $s \leqslant t$. Now formulae (5) and (6) are immediate consequences of (2): substitute $f(t, \omega) = \Theta$ and take into consideration that by the normality of $\mathsf{P}(\Theta \leqslant x \mid \mathscr{F}_t^\xi)$

$$\mathsf{E}[\Theta^2(\Theta - m_t) \mid \mathscr{F}_t^\xi] = 2m_t\gamma_t \qquad (\mathsf{P}\text{ a.e.})\,.$$

2. The above-mentioned case is a special case of the following more general situation.

We assume the "unobserved" process $\Theta = (\Theta_t)$ $t \geqslant 0$ and the observed one $\xi = (\xi_t)$ $t \geqslant 0$ satisfy the stochastic differential equations

$$\begin{aligned} d\Theta_t &= [a_0(t, \xi) + a_1(t, \xi)\Theta_t]dt + b_1(t, \xi)dw_1(t) + b_2(t, \xi)dw_2(t), \\ d\xi_t &= [A_0(t, \xi) + A_1(t, \xi)\Theta_t]dt + B(t, \xi)dw_2(t)\,, \end{aligned} \tag{9}$$

where $w_1 = (w_1(t))$, $w_2 = (w_2(t))$, $t \geqslant 0$, are independent Wiener processes.

Theorem 2. *Let the conditional distribution* $\mathsf{P}(\Theta_0 \leqslant x \mid \xi_0)$ *be normal* $N(m_0, \gamma)$, $|m_0| < \infty$, $0 \leqslant \gamma_0 < \infty$. *Assume the following conditions hold true:*

(i) $$\mathsf{P}\left(\int_0^t [a_0^2(s, \xi) + A_0^2(s, \xi) + b_1^2(s, \xi) + b_2^2(s, \xi)]ds < \infty\right) = 1\,,$$

(ii) $$|A_1(t, \xi)| \leqslant C < \infty \qquad \left|a_0(t, \xi) - \frac{b_2(t, \xi)}{B(t, \xi)}A_0(t, \xi)\right| \leqslant C < \infty\,,$$
$$\left|a_1(t, \xi) - \frac{b_2(t, \xi)}{B(t, \xi)}A_1(t, \xi)\right| \leqslant C < \infty\,,$$

(iii) $B^2(t, \xi) \geqslant C > 0$,

(iv) *arbitrary two continuous solutions of the equation*

$$\eta_t = \xi_0 + \int_0^t B(s, \eta)dw_2(s)\,, \qquad t \geqslant 0\,,$$

have the same finite dimensional distributions.

Then the stochastic process (Θ_t, ξ_t), $t \geqslant 0$ *is conditionally Gaussian, i.e. for arbitrary* $0 \leqslant t_1 \leqslant \ldots \leqslant t_n \leqslant t$ *this distribution* $\mathsf{P}(\Theta_{t_1} \leqslant x_1, \ldots, \Theta_{t_n} \leqslant x_n \mid \mathscr{F}_t^\xi)$ is (P a.e.) *normal. In addition* $m_t = \mathsf{E}(\Theta_t \mid \mathscr{F}_t^\xi)$ *and* $\gamma_t = \mathsf{E}((\Theta_t - m_t)^2 \mid \mathscr{F}_t^\xi)$ *satisfy the following system of equations:*

$$dm_t = [a_0 + a_1 m_t] + \frac{b_2 B + \gamma_t A_1}{B^2}[d\xi_t - (A_0 + A_1 m_t)dt], \qquad m_0 = m \tag{10}$$

$$\gamma_t = 2a_1\gamma_t + (b_1^2 + b_2^2) - \frac{(b_2 B + \gamma_t A_1)^2}{B^2}, \qquad \gamma_0 = \gamma \tag{11}$$

$$(a_0 = a_0(t, \xi),\ a_1 = a_1(t, \xi), \ldots).$$

This theorem contains, as special case, several former results in the theory of filtration. The most famous one is that of Kalman and Bucy [10]; they assume that every coefficient in (9) depends only on the time t. They deduced the equations for m_t, γ_t using the method of Wiener – Hopf and were able to consider non-stationary cases too, they did not allow, however, the dependence of coefficients on quantities observed in the past, which is extraordinary important if we want to control by observed past.

By Theorem 2 the a posteriori mean and a posteriori variance satisfy the system of equations (10), (11). The first equation is a stochastic differential equation and the second one is a Ricatti differential equation with random coefficients. The question is very important whether the solution of equations is unique and (11) unique in the class of all non-negative functions.

Theorem 3. *Under the conditions of Theorem 2 the system of equations* (10), (11) *has a unique solution.*

The complete proof of Theorem 2 and 3 will be given in the book [8] appearing in the near future. Here we remark only that the idea of the proof of Theorem 2 is very similar to that of the proof of Theorem 1 given before, i.e. the case $\Theta_t = \Theta$.

3. We use the result of Theorem 1 to give a simple deduction of a result in information-theory [11], [12], [13] about the structure of optimal coding and decoding, when a Gaussian random variable is transmitted through a channel with white noise and we may apply a noiseless feedback. Let $\Theta = \Theta(\omega)$ be a Gaussian random variable $N(m, \gamma)$ which must be transmitted through a channel with white noise. In addition, a feedback can be used. The corresponding model in continuous time can be built up in the following way. Let the receiver get the message $\xi = (\xi_t)$, $t \geqslant 0$,

$$d\xi_t = [A_0(t, \xi) + \Theta A_1(t, \xi)]dt + B \cdot dw_t, \qquad \xi_0 = 0 . \tag{12}$$

By choosing the functions (A_0, A_1) a coding is realized and its dependence on ξ means the possibility of a feedback without error.

The coding system is supposed to satisfy, in every moment, the condition

$$\mathsf{E}[A_0(t, \xi) + \Theta A_1(t, \xi)]^2 \leqslant P , \tag{13}$$

where P is a given constant characterizing the energetic possibilities of the transmitting system. In every moment $t \geqslant 0$ an estimation $\hat{\Theta}_t = \hat{\Theta}(t, \xi)$ must be given by the information ξ_s, $s \leqslant t$, in such a way that it minimize $\mathsf{E}(\Theta - \hat{\Theta}_t)^2$, the error of decoding. The problem is to find a coding (A_0^*, A_1^*) satisfying condition (13) and a decoding $\Theta_t^* = \Theta^*(t, \xi)$, $t \geqslant 0$ such that

$$\Delta(t) = \inf_{\hat{\Theta},(A_0, A_1)} \mathsf{E}(\Theta - \hat{\Theta}_t)^2$$

is achieved for every $t \geqslant 0$. If the coding (A_0, A_1) is already chosen, then obviously the best decoding would be $\Theta_t^* = m_t = \mathsf{E}(\Theta \mid \mathscr{F}_t^\xi)$. Thus the question about optimal decoding can be solved easily.

By virtue of (8), for every pair (A_0, A_1), $\gamma_t = \mathsf{E}[(\Theta - m_t)^2 \mid \mathscr{F}_t^\xi]$ can be expressed as follows:

$$\gamma_t = \frac{\gamma}{1 + \gamma \int_0^t \frac{A_1^2(s, \xi)}{B^2} ds} , \tag{14}$$

$$m_t = \gamma_t \left[\frac{m}{\gamma} + \int_0^t \frac{A_1(s, \xi)}{B} (d\xi_s - A_0(s, \xi) ds) \right]. \tag{15}$$

Rewrite (13)

$$P \geqslant \mathsf{E}[A_0(t, \xi) + \Theta A_1(t, \xi)]^2 =$$

$$= \mathsf{E}[(A_0(t, \xi) + m_t A_1(t, \xi)) + (\Theta - m_t) A_1(t, \xi)]^2 = \tag{16}$$

$$= \mathsf{E}[A_0(t, \xi) + m_t A_1(t, \xi)]^2 + \mathsf{E}[\gamma_t \cdot A_1^2(t, \xi)].$$

Hence

$$P \geqslant \mathsf{E}[\gamma_t \cdot A_1^2(t, \xi)]$$

which, together with (14), given inequality

$$P \geqslant \mathsf{E} \frac{\gamma A_1^2(t, \xi)}{1 + \int_0^t \frac{\gamma \cdot A_1^2(s, \xi)}{B^2} ds}. \tag{17}$$

Suppose now that the considered code $A_1(t, \xi)$ does not depend on ξ. Thus if $u(t) = \gamma A_1^2(t)$ then

$$u(t) \leqslant P + \frac{P}{B^2} \int_0^t u(s) ds. \tag{18}$$

By the Bellman – Gronwall inequality

$$u(t) \leqslant P \exp\left(\frac{P}{B^2} t \right),$$

i.e.

$$A_1^2(t) \leqslant \frac{P}{\gamma} \exp\left(\frac{P}{B^2} t \right). \tag{19}$$

Therefore

$$\mathsf{E}\gamma_t = \gamma_t = \frac{\gamma}{1 + \gamma \int_0^t \frac{A_1^2(s)}{B^2} ds} \geqslant \gamma e^{-\frac{P}{B^2} t} \tag{20}$$

So that

$$\Delta(t) = \inf_{\hat{\Theta}, (A_0, A_1)} \mathsf{E}(\Theta - \hat{\Theta}_t)^2 = \inf_{(A_0, A_1)} \mathsf{E}(\Theta - m_t)^2 =$$

(21)

$$= \inf_{(A_0, A_1)} \gamma_t \geqslant \gamma e^{-\frac{P}{B^t} t}$$

We remark now that equality can be reached in (19)-(21) everywhere choosing

$$A_1^2(t) = \frac{P}{\gamma} \exp\left(\frac{P}{B^2} t\right)$$

$$A_0(t, \xi) = - m_t A_1(t) .$$

Thus the following result has been proved (see [11], [12], [13]).

Theorem 4. *If* $A_1(t, \xi) = A_1(t)$ *(it does not depend on* ξ*), then the optimal coding can be given by the formulas*

$$A_1^*(t) = \sqrt{\frac{P}{\gamma}} \exp\left(\frac{P}{2B^2} t\right)$$

$$A_0^*(t, \xi) = - m_t A_1^*(t) .$$

The transmitted signal $\xi = (\xi_t)$, $t \geqslant 0$ *and the optimal decoding* $\Theta^* \equiv \equiv m = (m_t)$, $t \geqslant 0$ *are determined by the equations*

$$d\xi_t = \sqrt{\frac{P}{\gamma}} \exp\left(\frac{P}{2B^2} t\right)(\Theta - m_t)dt + Bdw_t ,$$

$$dm_t = \frac{\sqrt{P\gamma}}{B^2} \exp\left(-\frac{P}{2B^2} t\right) d\xi_t .$$

The error of the reproduction is

$$\Delta(t) = \gamma e^{-\frac{P}{B^2} t} .$$

Remark. One can prove in the same way that if the coding does not use feed-back, i.e. $A_0 = A_0(t)$, $A_1 = A_1(t)$, then the optimal solution is

$A_0^* = -mA_1^*$, $A_1^* = \sqrt{\dfrac{P}{\gamma}}$ which makes the error of reproduction as small as

$$\gamma\left(1 + \frac{P}{B^2}t\right)^{-1}.$$

§2. SEQUENTIAL ESTIMATION OF THE PARAMETERS IN DIFFUSION PROCESSES

1. Let us consider the diffusion process with differential

$$d\xi_t = \lambda A(t, \xi)dt + dw_t, \qquad \xi_0 = 0, \tag{22}$$

where the unknown parameter is λ, $-\infty < \lambda < \infty$, and $A(t, \xi)$ is $\mathscr{F}_t^\xi$-measurable for every $t \geq 0$. We assume that for every continuous function $x = (x_t)$, $t \geq 0$, $x_0 = 0$ there exists an $\epsilon = \epsilon(x) > 0$ such that $\int_0^{\epsilon(x)} A^2(t, x)dt < \infty$ and for every λ and $t \geq 0$

$$\mathsf{P}_\lambda\left(\int_0^t A^2(s, \xi)ds < \infty\right) = 1, \ \mathsf{P}_0\left(\int_0^t A^2(s, \xi)ds < \infty\right) = 1$$

where the index λ indicates that the process is considered for the given λ.

According to the results of [9], these conditions imply that the Wiener measure μ_w is equivalent for every $-\infty < \lambda < \infty$ to the measure μ_ξ^λ of process ξ, and

$$\frac{d\mu_\xi^\lambda}{d\mu_w} = \frac{d\mu_\xi^\lambda}{d\mu_\xi^0}(t, \xi) = \exp\left[\lambda\int_0^t A(s, \xi)d\xi_s - \frac{\lambda^2}{2}\int_0^t A^2(s, \xi)ds\right]. \tag{23}$$

The simplest and most known estimation of the parameter λ by the observation ξ_s, $s \leq t$, is the maximum likelihood estimation $\lambda_t(\xi)$, here

$$\lambda_t(\xi) = \frac{\int_0^t A(s, \xi)d\xi_s}{\int_0^t A^2(s, \xi)ds} \tag{24}$$

in strength of (23). We shall write E_λ for the mathematical expectation determined by the measure P_λ. Then, under natural conditions for $A(t, \xi)$ (e.g. $0 < c \leqslant |A(t, \xi)| \leqslant C < \infty$ is sufficient), it is not difficult to prove – using (23) – that

$$(25) \qquad \mathsf{E}_\lambda \lambda_t(\xi) = \lambda + \frac{\partial}{\partial \lambda} \mathsf{E}_\lambda \left[\int_0^t A^2(s, \xi) ds \right]^{-1} .$$

In consequence of this formula, the maximum likelihood estimation is biased and the bias, in general, equals to

$$\frac{\partial}{\partial \lambda} \mathsf{E}_\lambda \left[\int_0^t A^2(s, \xi) ds \right]^{-1}$$

R. Š. Lipcer and the author suggested to deal with sequential maximum likelihood estimation, which turned out to have several good properties. These estimations are made in the following way.

Let H be a non-negative number, whose role will be clear later. Define the stopping rule (Markov moment)

$$(26) \qquad \tau(H) = \inf \left\{ t \colon \int_0^t A^2(s, \xi) ds \geqslant H \right\} .$$

If for every λ, $\mathsf{P}_\lambda \left(\int_0^\infty A^2(s, \xi) ds < \infty \right) = 0$ then $\mathsf{P}_\lambda(\tau(H) < \infty) = 1$.

The estimation

$$\widetilde{\lambda}(H) = \lambda_{\tau(H)}(\xi), \qquad H \geqslant 0$$

is called the sequential maximum likelihood estimation.

Then, from (24) and (26)

$$(27) \qquad \widetilde{\lambda}(H) = \frac{1}{H} \int_0^{\tau(H)} A(s, \xi) d\xi_s .$$

One can see from (27) and also from (25) substituting $\tau(H)$ for t, that for every $H \geqslant 0$ the estimation is unbiased, $\mathsf{E}_\lambda \widetilde{\lambda}(H) = \lambda$. It is also clear that $\mathsf{D}_\lambda(\lambda(H)) = \frac{1}{H}$ and $\sqrt{H}(\tilde{\lambda}(H) - \lambda) = \frac{1}{H}\int_0^{\tau(H)} A(s, \xi)dw_s$. But the process $\beta(H) = \int_0^{\tau(H)} A(s, \xi)dw_s$, with $\tau(H)$ defined in (26), is well-known to be a Wiener process, consequently the quantity $H(\tilde{\lambda}(H) - \lambda)$ has a normal distribution $N(0, 1)$. Thus the following result can be stated:

Theorem 5. *If* $0 \leqslant c \leqslant |A(t, \xi)| \leqslant C < \infty$ *then the sequential maximum likelihood estimation* $\tilde{\lambda}(H)$

1) *is unbiased,* $\mathsf{E}_\lambda \tilde{\lambda}(H) \equiv \lambda$

2) *it has a constant variance* $\mathsf{D}_\lambda(\tilde{\lambda}(H)) = \frac{1}{H}$

3) *is normally distributed* $N\left(\lambda, \frac{1}{H}\right)$.

The above mentioned properties certainly do credit to the sequential maximum likelihood estimation. But the question arises very naturally, is it not a too long mean observation time, $\mathsf{E}_\lambda(\tau(H))$, the expense for these nice properties.

The study of this question for general $A(t, \xi)$ is very difficult. A. A. Novikov examined the quantity $\mathsf{E}_\lambda \tau(H)$ for $A(t, \xi) = -\xi_t$, i.e. when the process $\xi = (\xi_t)$, $t \geqslant 0$, has the differential

(28) $$d\xi_t = -\lambda\xi_t + dw_t .$$

He obtained the following result [14], [15].

Theorem 6. *If* $H = \text{const.}$, *then*

(29) $$\mathsf{E}_\lambda(\tau(H)) \sim \begin{cases} 2\lambda H, & \lambda \to \infty, \\ B_1\sqrt{H}, & \lambda \to 0, \quad B_1 \sim 2{,}09, \\ \dfrac{\ln 8\lambda^2 H}{2|\lambda|}, & \lambda \to -\infty. \end{cases}$$

Let us compare now $\mathsf{E}_\lambda(\lambda_\tau(\xi) - \lambda)^2$ and $\mathsf{E}_\lambda(\tilde{\lambda}(H) - \lambda)^2$ the quadratic errors for $\lambda_\tau(\xi)$, the ordinary, and $\tilde{\lambda}(H)$, the sequential maximum likelihood estimation. It is natural to require that if the time of observation in the ordinary case equals T, then for the comparing the estimation $\lambda_T(\xi)$ (with stopping-rule τ) must be take such that it satisfies the condition $\mathsf{E}_\lambda \tau \leqslant T$, $|H| < \infty$. The stopping rule τ introduced before has not this property.

Nevertheless, let $H = H(\lambda, T)$ be the value for which

$$\mathsf{E}_\lambda \tau(H(\lambda, T)) \equiv T$$

and

$$e(\lambda, T) = \frac{\mathsf{E}_\lambda (\lambda_T(\xi) - \lambda)^2}{\mathsf{E}_\lambda (\tilde{\lambda}(H(\lambda, T)) - \lambda)^2} .$$

The quantity $e(\lambda, T)$ characterizes the effectivity of the sequential method in comparison to the ordinary one and it gets some meaning, if it is a priori known that λ is in the neighbourhood of some known point λ_0. Then it is natural to choose $H = H(\lambda_0, T)$ as the "threshold" for H for the construction $\tilde{\lambda}(H)$, since for λ near to λ_0 we have

$$\frac{\mathsf{E}_\lambda (\lambda_T(\xi) - \lambda)^2}{\mathsf{E}_\lambda (\tilde{\lambda}(H(\lambda_0, T)) - \lambda)^2} \sim \frac{\mathsf{E}_\lambda (\lambda_T(\xi) - \lambda)^2}{\mathsf{E}_\lambda (\tilde{\lambda}(H(\lambda, T)) - \lambda)^2} = e(\lambda, T) .$$

The quantity $e(\lambda, T)$, the measure of effectivity, was studied for processes of type (28) by A. A. Novikov, who obtained the following result:

Theorem 7. *For the process* (28)

$$(30) \qquad e(\lambda, T) = \begin{cases} 1 + \dfrac{23}{4(\lambda T)} + O\left(\dfrac{1}{(\lambda T)^2}\right), & \lambda T \to \infty , \\ C_0 - C_1(\lambda T) + O((\lambda T)^2), & \lambda T \to \infty , \\ \dfrac{e^{|\lambda| T}}{4\sqrt{\pi}(|\lambda| T)^{\frac{1}{2}}} \left[1 + O\left(\dfrac{1}{|\lambda| T}\right)\right], & \lambda T \to \infty \end{cases}$$

where $C_0 \sim 3{,}04$; $C_1 \sim 2{,}13$.

This theorem shows that at fixed T the gain of using sequential maximum-likelihood estimation will be especially large, if $\lambda \leqslant 0$.

REFERENCES

[1] A.N. Širjaev, Study on the sequential analysis, (in Russian), *Math. Zametki,* 3 (1968), 739-754.

[2] P.Š. Lipcer – A.N. Širjaev, Non-linear filtration for diffusional Markov processes, (in Russian), *Trudy Mat. Inst. Steklov.,* 104 (1968), 135-180.

[3] G. Kallianpur – C. Striebel, Stochastic differential equations occurring in the estimation of continuous parameter of stochastic processes, *Teorija Verojatn. i Primenen.,* 14 (1970), 592-622.

[4] M.P. Eršov, Sequential estimation for diffusion process, (in Russian), *Teorija Verojatn. i Primenen.,* 15 (1970), 703-717.

[5] T. Kailath, An innovations approach to least-squares estimation, I, II, *IEEE Trans. on Autom. Control,* 13 (1968), 646-655, 655-660.

[6] P. Frost – T. Kailath, An innovations approach to least-squares estimation, III, *IEEE Trans. on Autom. Control,* 16 (1971), 217-226.

[7] M. Fujisaki – G. Kallianpur – H. Kunita, Stochastic differential equations for the non-linear filtering problem, Osaka Inst. of Math., 1972.

[8] P.Š. Lipcer – A.N. Širjaev, Statistics of stochastic processes (non-linear filtration and related topics), (in print).

[9] P.Š. Lipcer – A.N. Širjaev, On the absolut continuous measure with respect to the Wiener one corresponding to a diffusion-type process, (in Russian), *Izv. Akad. Nauk SSSR,* 36 (1972).

[10] R. Bucy – R. Kalman, New results in linear filtering and prediction theory, *J. of Basic Eng., Trans. of the ASME,* (1961), 95-108.

[11] J. Schalkwijk – T. Kailath, A coding scheme for additive noise channals with feedback, Part I: no band-width constraint, *IEEE Trans. on Inform. Theory,* 12 (1966), 172-182.

[12] K. Zigangirov, Transmission of communications on a duble Gauss channal with feedback, *Probl. Peredači Informacii,* 3 (1967), 98-101.

[13] P. Has'minskiy, Addendum (72 exercises) to the book of Dz. Turin: *Lessons on the combinatorics,* (in Russian), MIR, Moskau, 1972.

[14] A.A. Novikov, Sequential estimation of the parameters of diffusion processes, (in Russian), *Teorija Verojatn. i Primenen.,* 16 (1971), 394-396.

[15] A.A. Novikov, *Stochastical integrals and sequential estimations,* (in Russian), Dissertation, MIAN, 1972.

[11] J. P. M. Schalkwijk, T. Kailath, A coding scheme for additive noise channels with feedback. Part I: no band-width constraint, *IEEE Trans. Inform. Theory* 12 (1966), [illegible]-182.

[12] K. Sh. Zigangirov, Transmission of messages through a binary symmetric channel with feedback, *Probl. Peredachi Inform.* 3 (1967), 98-101.

[13] R. L. Dobrushin, Addendum (22 exercises) to the book of [illegible] *[illegible] of the combinatorics* (in Russian), MIR, Moscow, [illegible] (1972).

[14] A. [illegible] N[illegible]kov, Sequential estimation of the parameters of diffusion processes (in Russian), *Teoria Veroyat. i Primenen.* [illegible] (1971), [illegible].

[15] A. [illegible] N[illegible]kov, Stochastic [illegible] and [illegible] estimations (in Russian), Dissertation, MIAN, 1972.

COLLOQUIA MATHEMATICA SOCIETATIS JÁNOS BOLYAI
9. EUROPEAN MEETING OF STATISTICIANS, BUDAPEST (HUNGARY), 1972.

HIERARCHICAL PROCEDURES

J. STENE

SUMMARY

A new type of multiple decision procedures, hierarchical procedures, is introduced. These procedures are generalisations of stepwise procedures. The different test problems in a hierarchy of hypotheses are separated by systematic use of conditional testing. The test result is a detailed conclusion which usually is easy to interprete. The significance level of the whole procedure can be determined in advance.

1. LINEAR HIERARCHICAL MODELS

A linear hierarchical model is a model belonging to the Darmois – Koopman exponential class where the probability density can be written in the form

$$dF = A(\tau, \eta) \exp\{T(x_{ijk})\eta_{ijk} + U(x_{ijk})\tau\} dF_0 \tag{1}$$

where the parameter

(2) $\eta_{ijk} = \mu + \alpha_i + \beta_{ij} + \gamma_{ijk}$

$$(i = 1, \ldots, r; \; j = 1, \ldots, s_i; \; k = 1, \ldots, t_{ij})$$

and where τ is a nuisance parameter, in case of a three-stage hierarchy.

Example 1. Each of the probabilities for the different trials in a multinomial distribution may represent the probability for a simultaneous occurrence of several events. By the theorem on compound probabilities we get that such a probability p_{ijk} can be written in form

$$p_{ijk} = \mathsf{P}\{A_i, B_{ij}, C_{ijk}\} = \mathsf{P}\{A_i\} \cdot \mathsf{P}\{B_{ij} \mid A_i\} \cdot \mathsf{P}\{C_{ijk} \mid A_i, B_{ij}\} =$$

$$= \theta_i \cdot \theta_{ij} \cdot \theta_{ijk}$$

where

$$\theta_i = \mathsf{P}\{A_i\}, \; \theta_{ij} = \mathsf{P}\{B_{ij} \mid A_i\}, \; \theta_{ijk} = \mathsf{P}\{C_{ijk} \mid A_i, B_{ij}\}.$$

By a suitable reparametrization we get (1) and (2). In the context of [5], [7] A_i may mean the formation of a foetus of type i, B_{i1} that this foetus will be a liveborn child and B_{i2} that it will be aborted and C_{ijk} the type the child or miscarriage is classified to.

The parameters of (2) will usually not be uniquely determined unless we impose some restrictions on them. We introduce

$$\sum_{i=1}^{r} u_i \alpha_i = 0 \qquad \text{where} \qquad \sum_{i=1}^{r} u_i = 1$$

(3) $$\sum_{j=1}^{s_i} v_{ij} \beta_{ij} = 0 \qquad \text{where} \qquad \sum_{j=1}^{s_i} v_{ij} = 1 \qquad (i = 1, \ldots, r)$$

$$\sum_{k=1}^{t_{ij}} w_{ijk} \gamma_{ijk} = 0 \text{ where } \sum_{k=1}^{t_{ij}} w_{ijk} = 1 \qquad (j = 1, \ldots, s_i).$$

The u_i's, v_{ij}'s and w_{ijk}'s are known figures and satisfy some trivial algebraic conditions in order to determine the μ, α_i's, β_{ij}'s and γ_{ijk}'s uniquely.

The apriori conditions Ω are given by (1), (2) and (3).

2. STEPWISE PROCEDURES

The hypotheses which are usually tested in models of this type are (see e.g. [2], pp. 178-186)

$$(4)\quad \begin{aligned} &H_C: \text{ all } \gamma_{ijk} = 0, \quad H_B: \text{ all } \beta_{ij} = 0, \quad H_A: \text{ all } \alpha_i = 0, \\ &H': \mu = \mu^{(0)} . \end{aligned}$$

Very often these hypotheses are tested simultaneously by testing the hypothesis $\omega_0 = \Omega \cap H' \cap H_A \cap H_B \cap H_C$. For discrete distributions ω_0 is usually tested by a single χ^2-test. Although this test has some optimal properties, it has low power against all alternatives. In case of rejection the test result may be difficult to interprete.

The hypotheses may also be tested by a stepwise procedure. Introduce

$$(5)\quad \begin{aligned} \omega_1 &= H_C \cap \Omega \\ \omega_2 &= H_B \cap H_C \cap \Omega \\ \omega_3 &= H_A \cap H_B \cap H_C \cap \Omega \\ \omega_4 &= H' \cap H_A \cap H_B \cap H_C \cap \Omega . \end{aligned}$$

Here we test successively, ω_1 against $\Omega - \omega_1$, ω_2 against $\omega_1 - \omega_2$ if the preceding hypothesis is not rejected and ω_3 against $\omega_2 - \omega_3$ if the preceding hypotheses are not rejected and so on. Stepwise procedures of this kind are studied in [1]. The usual tests of the hierarchical or nested designs in analysis of variance belong to this group (see [2], pp. 178-186).

3. HIERARCHICAL PROCEDURES

Although the stepwise testing of the hypotheses (5) gives us a conclusion which is easier to interprete than the test of the hypothesis ω_0, there are a number of situations where it is not realistic to test the hypothesis H_C or the hypothesis H_B. Such situations arise when e.g. the hypothesis $\gamma_{ijk} = 0$ is likely to be false in a restricted, but unknown part of the material, but may hold in the rest of the material. In this case a stepwise procedure based on (5) will stop at the stage ω_1, although this hypothesis and the following ones do hold for the rest of the material. Materials of this type are discussed in [3], [4], [5], [7].

In order to be able to proceed in such situations for those parts of the material where the hypotheses seem to hold, we have introduced *hierarchical procedures.*

Instead of (4) we want to consider the following hypotheses

$$
(6)\quad
\begin{aligned}
&H_{C_{ij}}: \gamma_{ijk} = 0 \quad (k = 1, \ldots, t_{ij} \quad \text{for each } (i,j) \text{ separately}),\\
&H_{B_i}: \beta_{ij} = 0 \quad (j = 1, \ldots, s_i \quad \text{for each } i \text{ separately}),\\
&H_A: \alpha_i = 0 \quad (i = 1, \ldots, r),\\
&H': \mu = \mu^{(0)} .
\end{aligned}
$$

There is a hierarchical structure in this set of hypotheses. Let us further define the sets

$$J^{(i)} = \{j: H_{C_{ij}} \text{ is not rejected}, \quad j = 1, \ldots, s_i\}$$

$$I' = \{i: H_{C_{ij}} \text{ is not rejected for at least one } j, \quad i = 1, \ldots, r\}.$$

The hierarchical procedure, which is a generalization of (5), starts by testing $H_{C_{ij}}$ for each (i,j) separately. The next hypotheses to be tested are $H_{B(J^{(i)})}: \beta_{ij} = 0$ where $j \in J^{(i)}$, and they are tested for each $i \in I'$ separately. For this part of the material we assume that the $H_{C_{ij}}$'s hold, i.e. that all $\gamma_{ijk} = 0$. For all $i \notin I'$ and all $j \notin J^{(i)}$ we have obtained the terminal decision. We conclude that the model is the general model (1) and (2) without any simplifications of type (6) for this part of the material.

Let us then introduce the set

$$I = \{i: H_{B(J^{(i)})} \text{ is not rejected}, \quad i = 1, \ldots, r\}.$$

Obviously $I \subset I'$. Then we want to test the hypothesis

$$H_{A(I)}: \alpha_i = 0; \qquad i \in I$$

where we have excluded all α_i for which $i \notin I$. For all $i \notin I$ and $j \in J^{(i)}$ our terminal decision is the model (1) with parameter $\eta_{ijk} = \mu + \alpha_i + \beta_{ij}$.

If we reject the hypothesis $H_{A(I)}$ the terminal decision will be the model (1) with parameter $\eta_{ijk} = \mu + a_i$ for $i \in I$ and $j \in J^{(i)}$.

If we do not reject $H_{A(I)}$ we test the hypothesis

$$H'_{(I)}: \mu = \mu^{(0)} .$$

If we reject the hypothesis $H'_{(I)}$ the terminal decision will be the model (1) with parameter $\eta_{ijk} = \mu$ for $i \in I$ and $j \in J^{(i)}$.

If we do not reject $H'_{(I)}$ our terminal decision for $i \in I$ and $j \in J^{(i)}$ will be $\eta_{ijk} = \mu^{(0)}$.

The hierarchy of hypotheses may be written in the form

(7)
$$\begin{aligned}
&\omega_C(i,j): \Omega \cap H_{C_{ij}} \quad \text{separately for each } (i,j) , \\
&\omega_B(J^{(i)}): \Omega \cap H_{B(J^{(i)})} \bigcap_{j \in J^{(i)}} H_{C_{ij}} \quad \text{separately for each } i \in I' , \\
&\omega_A(I): \Omega \cap H_{A(I)} \bigcap_{i \in I} H_{B(J^{(i)})} \bigcap_{j \in J^{(i)}} H_{C_{ij}} , \\
&\omega'(I): \Omega \cap H'_{(I)} \cap H_{A(I)} \bigcap_{i \in J} H_{B(J^{(i)})} \bigcap_{j \in J^{(i)}} H_{C_{ij}}
\end{aligned}$$

which is a generalization of (5).

How the terminal decision for the whole material is constructed from the terminal decisions for the different parts of the material will be illustrated in Example 2 below.

When we test the hierarchy of hypotheses given in (7) the side conditions (3) can be kept unchanged. However, if we want to test a hypothesis about μ in (2), we have to impose another condition on the α_i's, e.g.

(8) $$\sum_{i \in I} u'_i \alpha_i = 0 \qquad \text{where} \qquad \sum_{i \in I} u'_i = 1 .$$

A hierarchical procedure differs from a stepwise one by considering different parts of the material separately instead of the whole material simultaneously.

A hierarchical procedure has been applied to the genetical problem discussed in [3], [4], [5]. In that case we had a five-stage hierarchy. The actual situation was that a chromosomal defect was inherited in different families. Such a chromosomal defect gives rise to malformed children with a probability which usually keeps constant throughout the family, but the probability may vary from family to family.

Owing to some biological or observational conditions this probability may vary within the family for a few families. However, families with this type of heterogeneity forms usually only a minor part of the material, and this part should be eliminated from further analysis.

4. EXAMPLES

Example 2. The number of different terminal decisions in a hierarchical procedure may be rather large. In order to demonstrate how to construct them, we will consider a simple example. We confine ourselves to a model with parameter $\gamma_{ij} = \mu + \alpha_i + \beta_{ij}$ and want to test the hypotheses in (6) about the β_{ij}'s, the α_i's and μ. In (2) let $r = 3$. s_i may vary with i.

In the genetical example mentioned above, index i indicates family number and j sibship number in that family. We want to test a single homogeneity hypothesis for each family separately, and eliminate from further analysis each family for which this homogeneity hypothesis is rejected. Let $\bar{\omega}$ denote rejection of a hypothesis and let ω denote non-rejection which we will call acceptance of a hypothesis. The different terminal decisions are given in Table 1.

Hypotheses		
H_{B_i}	$H_{A(I)}$	$\mu = \mu^{(0)}$
$\omega_{B_1} \cap \omega_{B_2} \cap \omega_{B_3}$	$\cap\, \omega_{A(1,2,3)}$	$\cap\, \omega'_{(1,2,3)}$
		$\cap\, \bar{\omega}'_{(1,2,3)}$
	$\cap\, \bar{\omega}_{A(1,2,3)}$	
$\omega_{B_1} \cap \omega_{B_2} \cap \bar{\omega}_{B_3}$	$\cap\, \omega_{A(1,2)}$	$\cap\, \omega'_{(1,2)}$
		$\cap\, \bar{\omega}'_{(1,2)}$
	$\cap\, \bar{\omega}_{A(1,2)}$	
$\omega_{B_1} \cap \bar{\omega}_{B_2} \cap \omega_{B_3}$	$\cap\, \omega_{A(1,3)}$	$\cap\, \omega'_{(1,3)}$
		$\cap\, \bar{\omega}'_{(1,3)}$
	$\cap\, \bar{\omega}_{A(1,3)}$	
$\bar{\omega}_{B_1} \cap \omega_{B_2} \cap \omega_{B_3}$	$\cap\, \omega_{A(2,3)}$	$\cap\, \omega'_{(2,3)}$
		$\cap\, \bar{\omega}'_{(2,3)}$
	$\cap\, \bar{\omega}_{A(2,3)}$	
$\omega_{B_1} \cap \bar{\omega}_{B_2} \cap \bar{\omega}_{B_3}$		$\cap\, \omega'_{(1)}$
		$\cap\, \bar{\omega}'_{(1)}$
$\bar{\omega}_{B_1} \cap \omega_{B_2} \cap \bar{\omega}_{B_3}$		$\cap\, \omega'_{(2)}$
		$\cap\, \bar{\omega}'_{(2)}$
$\bar{\omega}_{B_1} \cap \bar{\omega}_{B_2} \cap \omega_{B_3}$		$\cap\, \omega'_{(3)}$
		$\cap\, \bar{\omega}'_{(3)}$
$\bar{\omega}_{B_1} \cap \bar{\omega}_{B_2} \cap \bar{\omega}_{B_3}$		

Table 1.

Terminal decisions in Example 2.

E.g. the terminal decision $\omega_{B_1} \cap \bar{\omega}_{B_2} \cap \omega_{B_3} \cap \omega_{A(1,3)} \cap \omega'_{(1,3)}$ indicates that we have accepted the hypotheses H_{B_1} and H_{B_3} and rejected H_{B_2}. For $i = 2$ we conclude that the parameter is $\gamma_{2j} = \mu + \alpha_2 + \beta_{2j}$; $j = 1, \ldots, s_2$. In the further analysis the material belonging to $i = 2$ is excluded. We have further accepted $H_{A(1,3)}$, i.e. $\alpha_1 = \alpha_3 = 0$, but rejected the hypothesis $\mu = \mu^{(0)}$ for this part of the material. The terminal decision for the whole material is that

$$\eta_{1j_1} = \eta_{3j_3} = \mu(\neq \mu^{(0)}) \qquad (j_1 = 1, \ldots, s_1;\ j_3 = 1, \ldots, s_3),$$

$$\eta_{2j} = \mu + \alpha_2 + \beta_{2j} \qquad (j_2 = 1, \ldots, s_2).$$

The terminal decision $\bar{\omega}_{B_1} \cap \bar{\omega}_{B_2} \cap \omega_{B_3} \cap \omega'_{(3)}$ indicates that we have rejected H_{B_1} and H_{B_2}, but not H_{B_3}. The test of the hypothesis $\alpha_3 = 0$ cannot be tested independently of that of $\mu = \mu^{(0)}$, because of the side condition $\alpha_3 = 0$ from (8). The hypothesis $\mu = \mu^{(0)}$ is tested for the group $i = 3$ only and accepted. The terminal decision is here

$$\eta_{ij} = \mu + \alpha_i + \beta_{ij} \qquad (i = 1, 2;\ j = 1, \ldots, s_i),$$

$$n_{3j} = \mu^{(0)} \qquad (j = 1, \ldots, s_3).$$

Example 3. In the setup of Example 2 we want to demonstrate a hierarchical procedure for the hierarchical design in analysis of variance.

Let X_{ijm} be independently distributed normal variables with common unknown variance σ^2. Further is $\mathsf{E}(X_{ijm}) = \eta_{ij}$ and let $m = 1, \ldots$ $\ldots, M_{ij}$ be the number of observation in all (i, j). By suitable side conditions of the type (3) we get the following sums of squares

$$Q_0 = \sum_{i=1}^{r} \sum_{j=1}^{s_i} \sum_{m=1}^{M_{ij}} (X_{ijm} - \bar{X}_{ij.})^2,$$

$$Q_1(i) = \sum_{j=1}^{s_i} M_{ij}(\bar{X}_{ij.} - \bar{X}_{i..})^2.$$

The hypothesis H_{B_i} is tested by means of the F-statistic

$$\frac{Q_1(i)}{Q_0} \frac{\sum_{i=1}^{r} \sum_{j=1}^{s_i} (M_{ij} - 1)}{s_i - 1}$$

for each $\cdot\, i$.

In order to test the hypothesis $H_{A(I)}$ we need

$$Q_2 = \sum_{i \in I} n_i (\bar{X}_{i\,..} - \bar{X}_{...})^2$$

where

$$n_i = \sum_{j=1}^{s_i} M_{ij} \quad \text{and} \quad \bar{X}_{...} = \frac{\sum_{i \in I} n_i \bar{X}_{i\,..}}{\sum_{i \in I} n_i}.$$

The test is carried out by means of the F-statistic

$$\frac{Q_2}{Q_0 + \sum_{i \in I'} Q_1(i)} \cdot \frac{\sum_{i=1}^{r} \sum_{j=1}^{s_i} (M_{ij} - 1) + \sum_{i \in I} (s_i - 1)}{n_I - 1}.$$

Further details about this example will be given elsewhere.

Hierarchical procedures consisting of up to five stages are given in [3], [4], [5], [7]. In these cases the distributions are binomial ones, and the tests are χ^2-tests, often small-sample versions. A more general discussion of the hierarchical procedures for binomial distributions has been given in [6]. In [5], [7] there are demonstrated how detailed the conclusions of hierarchical procedures can be.

5. SIGNIFICANCE LEVELS

Let s_i', $n_{I'}$ and n_I be the number of elements in the sets $J^{(i)}$, I' and I respectively.

It will be shown elsewhere that when the different tests are similar test for the respective hypotheses and depend on the observations through complete sufficient statistics and all the tests about the γ_{ijk}'s have level

ϵ_1, all tests about the β_{ij}'s level. ϵ_2, the test about the α_i's level ϵ_3 and the test about $\mu = \mu^{(0)}$, level ϵ_4, then the hierarchical procedure has multiple level (p_1, p_2, p_3, p_4) where the $p_1, \ldots, p_4$ are determined from the different test levels by

$$
(9) \quad
\begin{aligned}
p_1 &= 1 - (1 - \epsilon_1)^{\sum_{i=1}^{r} s_i}, \\
p_2 &= (1 - \epsilon_1)^{\sum_{i \in I'} s_i'} (1 - (1 - \epsilon_2)^{n_{I'}}), \\
p_3 &= (1 - \epsilon_1)^{\sum_{i \in I} s_i'} (1 - \epsilon_2)^{n_I} \cdot \epsilon_3, \\
p_4 &= (1 - \epsilon_1)^{\sum_{i \in I} s_i'} (1 - \epsilon_2)^{n_I} (1 - \epsilon_3) \cdot \epsilon_4 .
\end{aligned}
$$

If the hypothesis $H_{A(I)}$ cannot be tested, because $n_I = 1$, but $\mu = \mu^{(0)}$ can be tested, then ϵ_3 is substituted by ϵ_4 in the third equation in (9) and the fourth equation is omitted.

If, on the other hand, we determine the level ϵ of the whole procedure, i.e. the probability for at least one false rejection, in advance, then we determine the p_i's such that $p_1 + p_2 + p_3 + p_4 = \epsilon$ and the levels of the different tests can be found from (9).

REFERENCES

[1] S. Das Gupta, Step-down multiple decision rules, *R.C. Bose, et. al. (ed.): Essays in Probability and Statistics,* The University of North Carolina Press, Chapel Hill, 1970.

[2] H. Scheffé, *The Analysis of Variance,* J. Wiley et Sons, New York, 1959.

[3] J. Stene, Analysis of segregation patterns between sibships within families ascertained in different ways, *Annals of Human Genetics,* 33 (1970), 261-283.

[4] J. Stene, Comparison of segregation rations for families assertained in different ways, *Annals of Human Genetics,* 33 (1970), 395-412.

[5] J. Stene, Statistical inference on segregation ratios for *D/G*-translocations, when the families are ascertained in different ways, *Annals of Human Genetics,* 34 (1970), 93-115.

[6] J. Stene, Nested step-down multiple decision procedures for testing homogeneity in Bernoulli trials, *Skandinavisk Aktuarietidsskrift,* 55 (1972), 92-106.

[7] J. Stene, Ascertainment of families with inherited characters. Preprint of University of Copenhagen, 1971.

COLLOQUIA MATHEMATICA SOCIETATIS JÁNOS BOLYAI
9. EUROPEAN MEETING OF STATISTICIANS, BUDAPEST (HUNGARY), 1972.

PROBABILITY LIMIT IDENTIFICATION FUNCTION

J. ŠTĚPÁN

The classical situation in the estimation theory can be described by means of a parametrical space Θ and a family of probabilities $\{P_\theta, \theta \in \in \Theta\}$ defined on a measurable space $(\Omega, \mathscr{A})$. Estimating a parametrical function $l(\theta)$: $\Theta \longrightarrow R^1$ we are often satisfied with a sequence of estimators h_n: $\Omega \longrightarrow R^1$ which is convergent to the $l(\theta)$ in probability P_θ for all $\theta \in \Theta$. The sequence h_n is usually called consistent estimator of the l. It seems that for the statistician to be satisfied with such estimator is necessary to have some evidence for the fact that the values of h_n identify the values of l. In other words we have the problem:

Is there a function f: $R^\infty \longrightarrow R^1$ such that $f(h_1, h_2, \ldots) = l(\theta)$ a.s. P_θ for all $\theta \in \Theta$? It is clear that we cannot simply construct the f by subsequencing the h_n to an almost surely convergent subsequence as this process depends upon the unknown parameter θ.

G. Simons (1971) in [1] put the problem into the following more general setting:

Consider a probability space $(\Omega, \mathscr{A}, P)$ and denote by $\mathscr{E}$ the space of random sequences X defined here which are convergent in probability. A function $f\colon R^\infty \longrightarrow R^1$ is called a probability limit identifying function (PLIF) on $\mathscr{E}' \subset \mathscr{E}$ if for every $X \in \mathscr{E}'$ the set $\{f(X) \neq p(X)\}$ is contained in a P-null set of $\mathscr{A}$. (The $p(X)$ denotes the probability limit of the sequence X.) Simons has menaged in proving that there is a PLIF on $\mathscr{E}$ if and only if there is a PLIF on $\mathscr{E}^*$ the set of $X \in \mathscr{E}$ whose coordinates are 0-1 variables and whose probability limit $p(X)$ is a constant almost surely. It is easy to construct the PLIF's on finite or countable parts of $\mathscr{E}$. The inability to use some form of the axiom of choice (Zorn's lemma, for example) represents the main difficulty in the construction of PLIF on $\mathscr{E}$. We are going to employ the continuum hypothesis (the axiom of choice, too) to construct the PLIF on $\mathscr{E}$.

We shall prefer the following presentation of the problem: Denote by $\mathscr{B}^\infty$ the Borel σ-algebra of the metric space R^∞ and by $\mathscr{P}$ the set of probabilities μ on $\mathscr{B}^\infty$ such that the coordinates x_n are convergent in probability μ. We can write

$$\mathscr{P} = \{\mu\colon \mu\{x \in R^\infty\colon |x_n - x_m| > \epsilon\} \underset{\substack{n \to \infty \\ m \to \infty}}{\longrightarrow} 0 \text{ for all } \epsilon > 0\}\,.$$

where $x = (x_1, x_2, \ldots)$ denotes the elements of R^∞.

Denote by $(\mathscr{B}^\infty_\mu, \bar{\mu})$ the μ-completion of the probability space $(\mathscr{B}^\infty, \mu)$. The function $f\colon R^\infty \longrightarrow R^1$ which is measurable with respect to the σ-algebra $\cap \{\mathscr{B}^\infty_\mu, \mu \in \mathscr{P}\}$ such that $\bar{\mu}\{|x_n - f| > \epsilon\} \underset{n \to \infty}{\longrightarrow} 0$ for all $\epsilon > 0$ and $\mu \in \mathscr{P}$ will be called the PLIF (on $\mathscr{P}$).

It is easy to see that the PLIF's in the sense of this definition are the PLIF's on $\mathscr{E}$ for all probability spaces $(\Omega, \mathscr{A}, \mathrm{P})$ in the sense of [1].

One can make some simple remarks on the PLIF's. For example if $x = (x_1, x_2, \ldots)$ is a convergent element of R^∞ then $f(x) = \lim\limits_{n \to \infty} x_n$. Our main result is

Theorem ([2]). *There is a PLIF on $\mathscr{P}$ under the continuum hypothesis.*

The proof is based on the following facts:

Lemma. *The set* $\mathscr{P}$ *has the cardinality of continuum.*

It follows from the continuum hypothesis that we can order the $\mathscr{P}$ by the countable ordinal numbers: $\mathscr{P} = \{\mu_\alpha, \alpha < \Omega\}$, here Ω denotes the first uncauntable ordinal. Let us choose $\mathscr{B}^\infty$-measurable functions g_α, such that the coordinates x_n converge to the g_α in probability μ_α. The proof will be completed when constructing the function $f\colon R^\infty \longrightarrow R^1$ such that

$$f = g_\alpha \quad \text{a.s.} \quad [\overline{\mu_\alpha}] \quad \text{for all} \quad \alpha < \Omega .$$

One can do it using the transfinite construction. Riesz's theorem on the relation between convergence in probability and a.s. convergence is substantially employed in the course of the construction.

At least two problems seem to be open:

1. Is there any $\mathscr{B}^\infty$-measurable PLIF?

2. Is the continuum hypothesis necessary for the existence of the PLIF's?

REFERENCES

[1] G. Simons, Identifying Probability Limits, *Ann. Math. Statist.*, 42 (1971), 1429-1433.

[2] J. Štěpán, Probability limit identification function exists under the continuum hypothesis. *Annals of Prob.*, 1 (1973), 712-715.

COLLOQUIA MATHEMATICA SOCIETATIS JANOS BOLYAI
9. EUROPEAN MEETING OF STATISTICIANS, BUDAPEST (HUNGARY), 1972.

LARGE DEVIATION CONNECTIONS

M. STONE

The provocative findings of B r o w n [2] relating to tests, based on a random sample $x^{(n)}$ of size n, of a composite null hypothesis Θ_0 against a composite alternative Θ_1 include the following:

There exists a test statistic λ_n^* (a likelihood-ratio test statistic of Θ_0 versus $\Theta_1^* \supset \Theta_1$) such that, under regularity conditions, if α_n is the size and $1 - \beta_n(\theta_1)$ the power at θ_1 of any test with $\limsup_{n\to\infty} \alpha_n < 1$, there are tests based on λ_n^* with $\alpha_n^* \leqslant \alpha_n$ and $\liminf_{n\to\infty} \frac{1}{n} (\log \beta_n(\theta_1) - - \log \beta_n^*(\theta_1)) \geqslant 0$ for all $\theta_1 \in \Theta_1$. *There exist examples with strict inequalities even when* $(\alpha_n, \beta_n) = (\alpha_n^\lambda, \beta_n^\lambda)$, *the error rates for the likelihood-ratio test* (λ_n) *of* Θ_0 *versus* Θ_1.

When all critical regions are considered simultaneously the following picture illustrates the possibility revealed by B r o w n (B_1 and B_2 mark the relevant features):

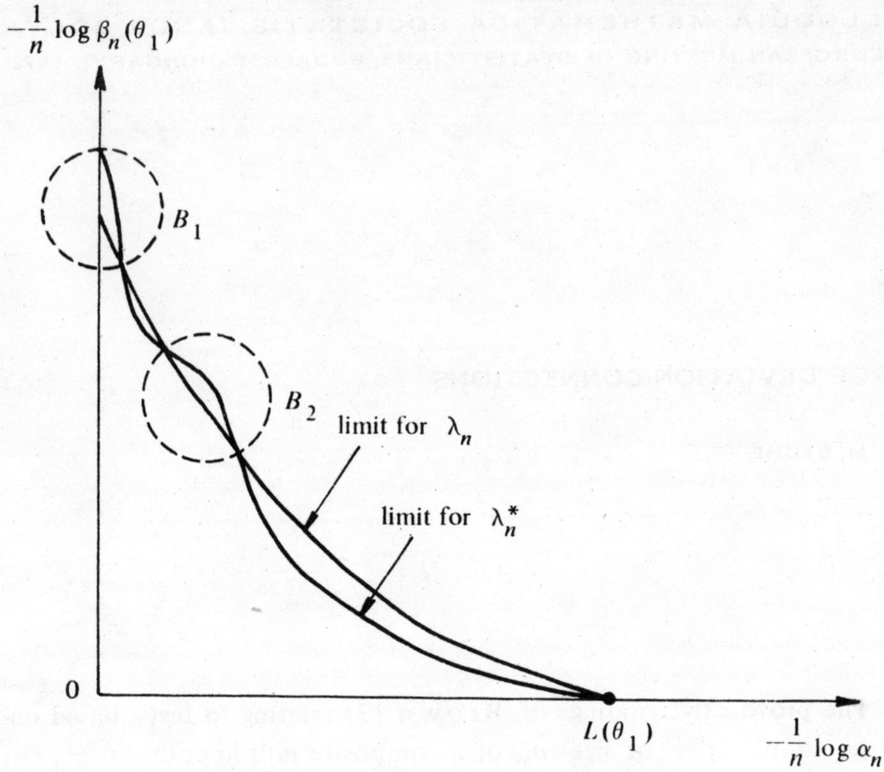

Brown's work is strictly Neyman – Pearsonian. In contrast, Bahadur [1] employs the (stochastic) *observed significance level* given by $\alpha(x^{(n)}) = \sup_{\theta_0 \in \Theta_0} \mathsf{P}_{\theta_0}(\widetilde{x}^{(n)} \gtrsim x^{(n)})$ where $\gtrsim$ is the weak ordering corresponding to the test statistic. Bahadur misrepresents his own findings but shows that, under regularity conditions, $\operatorname*{plim}_{n\to\infty} -\frac{1}{n}\log\alpha(x^{(n)})$ exists and is, under θ_1, maximized at $L(\theta_1)$ by $\{\lambda_n\}$. We show $L(\theta_1)$ on the diagram.

An *overall* assessment of the situation is given by an examination of *expected significance level* (Stone [3]) defined here as the function of θ_0 and θ_1

$$\bar{\alpha}_n(\theta_0, \theta_1) = \int \mathsf{P}_{\theta_0}(\widetilde{x}^{(n)} \gtrsim x^{(n)})\, d\mathsf{P}_{\theta_1}(x^{(n)})\,.$$

(When the distributions are continuous, $\bar{\alpha}_n(\theta_0, \theta_1)$ is the area under the

(α, β)-curve for the test for θ_0 versus θ_1.)

Theorem (the asymptotic minimaxity of the λ-test). *Under regularity conditions which include continuity*

$$(1) \qquad \inf_{\theta_0, \theta_1} \lim_{n \to \infty} -\frac{1}{n} \log \bar{\alpha}_n(\theta_0, \theta_1)$$

is maximized by the likelihood-ratio test.

The quantity (1) is not, in general, equal to $L(\theta_1)$. Informally, we can say that Brown's bumps can be averaged and infimized out.

REFERENCES

[1] R.R. Bahadur, Rates of convergence of estimates and test statistics, *Ann. Math. Statist.*, 38 (1967), 303-324.

[2] L.D. Brown, Non-local asymptotic optimality of appropriate likelihood-ratio tests, *Ann. Math. Statist.*, 42 (1971), 1206-1240.

[3] M. Stone, The role of significance testing: some date with a message, *Biometrika*, 56 (1969), 485-493.

COLLOQUIA MATHEMATICA SOCIETATIS JÁNOS BOLYAI
9. EUROPEAN MEETING OF STATISTICIANS, BUDAPEST (HUNGARY), 1972.

OPTIMAL ESTIMATION OF SOME STOCHASTIC PROCESSES BY INCOMPLETE DATA

J.M. STOYANOV

On the probability space $(\Omega, \mathscr{F}, \mathsf{P})$ the stochastic processes θ_t and ξ_t are given, adapted with the family of increasing σ-algebras $\{\mathscr{F}_t\}$, $\mathscr{F}_t \subset \mathscr{F}$. We suppose that the process ξ_t depends on the process θ_t in a definite manner. The time parameter t may be discrete as well as continuous but in both cases will be writen $t \in [0, T]$ $(0 < T < \infty)$.

Let the process θ_t be non-observable and all the information on it is contained in the observable process ξ_t. Let ξ_0^t denote the results of the observations on the process ξ_τ in the time-interval $[0, t]$, and $\mathscr{F}^{\xi}_{[0,t]} = \sigma\{\omega\colon \xi_\tau(\omega),\ \tau \in [0, t]\}$. By the observations ξ_0^t we must estimate in a definite sense and in a optimal manner a certain function $\varphi(\theta_s)$ at the moment s. In this sense we will speak about estimation of a stochastic process by incomplete data, and the optimality is given with the help of a certain criterion at every time. We assume that if $\hat{\psi}_t(s)$ is any estimator of $\varphi(\theta_s)$ by the observations ξ_0^t then it must satisfy the conditions:

1. $\hat{\psi}_t(s)$ is measurable with respect to $\mathscr{F}^{\xi}_{[0,t]}$ for all $s, t \in [0, T]$;

2. $\mathsf{E}\{\hat{\psi}_t(s) - \varphi(\theta_s)\} = 0$.

The quality of the estimator $\hat{\psi}_t(s)$ will be measured with the help of a mean-quadratic criterion, that is, we require that

3. $\mathsf{E}\{|\hat{\psi}_t(s) - \varphi(\theta_s)|^2\} = \min$.

The estimator $\widetilde{\psi}_t(s)$ for which the conditions 1 and 2 are valid and

$$\mathsf{E}\{|\widetilde{\psi}_t(s) - \varphi(\theta_s)|^2\} = \min_{\hat{\psi}_t(s)} \mathsf{E}\{|\hat{\psi}_t(s) - \varphi(\theta_s)|^2\}$$

will be called optimal.

It is well known that in the class (which corresponds to the criterion we use) of the estimators with finite second moments the optimal estimator exists and is equal to the conditional expectation $\mathsf{E}\{\varphi(\theta_s) | \mathscr{F}^{\xi}_{[0,t]}\}$.

One of the first authors who studied the Markovian case of stochastic processes (θ_t, ξ_t) in different situations was A. N. Širjaev in a series of papers, published during the period 1961-1965 (see [4] and the bibliography mentioned there). Another paper which deals with a special case of estimation is W. M. Wonham [7]. After the above results, with the help of various methods, a number of new results have been obtained ([2], [3], [6] etc.).

Following the general classification, belonging to A. N. Širjaev, the estimation problem of $\varphi(\theta_s)$ at the moment s by observations ξ_0^t with $s < t$, $s = t$ and $s > t$, we shall call the interpolation, filtration and extrapolation problems respectively.

In the present paper we examine the case when θ_t is a general jump-like Markov process, given with its transitional probabilities and is taking values in the measurable space $(X, \mathscr{X})$, where X is complete separable metric space and the observable process ξ_i satisfies the following stochastic differential equation:

$$\xi_t = \xi_0 + \int_0^t a(\theta_s, \xi_s, s)ds + \int_0^t \sigma(\xi_s, s)dW_s \, . \tag{1}$$

Here W_s, $s \in [0, T]$ is a standard Wiener process (independent of the Markov process θ_t for every t), with respect to the family $\{\mathscr{F}_t\}$, and $\int_0^t \sigma(\cdot)dW_s$ is a stochastic integral in K. Ito's sense. We assume that the functions $a(\cdot)$ and $\sigma(\cdot)$, called respectively coefficients of drift and diffusion are such that the solution ξ_t, $t \in [0, T]$ of the equation (1) exists and is unique. Moreover, it is true that the pair (θ_t, ξ_t) forms a two-dimensional Markov process.

Let $\mathscr{P} = \{p_0(A), p(s, x; t, A),\ s, t \in [0, T], s < t, x \in X, A \in \mathscr{X}\}$ be the family of transition probabilities of the process θ_t. We require that the following property should be valid [1]:

There exists a function $q(s, x, A)$ which is continuous in s and is a measure with respect to A on the σ-algebra $\mathscr{X}$ and when $t \downarrow s$

$$\frac{p(s, x; t, A) - I_A(x)}{t - s} \to q(s, x, A)$$

uniformly in (s, x, A).

We aim to find the optimal estimators for the function $\varphi(\theta_s)$ by the observations ξ_0^t (which is the same as by $\mathscr{F}^{\xi}_{[0,t]}$) and the family $\mathscr{P}$ in each of the cases $s < t$, $s = t$ and $s > t$.

The fact that we know the family $\mathscr{P}$ in advance admits the Bayesian approach to examine the problems.

If the moment t in which we stop the observations and the moment of estimation s are fixed, then the optimal estimator is $\psi_t(s) = E\{\varphi(\theta_s) | \mathscr{F}^{\xi}_{[0,t]}\}$ and the estimation problem may be considered solved. In fact we are interested in the behavior of these optimal estimators with various data t, with the fixed moment of estimation s and vice versa. The Markovian character of the two-dimensional process (θ_t, ξ_t) and the type (1) of the observable process ξ_t allow us to hope that $\psi_t(s)$ in all cases will satisfy ordinary equations or stochastic differential equations in K. Ito's sense.

For an arbitrary measurable set $A \in \mathscr{X}$ we put

$$\pi_t(s, A) = \mathsf{P}\{\omega\colon \theta_s(\omega) \in A \mid \mathscr{F}^{\xi}_{[0,t]}\}. \tag{2}$$

It is easy to see that between $\pi_t(s, A)$ and the optimal estimator $\psi_t(s)$ there exists a determined connection. On the one hand if $\varphi(x) = I_A(x)$ is an indicator function of the set $A \in \mathscr{X}$, $\psi_t(s)$ goes into $\pi_t(s, A)$. On the other hand according to Skorohod's theorem [1]

$$\psi_t(s) = \int_X \varphi(x)\pi_t(s, dx)$$

with probability 1.

From here it follows that the finding the optimal estimators $\psi_t(s)$ is similar to finding the a posterior condition probabilities $\pi_t(s, A)$, i.e. knowing the stochastic equations for $\pi_t(s, A)$ after an integration we will obtain $\psi_t(s)$.

Before formulating some of the results we will give the following lemma (see [3]). Without restricting the generality we may put $\sigma(\cdot) \equiv 1$.

Lemma 1. *There exists a standard Wiener process* $\bar{W}_t$, $t \in [0, T]$ *with respect to the family of* σ*-algebras* $\mathscr{F}^{\xi}_{[0,t]}$ *such that the process* ξ_t *admits the representation*

$$\xi_t = \xi_0 + \int_0^t \bar{a}(s)ds + \bar{W}_t, \qquad t \in [0, T]$$

where $\bar{a}(s) = \mathsf{E}\{a(\theta_s, \xi_s, s) \mid \mathscr{F}^{\xi}_{[0,s]}\}$.

Let $\varphi(x)$ be an $\mathscr{X}$-measurable function for which there exist $\mathsf{E}\{|\varphi(\theta_s)|^2\}$ and $\mathsf{E}\{L_s\varphi(\theta_s)\}$ for every $s \in [0, T]$, $L_s\varphi(x) = \int_X \varphi(z)q(s, x, dz)$.

We will introduce also the following notations:

$$\bar{a}_x(s, t) = \mathsf{E}\{a(\theta_t, \xi_t, t) \mid \mathscr{F}^{\xi}_{[0,t]}, \theta_s = x\}, \qquad s < t,$$

and

$$\Phi_t^s(x) = \int_X \varphi(z)p(t, x; s, dz) \qquad \text{for} \qquad s > t\,.$$

The following theorems are hold.

Theorem 1. *In the filtration problem (that is if* $s = t$*)* $\psi_t = \psi_t(t)$, $t \in [0, T]$ *admits* $[\mathsf{P}\ a.e.]$ *a stochastic differential*

$$(3) \qquad d\psi_t = \alpha(t, \omega)dt + \beta(t, \omega)(d\xi_t - \bar{a}(t)dt)$$

with initial condition $\psi_0 = \mathsf{E}\{\varphi(\theta_0)\} = \int_X \varphi(x)p_0(dx)$, *where* $\alpha(t, \omega) = \mathsf{E}\{L_t\varphi(\theta_t) \mid \mathcal{F}^{\xi}_{[0,t]}\}$, $\beta(t, \omega) = \mathsf{E}\{\varphi(\theta_t)[a(\theta_t, \xi_t, t) - \bar{a}(t)] \mid \mathcal{F}^{\xi}_{[0,t]}\}$.

Theorem 2. *In the interpolation problem with a fixed* s $\psi_t(s)$ *admits* $[\mathsf{P}\ a.e.]$ *in* t, $t > s$, *stochastic differential*

$$(4) \qquad d_t\psi_t(s) = \gamma_s^t(\omega)(d\xi_t - \bar{a}(t)dt)$$

with initial condition $t = s$, $\psi_s(s) = \psi_s$ *which is calculated from* (3), *and* $\gamma_s^t(\omega) = \mathsf{E}\{\varphi(\theta_s)[\bar{a}_{\theta_s}(s, t) - \bar{a}(t)] \mid \mathcal{F}^{\xi}_{[s,t]}\}$.

Theorem 3. *In the extrapolation problem with fixed* t $[\mathsf{P}\ a.e.]$ *in* s, $s > t$, *for* $\psi_t(s)$ *the equation*

$$(5) \qquad \frac{\partial \psi_t(s)}{\partial s} = \mathsf{E}\{L_s\varphi(\theta_s) \mid \mathcal{F}^{\xi}_{[t,s]}\}$$

is valid, with initial condition, $s = t$, $\psi_t(t) = \psi_t$, *calculated from* (3).

Moreover, when s *is fixed* $\psi_t(s)$ *admits* $[\mathsf{P}\ a.e.]$ *in* t, $0 \leqslant t < s$, *stochastic differential*

$$(6) \qquad d_t\psi_t(s) = V_t^s(\omega)(d\xi_t - \bar{a}(t)dt)$$

with initial condition $t = 0$, $\psi_0(s) = \int_X \varphi(x)p_s(dx)$, *and* $V_t^s(\omega) = \mathsf{E}\{\Phi_t^s(\theta_t)[a(\theta_t, \xi_t, t) - \bar{a}(t)] \mid \mathcal{F}^{\xi}_{[0,t]}\}$. *(The measure* $p_s(A)$ *in the initial condition of function* $\psi_0(s)$ *is the non-conditional distribution of the process* θ_t *at the moment* s *and it is calculated from the Kolmogorov equation*

$$p_s(A) = p_0(A) + \int_0^t L^* p_s(A)ds ,$$

see [1].)

The reader can find detailed proofs of Theorems 1-3 in the papers [5] and [6]. We would like to point out the following important features:

a) from Lemma 1 it follows that (3), (4) and (6) are stochastic differentials in K. Ito's sense;

b) from theorems 1-3 it is clear that the filtration problem is formulated and solved independently, while the interpolation and extrapolation problems use essentialy the optimal estimator of the filtration problem;

c) it is easy to see that the optimal estimators found are non-linear, that is, they are not linear functions of the observations ξ_0^t;

d) when finding the stochastic equations for optimal estimators, we act in the following manner (see [6]): first we obtain stochastic equations for a posterior conditional probabilities $\pi_t(s, A)$, and after integration we obtain stochastic equations for $\psi_t(s)$. This enables us to draw the following conclusion:

The a posterior conditional probabilities $\pi_t(s, A)$ are sufficient statistics in the optimal non-linear interpolation, filtration and extrapolation problems.

At the end the author would like to express his gratitude to Professor A.N. Širjaev for has kind suggestions and help.

REFERENCES

[1] I.I. Gihman – A.V. Skorohod, *Introduction to the theory of random processes,* (in Russian), Moscow, 1965.

[2] G. Kallianpur – C. Striebel, Stochastic differential equations occurring in the estimation of continuous parameter stochastic processes, *Teorija Verojatn. i Primenen.*, 14 (1969), 597-622.

[3] R.S. Lipcer – A.N. Širjaev, Interpolation and filtration of jump-like component of Markov process, *Izv. Akad. Nauk SSSR*, 33 (1969), 901-914.

[4] A.N. Širjaev, Some new results in the theory of the controlled random processes, *Trans. Fourth Prague Conference on Inform. Theory*, (1967), 131-203.

[5] J.M. Stoyanov, Filtering of general jump-like Markov processes, *Izv. Math. Inst. Bulg. Acad. Sci.*, 14 (1973), 73-89.

[6] J.M. Stoyanov, On the estimation of partially-observable stochastic processes, *Mathematica Balkanica*, 2 (1972), 235-250.

[7] W.M. Wonham, Some applications of stochastic differential equations to optimal non-linear filtering, *SIAM J. Control*, 2 (1965), 347-369.

COLLOQUIA MATHEMATICA SOCIETATIS JÁNOS BOLYAI
9. EUROPEAN MEETING OF STATISTICIANS, BUDAPEST (HUNGARY), 1972.

ON THE RATE OF CONVERGENCE IN LEVY'S METRIC FOR RANDOM INDICED SUMS

D.O.H. SZÁSZ

Since 1937, when it was defined by Lévy [1], Lévy's metric has been generally regarded as a curiosity. It turned out only in the last years that the Lévy-distance may be useful in various investigations and has become the subject of more detailed analyses (see e.g. [2]). In the recent paper a method is developed for the estimation of the rate of convergence of random indiced sums in Lévy's metric.

Recall that the Lévy-distance between two distribution functions F and G is defined by the equality

$$(1) \qquad L(F, G) = \inf \{h: F(x-h) - h \leqslant G(x) \leqslant F(x+h) + h \text{ for all } x\} .$$

Let $\xi_{n1}, \ldots, \xi_{nk}, \ldots$ be a sequence of independent identically distributed random variables and ν_n be a random index, independent of the sequence above $(n = 1, 2, \ldots)$. Write

$$S_k^{(n)} = \xi_{n1} + \ldots + \xi_{nk} .$$

We shall denote the distribution function of the random variable X by $\mathscr{L}(X)$.

In 1969 Gnedenko and Fahim [5] proved that if there exists a sequence $\{k_n\}$ of natural numbers $(\lim_{n\to\infty} k_n = \infty)$ such that

(A) $\mathscr{L}(S_{k_n}^{(n)}) \to \Phi$

and

(B) $\mathscr{L}\left(\frac{\nu_n}{k_n}\right) \to A\,,$

where Φ and A are distribution functions and the convergence is understood in the weak sense, then

(C) $\mathscr{L}(S_{\nu_n}^{(n)}) \to \Psi\,,$

where the characteristic function of Ψ is determined by the equality

$$\psi(t) = \int_0^\infty [\varphi(t)]^y \, dA(y)\,.$$

Here $\varphi(t)$ is the characteristic function of the (necessarily infinitely divisible) distribution Φ.

Our aim is to estimate the Lévy-distance

$$L(\mathscr{L}(S_{\nu_n}^{(n)}), \Psi)$$

with the aid of the Lévy-distances

$$L(\mathscr{L}(S_{k_n}^{(n)}), \Phi) \qquad \text{and } L\left(\mathscr{L}\left(\frac{\nu_n}{k_n}\right), A\right).$$

In this paper we deal only with the special case $A(x) = E(x-1)$, where E is the unit law $(E(x) = 0,$ if $x \leqslant 0;$ $E(x) = 1,$ if $x > 0)$.

We need some lemmas.

Lemma 1. *If* $p_i \geqslant 0,$ $\sum_1^n p_i = 1,$ *then*

$$L\left(\sum_1^n p_i F_i, \sum_1^n p_i G_i\right) \leqslant \max_i L(F_i, G_i)\,.$$

Lemma 2.

$$L(pF_1 + (1-p)F_2,\ pG_1 + (1-p)G_2) \leqslant L(F_1, G_1) + (1-p)\,.$$

The proof of these lemmas is straightforward on the basis of the definition (1). We remark that Lemma 1 expresses the local convexity of Lévy's metric, while in some cases Lemma 2 gives a stronger bound for the Lévy-distance of mixed laws.

Lemma 3. (V.M. Zolotarev [3]). *Let*

$$c(x) = \begin{cases} x^2 & \text{for} \quad x \leqslant c \\ c^2 & \text{for} \quad x > c\,, \end{cases}$$

then for $L(E, F) \leqslant c$

$$L(E, F) \leqslant \left[\int c(x) F(dx)\right]^{\frac{1}{3}}$$

and for $L(E, F) > c$

$$L(E, F) \leqslant \frac{1}{c^2} \int c(x) F(dx)\,.$$

Lemma 4. *Let* Φ *be a fixed distribution function, then to any* α $\left(0 < \alpha < \frac{1}{7}\right)$ *there exists a positive* Δ *such that whenever the distribution function* F *satisfies the inequality* $L(\Phi, F) < \alpha$, *then in the interval* $|t| \leqslant \Delta$ *the characteristic function* $f(t)$ *of* F *satisfies the inequality*

$$|1 - f(t)| \leqslant 7\alpha$$

(Δ *may depend on* Φ *and* α *only*).

Proof. Choose the interval $(-a, a)$ so that $\Phi(a) - \Phi(-a) > 1 - \alpha$.

On the basis of the inequality $L(\Phi, F) < \alpha$ we have

$$F(a+\alpha) - F(-a-\alpha) \geqslant \Phi(a) - \Phi(-a) - 2\alpha > 1 - 3\alpha\,.$$

and so

$$|1-f(t)| \leqslant \int_{-a-\alpha}^{a+\alpha} (1-\cos tx)F(dx) + \int_{-a-\alpha}^{a+\alpha} |\sin tx|\, F(dx) + 2 \cdot 3\alpha .$$

Using the estimates

$$1-\cos tx \leqslant \frac{t^2x^2}{2}, \qquad |\sin tx| \leqslant |tx| ,$$

we obtain that in an interval $|t| \leqslant \Delta$

$$|1-f(t)| \leqslant \frac{\Delta^2(a+\alpha)^2}{2} + \Delta(a+\alpha) + 2 \cdot 3\alpha \tag{2}$$

holds. Putting $\Delta = \dfrac{\alpha}{2(a+\alpha)}$, (2) implies that

$$|1-f(t)| \leqslant 7\alpha \qquad (|t| \leqslant \Delta) .$$

Let us denote simply by F^k the kth convolution power of the distribution function F.

Lemma 5. *Let Φ be a fixed distribution function. There exists a constant C_1, depending only on Φ such that if*

$$L(\Phi, F^K) < \frac{1}{14},$$

then for all k

$$L(E, F^k) \leqslant C_1 \left(\frac{k}{K}\right)^{\frac{1}{3}} .$$

Proof. We can suppose that $k \leqslant K$. It is easy to see that if for the complex number z

$$|z-1| \leqslant \frac{1}{2},$$

then for any $\beta \in [0, 1]$

$$|1-z^\beta| \leqslant 4\beta |1-z| . \tag{3}$$

Now, applying the assertion of Lemma 4 with $\alpha = \frac{1}{14}$ for the distribution F^K, it follows that in the interval $|t| \leqslant \Delta$

(4) $$|1 - f^K(t)| \leqslant \frac{1}{2}.$$

($f(t)$ is the characteristic function of F). Chose $c = \frac{1}{\Delta}$ and apply Lemma 3

(5) $$L(E, F^k) \leqslant \max\left\{\left(\int c(x)F^k(dx)\right)^{\frac{1}{3}}, \frac{1}{c^2}\int c(x)F^k(dx)\right\}.$$

Recall the well-known truncation inequalities (see [4], Ch. 4. 12.4. B'): for $c > 0$

$$\int_{|x|<c} x^2 F^k(dx) \leqslant 3c^2\left(1 - \operatorname{Re} f^k\left(\frac{1}{c}\right)\right)$$

$$\int_{|x|\geqslant c} F^k(dz) \leqslant 7c\int_0^{\frac{1}{c}}(1 - \operatorname{Re} f^k(t))dt.$$

Thus

$$\int c(x)F^k(dx) \leqslant 3c^2\left(1 - \operatorname{Re} f^k\left(\frac{1}{c}\right)\right) + 7c^3\int_0^{\frac{1}{c}}(1 - \operatorname{Re} f^k(t))dt \leqslant$$

$$\leqslant 3c^2\left|1 - f^k\left(\frac{1}{c}\right)\right| + 7c^3\int_0^{\frac{1}{c}}|1 - f^k(t)|\,dt$$

and by (3) and (4)

(6) $$\int c(x)F^k(dx) \leqslant 12c^2\frac{k}{K}\left|1 - f^K\left(\frac{1}{c}\right)\right| + 28c^3\frac{k}{K}\int_0^{\frac{1}{c}}|1 - f^K(t)|\,dt \leqslant$$

$$\leqslant C_2\frac{k}{K}.$$

This estimate and (5) show that if $\frac{k}{K}$ is small enough (say $\frac{k}{K} \leqslant \delta$) then $L(E, F^k) \leqslant c$, and in this case we can see that by applying Lemma 3 it is sufficient to take

(7) $$L(E, F^k) \leqslant \left(\int c(z) F^k(dz)\right)^{\frac{1}{3}}$$

instead of (5). If $\frac{k}{K} > \delta$, then the assertion of the Lemma is true with a C_1 chosen large enough. In the case $\frac{k}{K} \leqslant \delta$ the lemma is the consequence of (7) and (6).

We remark that from the proof it is not clear whether $\frac{1}{3}$ is the best order of estimation in the inequality given by the lemma.

Theorem. *If*

$$L(\mathscr{L}(S_{k_n}^{(n)}), \Phi) = \alpha \quad \textit{and} \quad L\left(\mathscr{L}\left(\frac{\nu_n}{k_n}\right), E_1\right) = \beta ,$$

then there exist constants C_3 *and* C_4, *depending on* Φ, *but not on* α *and* β, *such that*

$$L(\mathscr{L}(S_{\nu_n}^{(n)}), \Phi) \leqslant C_3 \alpha + C_4 \beta^{\frac{1}{3}}$$

(E_1 *denotes the distribution function* $E(x-1)$).

Proof. Suppose that $\alpha < \frac{1}{14}$. Denote by H_β the event $\left\{\left|\frac{\nu_n}{k_n} - 1\right| \leqslant \beta\right\}$. By the definition of the Lévy-distance

$$\mathsf{P}(\bar{H}_\beta) \leqslant 2\beta .$$

Denote the conditional distribution function of $S_{\nu_n}^{(n)}$ under the conditions H_β and $\bar{H}_\beta$ by F_1 and F_2, resp., then

$$\mathscr{L}(S_{\nu_n}^{(n)}) = \mathsf{P}(H_\beta) \cdot F_1 + \mathsf{P}(\bar{H}_\beta) \cdot F_2$$

and by Lemma 2

(8) $$L(\mathscr{L}(S_{\nu_n}^{(n)}), \Phi) \leqslant L(F_1, \Phi) + \mathsf{P}(\bar{H}_\beta) \leqslant L(F_1, \Phi) + 2\beta .$$

But by Lemma 1

$$L(F_1, \Phi) \leqslant \max_{\left\{k: \left|\frac{k}{k_n} - 1\right| \leqslant \beta\right\}} L(\mathscr{L}(S_k^{(n)}), \Phi) .$$

Here, using the triangle inequality and the property $L(F * H, G * H) \leqslant \leqslant L(F, G)$, we obtain

$$L(\mathscr{L}(S_k^{(n)}), \Phi) \leqslant L(\mathscr{L}(S_k^{(n)}), \mathscr{L}(S_{k_n}^{(n)})) + L(\mathscr{L}(S_{k_n}^{(n)}), \Phi) \leqslant$$
$$\leqslant L(\mathscr{L}(S_{|k - k_n|}^{(n)}), E) + \alpha ,$$

and so by Lemma 5

$$L(\mathscr{L}(S_k^{(n)}), \Phi) \leqslant C_1 \left|\frac{k - k_n}{k_n}\right|^{\frac{1}{3}} + \alpha .$$

Taking into account our results following (8) we obtain

$$L(\mathscr{L}(S_{\nu_n}^{(n)}), \Phi) \leqslant C_1 \beta^{\frac{1}{3}} + \alpha + 2\beta = \alpha + C_4 \beta^{\frac{1}{3}} .$$

If $\alpha \geqslant \frac{1}{14}$ then the lemma is necessarily true with $C_3 \geqslant 14$. Q.e.d.

REFERENCES

[1] P. Lévy, *Théorie de l'addition des variables aléatoires,* Gauthier-Villars, Paris, 1937.

[2] V.M. Zolotarev, Some new inequalities in probability connected with Lévy's metric, *Dokl. Akad. Nauk SSSR,* 190 (1970), 1019-1021.

[3] V.M. Zolotarev, *Lecture notes on the general theory of limit theorems for sums of independent random variables,* Sydney, 1971, (manuscript).

[4] M. Loeve, *Probability theory,* Van Nostrand, Princeton, New York, 1955.

[5] B.V. Gnedenko – H. Fahim, On a transfer theorem, *Dokl. Akad. Nauk SSSR,* 187 (1969), 15-17.

[illegible]

Here using the [illegible] inequality and the property [illegible] we obtain

[illegible]

and so by Lemma [illegible]

[illegible]

Taking into account our results following we obtain

[illegible]

[illegible]

REFERENCES

[illegible]

COLLOQUIA MATHEMATICA SOCIETATIS JÁNOS BOLYAI
9. EUROPEAN MEETING OF STATISTICIANS, BUDAPEST (HUNGARY), 1972.

STATISTICAL THEORY OF TOPOLOGICAL GROUPS

G.J. SZÉKELY

In this paper we generalize a group theoretical result of Erdős and Rényi [1]. It is also possible to generalize other results of the statistical theory of finite groups and to develop the statistical theory of topological groups.

Let G be a compact topological Abelian group. (The topological space of G is a Hausdorff space. The group operation will be written as addition.) We shall denote by λ the Haar measure of G ($\lambda(G) = 1$). Let $g, g_1, g_2, \ldots, g_k$ be $k + 1$ elements of G chosen at random, independently of each other, so that the probabilities of the events $\{g \in V\}$, $\{g_i \in V\}$ $(i = 1, 2, \ldots, k)$ be equal to $\lambda(V)$, where V is an arbitrary λ-measurable set. Finally let us denote by V_k the number of occurences of the relation $\sum_{i=1}^{k} \epsilon_i g_i \in g + V$, where each of the numbers ϵ_i may have the value 0 or 1. We shall prove the following

Theorem. *If* $\epsilon > 0$, $\delta > 0$ *are arbitrary small positive numbers,* $\lambda(V) > 0$, *and*

(1) $$k \geqslant \frac{\log\frac{1}{\lambda(V)} + 2\log\frac{1}{\epsilon} + \log\frac{1}{\delta}}{\log 2}$$

then

$$\mathsf{P}(|V_k - 2^k\lambda(V)| \leqslant \epsilon 2^k \lambda(V)) > 1 - \delta .$$

Proof. It is not difficult to prove, that the mean value

$$\mathsf{E}[(V_k - 2^k\lambda(V)^2] = 2^k\lambda(V)(1 - \lambda(V))$$

and thus by the Markov inequality

$$\mathsf{P}((V_k - 2^k\lambda(V))^2 > \epsilon^2 2^{2k}\lambda^2(V)) < \delta$$

if (1) holds.

Remark. If we choose at random only $g_1, g_2, \ldots, g_k$ and $\sum_{i=1}^{k} \epsilon_i g_i \in$ $\in g + V$ holds $V_k(g)$ times, then

$$\mathsf{P}(\max_{h\in G}\ \min_{g\in h+V} |V_k(g) - 2^k\lambda(V)| < \epsilon 2^k \lambda(V)) > 1 - \delta ,$$

if

$$k \geqslant \frac{2\log\frac{1}{\lambda(V)} + 2\log\frac{1}{\epsilon} + \log\frac{1}{\delta}}{\log 2} .$$

This is also a consequence of the Markov inequality, because

$$\mathsf{E}\left(\frac{1}{\lambda(V)}\int_G (V_k(g) - 2^k\lambda(V))^2\lambda(dg)\right) = 2^k(1 - \lambda(V))$$

and

$$\frac{1}{\lambda(V)}\int_G (V_k(g) - 2^k\lambda(V))^2\lambda(dg) \geqslant \max_{h\in G}\ \min_{g\in h+V} (V_k(g) - 2^k\lambda(V))^2 .$$

The theorem of Erdős and Rényi is a special case of this theorem, when G is a finite Abelian group (with the discrete topology) and V is the zero element of G.

Let us mention a further result of the statistical theory of topological groups. Let V be an arbitrary λ-measurable $(\lambda(V) > 0)$ subset of G

and let us choose k elements $(g_1, g_2, \ldots, g_k)$ of G at random independently of each other, so that the probabilities of the events $\{g_i \in V\}$ $(i = 1, 2, \ldots, k)$ be equal to $\lambda(V)$.

The question is now, what is the mean value of $\lambda(A_j)$, where A_j is the set of points belonging to just j of the sets $g_i + V$ $(i = 1, 2, \ldots, k;$ $j = 0, 1, \ldots, k)$?

$$\mathsf{E}(\lambda(\{g_1 + V\} \cap \{g_2 + V\} \cap \ldots \cap \{g_h + V\}) = \lambda(V)^h$$
$$(h = 1, 2, \ldots)$$

and thus

$$\mathsf{E}(\lambda(A_j)) = \sum_{h=j}^{k} \frac{(-1)^{h-k} h!}{j!(h-j)!} \frac{k!}{h!(k-h)!} \lambda(V)^h =$$

$$= \frac{k!}{j!(k-j)!} \lambda(V)^j (1 - \lambda(V))^{k-j} \to \frac{c^j e^{-c}}{j!}$$

if $\lambda(V) \to 0$ and $k = \left[\frac{c}{\lambda(V)}\right]$ ([ ] is the entire function), where c is an arbitrary positive constant.

REFERENCE

[1] P. Erdős – A. Rényi, Probabilistic methods in group theory, *J. Analyse Math.*, 14 (1965), 127-138.

COLLOQUIA MATHEMATICA SOCIETATIS JÁNOS BOLYAI
9. EUROPEAN MEETING OF STATISTICIANS, BUDAPEST (HUNGARY), 1972.

A BAYESIAN APPROACH TO THE COMPARISON OF SENSITIVITIES OF TWO EXPERIMENTS

W.Y. TAN — I. GUTTMAN

1. SUMMARY

In many instances, the comparison of two experiments is effected by making inferences about the ratio of the *sensitivities* of the experiments. To get at a definition of sensitivity, we have in mind the following model for the two experiments: We suppose that independent observations y_{ist} made in the ith experiment are such that

$$y_{ist} = \mu_i + \alpha_{is} + \epsilon_{ist} \qquad (i = 1, 2) \tag{1.1}$$

with $s = 1, \ldots, k_i$ and $t = 1, \ldots, n_i$, and where the random disturbances ϵ_{ist} are independent normally distributed $N(0, \sigma_i^2)$ random variables. In addition, it is supposed that the random effects α_{is} are independent $N(0, \sigma_{\alpha i}^2)$ random variables, and independent of the ϵ_{ist}'s. The sensitivity of the ith experiment is defined as

$$\theta_i = \frac{\sigma_{\alpha i}^2}{\sigma_i^2} \qquad (i = 1, 2), \tag{1.2}$$

and we will be interested in the ratio

(1.3) $$r = \frac{\theta_1}{\theta_2} .$$

Moreover, our approach, unlike the approach in the previous literature, is Bayesian in outlook. Hence, we will first find the posterior of τ (Section 2) and give some examples of its use (Section 3). We will also examine the behaviour, for specific examples, of the posterior as sample sizes grow large.

As implied in the above, this problem has been treated in the literature from the classical sampling theory approach – see Dar [4], Schumann – Bradley [2], amongst others. For our model, their approach would be as follows.

Construct

(1.4) $$F_i = \frac{MS_{2i}}{MS_{1i}}$$

where

(1.4a) $$MS_{2i} = \frac{n_i}{k_i - 1} \sum_{s=1}^{k_i} (\bar{y}_{is.} - \bar{y}_{i..})^2 = \frac{1}{k_i - 1} S_{2i} .$$

(1.4b) $$MS_{1i} = \frac{1}{k_i(n_i - 1)} \sum_{s=1}^{k_i} \sum_{t=1}^{n_i} (y_{ist} - \bar{y}_{is.})^2 = \frac{1}{k_i(n_i - 1)} S_{1i}$$

with

$$\bar{y}_{is.} = \frac{1}{n_i} \sum_{t=1}^{n_i} y_{ist} , \qquad \text{and}$$

(1.4c)

$$\bar{y}_{i..} = \frac{1}{k_i n_i} \sum_{s=1}^{k_i} \sum_{t=1}^{n_i} y_{ist} = \frac{1}{k_i} \sum_{s=1}^{k_i} \bar{y}_{is.} .$$

The comparison of the two experiments is then made though use of the ratio

$$W = \frac{F_1}{F_2} .$$

This is used to test the hypothesis that $r = 1$, that is, $\theta_1 = \theta_2$. However, the (sampling) distribution of W is "messy" and rather hard to work with (indeed needed approximations to the distribution are discussed by Dar [4]) unless the experiments are *similar*, i.e. $k_1 = k_2$ or *identical* ($k_1 = k_2$, $n_1 = n_2$), which more often than not does not hold in practice. In the Bayesian approach, no such restriction need be applied.

2. THE POSTERIOR OF τ

Under the conditions intendent in (1.1), it has been shown by Tan [6] that the posterior distribution of θ_i is given by

$$(2.1) \qquad p(\theta_i \mid y_i) = c_i (1 + n_i\theta_i)^{\frac{k_i(n_i-1)}{2} - 1} \left[1 + \frac{1 + n_i\theta_i}{\Phi_i}\right]^{-\frac{kn_i - 1}{2}}$$

where

$$(2.1a) \qquad \Phi_i = \frac{S_{2i}}{S_{1i}} = \frac{k_i - 1}{k_i(n_i - 1)} F_i ,$$

with F_i, S_{2i}, S_{1i} defined as in (1.4), (1.4a), (1.4b), and where c_i is such that

$$(2.1b) \qquad \frac{1}{c_i} = \Phi_i^{\frac{k_i(n_i-1)}{2}} \int_0^{\frac{\Phi_i}{1+\Phi_i}} u^{\frac{k_i - 1}{2} - 1} (1 - u)^{\frac{k_i(n_i-1)}{2} - 1} du .$$

The prior taken by Tan [6] to find (2.1) was found using Jeffreys' invariance argument and given by

$$(2.1c) \qquad p(\mu_i, \sigma_{\alpha i}^2, \sigma_i^2) \propto (\sigma_i^2)^{-2} (\sigma^2 + n\sigma_{\alpha i}^2)^{-1} .$$

Since we have assumed independence of experiments, we of course have that

$$(2.2) \qquad p(\theta_1, \theta_2 \mid y_1, y_2) = \prod_{i=1}^{2} p(\theta_i \mid y_i) .$$

Hence, on making the transformation $\tau = \frac{\theta_1}{\theta_2}$, $\gamma = \theta_2$, it is easy to see that

$$p(\tau, \gamma | y_1, y_2) =$$

(2.3) $$= \frac{c_1 c_2 \gamma (1+n_2\gamma)^{\frac{k_2(n_2-1)}{2}-1} (1+n_1\gamma\tau)^{\frac{k_1(n_1-1)}{2}-1}}{\left(1+\frac{1+n_2\gamma}{\Phi_2}\right)^{\frac{k_2 n_2 - 1}{2}} \left(1+\frac{1+n_1\gamma\tau}{\Phi_1}\right)^{\frac{k_1 n_1 - 1}{2}}} .$$

Hence, the posterior of τ is simply given as

(2.4) $$p(\tau | y_1, y_2) = \int_0^\infty p(\tau, \gamma | y_1, y_2) d\gamma .$$

3. EXAMINATION OF THE POSTERIOR OF τ

As is well known, the posterior distribution summarizes all the information we have about the parameter τ. We would, then, obtain a $100(1-\alpha)$ % posterior interval for τ, say of the form

(3.1) $$I = (\tau_l, \tau_u) = [\tau | \tau_l < \tau \leqslant \tau_u]$$

where

(3.1a) $$\int_0^{\tau_l} p(\tau | y_1, y_2) = \int_{\tau_u}^\infty p(\tau | y_1, y_2) = \frac{\alpha}{2} .$$

Hence, the posterior probability of τ falling in I is

(3.2) $$\mathrm{P}(\tau_l < \tau \leqslant \tau_u | y_1, y_2) = 1 - \mathrm{P}(\tau \leqslant \tau_l) - \mathrm{P}(\tau > \tau_u) = 1 - \alpha .$$

Numerically, this is a simple routine affair, and we have carried out the computation for three sorts of situations. We note that $p(\tau = 0 | y_1, y_2) < < \infty$ only if $k_2 > 3$, as can easily be seen from (2.3) and (2.4).

Example 3.1. Determinations of melting temperatures of hydroquinine with four different types of thermometers have been made in two separate experiments, with the following results.

Experiment 1

$n_1 = 2, \quad k_1 = 4$

Thermometer type	1	2	3	4
Metting points (coded)	2.0 1.5	1.0 1.5	−0.5 0.5	1.5 1.5

Experiment 2

$n_2 = 4, \quad k_2 = 4$

Thermometer type	1	2	3	4
Metting points (coded)	1.0, 1.0 1.5, 1.0	0.0, 1.0 1.0, 1.5	−1.0, 0.0 1.0, 1.0	−1.0, 0.0 0.5, 1.0

Here, the experiments are similar ($k_1 = k_2 = 4$). It is easy to calculate that

$$S_{21} = \frac{29}{8}, \quad S_{11} = \frac{6}{8}, \quad S_{22} = \frac{44.75}{16}, \quad S_{12} = \frac{21.25}{4},$$

so that

$$\varphi_1 = \frac{29}{6}, \quad \varphi_2 = \frac{44.75}{101}.$$

Using this information, we are able to find τ_l and τ_u defined in (3.1a) for various values of $1 - \alpha$, and tabulate these in the first panel of Table 3.1. A graph of $p(\tau \mid y_1, y_2)$ for this example is given in Figure 3.1.

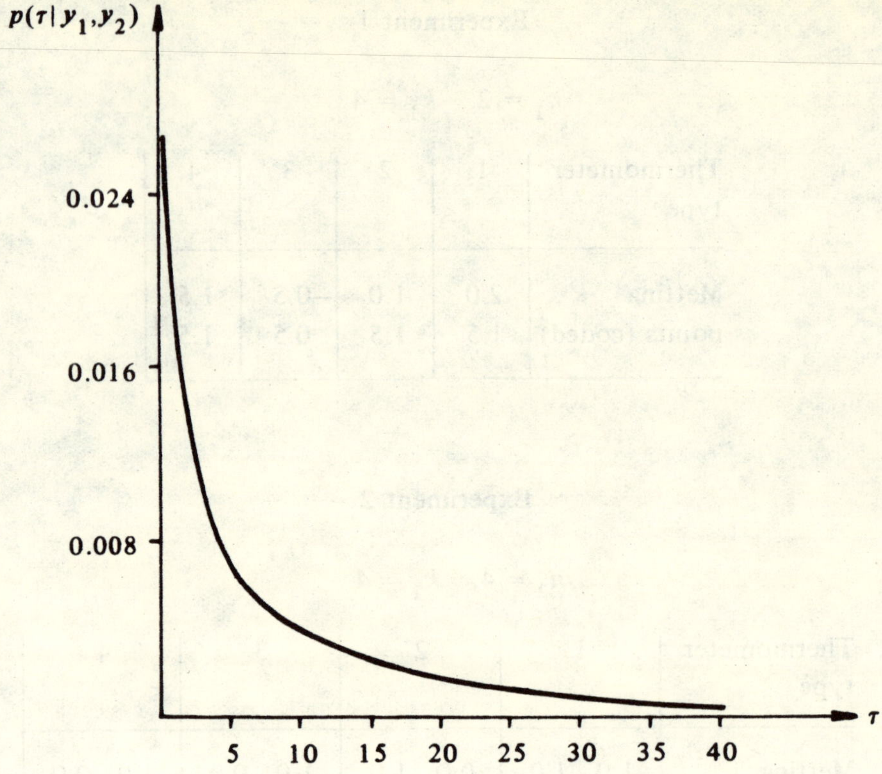

Figure 3.1

A plot if $p(\tau|y_1,y_2)$ of Example 3.1

It is interesting to look at the behaviour of τ_l and τ_u for different values of φ_1 and φ_2. For this purpose, we set

$$\varphi_1 = b\ ,\ \ \varphi_2 = 1$$

and plot $\tau_l(\alpha = 0.025)$ and $\tau_u(\alpha = 0.025)$ against b for the conditions of this example, that is, $n_1 = 2$, $k_1 = 4$, $k_2 = 4$. The plots are "highly linear" and are given in Figure 3.2.

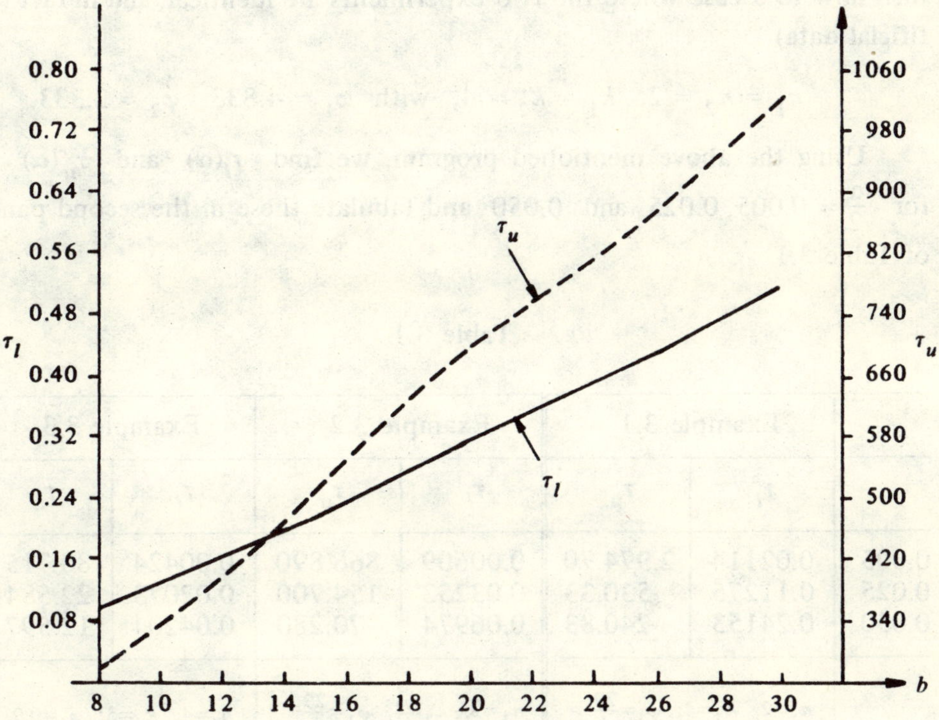

Figure 3.2

A plot of τ_l and τ_u fur $\frac{\alpha}{2} = 0.025$, when
$n_1 = 2,\ n_2 = 4;\ k_1 = k_2 = 4,$
with $\varphi_2 = 1$ and $\varphi_1 = b$

The author wishes to thank G. Desrocher, Université de Montréal, for writing the necessary program which will take observations of two-one-way analysis of variance experiments and produce a table of $p(\tau | y_1, y_2)$ and (τ_l, τ_u) for specified α. The program is very flexible, and for example, if φ_1 and φ_2 are computed elsewhere, this information along with the values of n_1, n_2, k_1 and k_2 will produce $p(\tau | y_1, y_2)$ and (τ_l, τ_u).

Copies are available on request from the author.

Example 3.2. The above example dealt with *similar* experiments. We turn now to a case where the two experiments are identical, and in fact (artificial data)

$$n_1 = n_2 = 2, \ \ k_1 = k_2 = 4, \ \ \text{with} \ \ \varphi_1 = 4.833, \ \ \varphi_2 = 2.333.$$

Using the above mentioned program, we find $\tau_l(\alpha)$ and $\tau_u(\alpha)$ for $\frac{\alpha}{2} = 0.005, 0.025$ and 0.050 and tabulate these in the second panel of Table 3.1.

Table 3.1

	Example 3.1		Example 3.2		Example 3.3	
	τ_l	τ_u	τ_l	τ_u	τ_l	τ_u
0.005	0.02114	2.974.70	0.00609	866.890	0.00424	80.715
0.025	0.11275	530.33	0.03253	154.700	0.02073	22.584
0.050	0.24153	240.83	0.06974	70.280	0.04204	12.597
	$n_1=2,\ k_1=4;\ \varphi_1=\frac{29}{4}$		$n_1=2,\ k_1=4;\ \varphi_1=\frac{29}{6}$		$n_1=4,\ k_1=5;\ \varphi_1=0.9$	
	$n_2=4,\ k_2=4;\ \varphi_2=0.443$		$n_2=2,\ k_2=4;\ \varphi_2=\frac{27}{12}$		$n_2=7,\ k_2=4;\ \varphi_2=0.7$	
	Similar		Identical		—	

Example 3.3. We turn now to an example in which experiments are neither similar nor identical. For two experiments, the pertinent data is

$$n_1 = 4, \ \ k_1 = 5; \ \ \varphi_1 = 0.9,$$
$$n_2 = 7, \ \ k_2 = 4; \ \ \varphi_2 = 0.7.$$

Once again, we find the lower and upper $100(1-\alpha)\ \%$ limits τ_l

and τ_u for $\frac{\alpha}{2} = 0.005$, 0.025 and 0.050 and tabulate these in the third panel of Table 3.1. For example, consulting Table 3.1, we immediately see that for the data of Example 3.1, the 95 % posterior interval for τ is $1 - \alpha = 0.95$ or $\frac{\alpha}{2} = 0.025$,

(3.3) $\qquad I = (\tau_l, \tau_u) = (0.11275,\ 530.33)\,.$

It is of interest to inquire into the behaviour of τ_l and τ_u as the n_i increase, for fixed φ_1 and φ_2. To gather some insight for this purpose, we have selected

(3.4) $\qquad n_1 = m,\ n_2 = 2m,\ k_1 = k_2 = 4,\ \varphi_1 = 4.833,\ \varphi_2 = 0.443$

and we let $m = 8(8)24{,}48$ and 60. This forms a set of similar experiments. (The case $m = 2$ is of course, the situation given in Example 3.1.) We tabulate τ_l and τ_u for the usual values of $\frac{\alpha}{2}$ in Table 3.2.

Table 3.2

Values of τ_l and τ_u for the set of similar experiments (3.4)
$n_1 = m,\ n_2 = 2m;\ k_1 = k_2 = 4;\ \varphi_1 : \varphi_2 = 4.833 : 0.443$

	$\frac{\alpha}{2} = .005$		$\frac{\alpha}{2} = .025$		$\frac{\alpha}{2} = .05$	
m	τ_l	τ_u	τ_l	τ_u	τ_l	τ_u
8	0.18493	722.47	0.58073	222.47	0.98447	129.17
16	0.20839	583.32	0.64642	187.59	1.0845	111.62
24	0.21573	557.44	0.66702	180.14	1.1158	107.61
48	0.27287	535.98	0.68706	173.83	1.1463	104.17
60	0.22427	532.10	0.69100	172.68	1.1523	103.54

We note from Table 3.2, that for $n_1 = 16$, $n_2 = 32$, $k_1 = 4 = k_2$, that if $\varphi_1 = 4.833$ and $\varphi_2 = 0.443$, that the 90 % posterior interval for τ would be (1.0845, 111.62), that is, would not include the value 1.

4. ACKNOWLEDGEMENT

Thanks are due to G. Desrochers, Université de Montréal, for permission to use his program (see Desrochers [5]) for the needed calculations of this paper.

REFERENCES

[1] R.A. Bradley – D.E.W. Schumann, The comparison of the sensitivities of similar experiments: Applications, *Biometrics,* 13 (1957), 496-510.

[2] D.E.W. Schumann – R.A. Bradley, The comparison of the sensitivities of similar experiments: Theory, *Ann. Math. Statist.*, 28 (1957), 902-920.

[3] D.E.W. Schumann – R.A. Bradley, The comparison of the sensitivities of similar experiments: Model II of the Analysis of Variance, *Biometrics,* 15 (1959), 405-416.

[4] S.N. Dar, On the Comparison of the Sensitivities of Experiments, *J.R.S.S.*, 24 (1962), 447-453.

[5] G. Desrochers, A program for the determination of lower and upper posterior limits for the sensitivity ratio $\tau = \frac{\theta_1}{\theta_2}$, (unpublished – copies available on request).

[6] W.Y. Tan, Bayesian Analysis of Random Effect Models in the Analysis of Variance, Ph. D. Thesis, University of Wisconsin; Madi-Wisconsin, 1964.

COLLOQUIA MATHEMATICA SOCIETATIS JÁNOS BOLYAI
9. EUROPEAN MEETING OF STATISTICIANS, BUDAPEST (HUNGARY), 1972.

A SIMPLE ALGORITHM TO DETERMINE THE ERGODIC CLASSES OF A MARKOV CHAIN

J. TANKÓ

It may often be necessary to determine serially the ergodic classes of Markov chains. It is indispensable e.g. for the control of a controllable Markov chain. I met the problem when I had to make a program for the Howard-algorithm to determine the optimal control for Markov chains with rewards.

I report, in the following, on a simple algorithm found by me to determine the ergodic classes of a finite-state Markov chain. In case of few states the algorithm can be executed in a simple graphic form. When the number of states is greater, then a simple computer program enables the algorithm to be executed by a computer.

The result of the algorithm can be represented by a marking vector M from which it can be read which states belong to the same ergodic class and which states are unessential. On this basis rearrangement of the one-step transition matrix to a quasi-diagonal supermatrix by the renumbering of the states of the Markov chain can be easily done.

We set forth the algorithm in line with a presentation by an example. Its detailed description and demonstration can be found in No. 7 of the Communications of the Computing Centre of the Hungarian Academy of Sciences in Hungarian language.

Let us characterize the Markov chain by its one-step transition matrix $\boldsymbol{P} = \{p_{ij},\ \ i,j = 1, \ldots, N\}$ and denote $M = \{m_i,\ \ i = 1, \ldots, N\}$ the marking vector to be defined later.

Let us consider e.g. the stochastic matrix to be seen in Fig. 1. This is a one-step transition matrix of a Markov chain with 12 states. We have chosen such a simple matrix for the sake of easier survey.

	1	2	3	4	5	6	7	8	9	10	11	12
1							1.0					
2			0.4		0.3							0.3
3						0.4				0.6		
4	1.0											
5				1.0								
6									1.0			
7					1.0							
8	1.0											
9											1.0	
10		1.0										
11						1.0						
12						1.0						

Fig. 1.

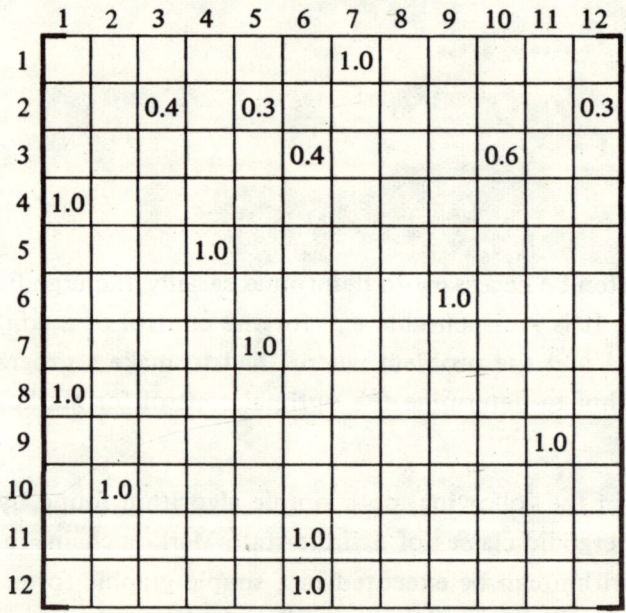

For illustration's sake we can picture the Markov chain also in the form of a directed graph where the vertices correspond to the states of the Markov chain and the arrows represent the positive one-step transition probabilities.

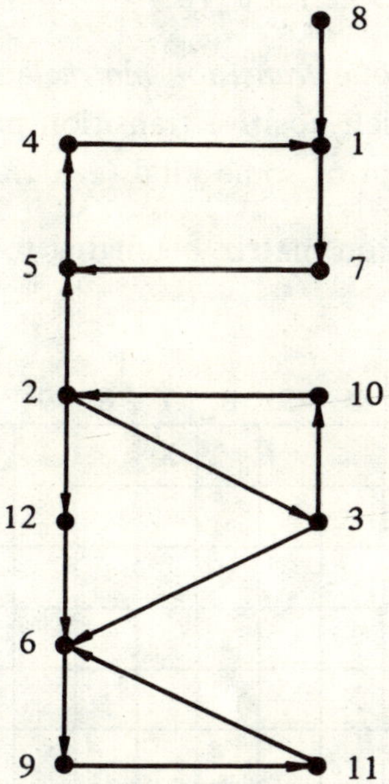

Fig. 2.

In our example the graph takes the shape presented in Fig. 2.

We purposefully drew the graph in a way to facilitate the recognition of the ergodic classes. The arrows show that the states (1, 4, 5, 7) and (6, 9, 11) make each an ergodic class, the other states are transient states.

A state is called greater than the other if its serial number is greater.

The *indicator* of an ergodic class is to be the serial number of its greatest state. E.g. the indicator of the ergodic class (1, 4, 5, 7) is 7. An ergodic class is called greater than the other if its indicator is greater.

In a directed graph the branches of a "cluster" belonging to a vertex are the vertices at which arrows are pointing from the respective vertex, and the roots are the vertices starting from which arrows point at the respective vertex. E.g. the cluster belonging to the vertex 2 has three branches: 3, 5, 12, and a sole root: 10.

We mean by a *symbolic transition matrix* an $n \times n$ matrix in which those (i, j) fields, to which positive transition probabilities $p_{ij} > 0$ belong, are denoted by a sign of some kind (e.g. X).

The symbolic transition matrix belonging to our example is shown in Figure 3.

	1	2	3	4	5	6	7	8	9	10	11	12
1							X					
2			X		X							X
3						X				X		
4	X											
5				X								
6									X			
7					X							
8	X											
9											X	
10		X										
11						X						
12						X						

Figure 3.

The algorithm consists in performing, proceeding in an increasing order along the principal diagonal fields of the symbolic transition matrix, the following procedure, and determining thereafter the ergodic classes.

The procedure is the following: We sign the intersection field of the columns corresponding to the signed fields of the row of the principal diagonal field and the rows containing signed fields in the column of the principal diagonal field below the principal diagonal. Briefly, we sign directly the transitions which are possible through the state corresponding to the principal diagonal element. In order of later construction, we may use another sign (e.g. +) or we may enter the number indicating the step at which the signing takes place. As a result of the procedure we obtain the symbolic matrix to be seen in Fig. 4.

	1	2	3	4	5	6	7	8	9	10	11	12	
1							X						7
2			X		X							X	0
3						X				X			0
4	X						①						7
5	④			X			④						7
6									X				11
7	⑤			⑤	X		⑤						7
8	X			⑦	⑦		①						0
9											X		11
10	⑤	X	②	⑤	②	③	⑤		⑥	③	⑨	②	0
11						X			⑥		⑨		11
12						X			⑥		⑨		0

Figure 4.

This procedure represented in the graph means that we connect the greater vertices of the root of the cluster with the branches of the cluster by arrows showing at the latter. (In order of subsequent construction we may indicate the arrows another way, (e.g. by a thinner line) or quote with the number corresponding to the place of the step at which they originated). The result of the procedure in our example can be seen in Figure 5.

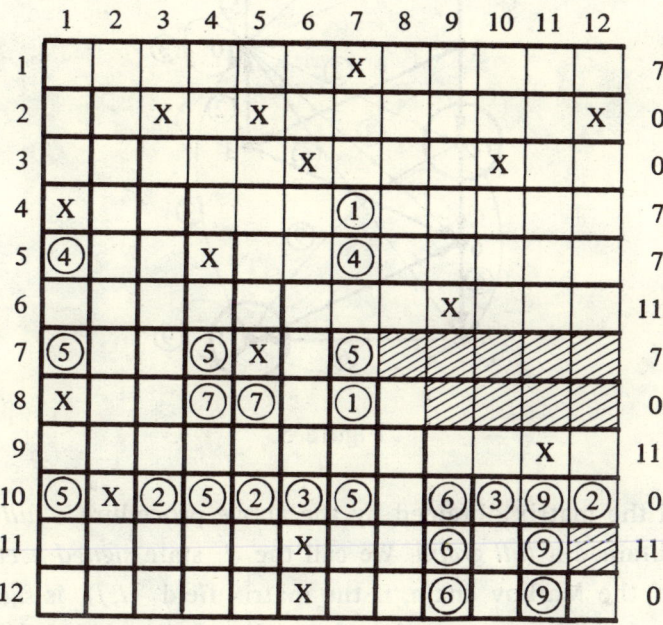

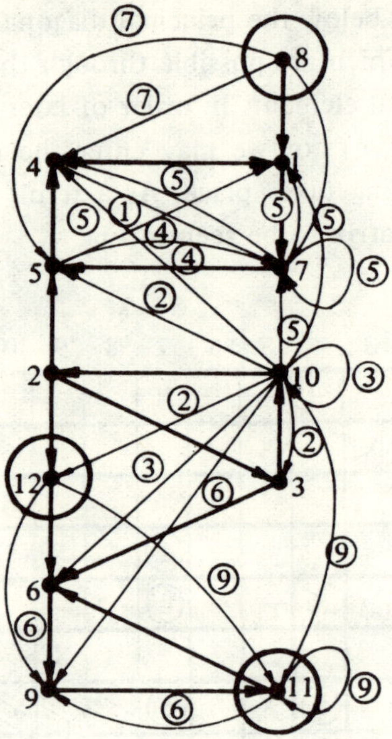

Figure 5.

We call the matrix obtained by the above procedure a *full matrix*, the graph obtained a *full graph*. We call the j state *signed accessible* from a state i of the Markov chain, if the matrix field (i, j) is signed in the full matrix, resp. if an arrow points from vertex i at j in the full graph.

The signed fields of the full matrix and the arrows of the full graph correspond to each other. It follows from the Procedure that all signed transitions are possible transitions in the Markov chain. In the full matrix, resp. full graph not every possible transition appears as a signed transition. The appearence depends on the numbering order of the states.

Those states, rows corresponding to which are void right of the principal diagonal in the full matrix, resp. from the graph-vertex corresponding to which no arrows point at the greater vertices in the full graph are called *states to be examined.* The greatest state of the Markov chain is always a state to be examined. In our example, the states to be examined are therefore 7, 8, 11 and 12.

We mean by the *marking vector* of a Markov chain a vector of n components, the kth component of which is 0 if the k state of the chain is a transient state or is the indicator of the ergodic class to which the k state belongs, respectively. The components of the marking vector in our case are indicated in Figure 4 on the right side of the matrix. Two theorems follow:

Theorem 1. *The indicator of each ergodic class is a state to be examined.*

Theorem 2. *One of the following two statements is true for a state to be examined:*

either the serial number of the state to be examined is the indicator of an ergodic class and the ergodic class is constituted of states accessible from it on a signed way,

or the state to be examined is transient (unessential) depending on that whether none or any of the states accessible from it on a signed way belongs to a smaller ergodic class, respectively.

The proof will be here omitted. It can be found in the paper referred to. Instead of it we execute the first two steps of the algorithm in our example. We start from the symbolic matrix of Fig. 3.

From *state* 1 one can arrive to 7. This is the branch of the cluster belonging to the state 1. There is a transition to state 1 from states 4 and 8. These are the roots of the cluster. The latter must be connected with the branches. This means for the matrix that fields corresponding to the signed fields of column 1 must be signed also in column 7.

There is a transition from *state* 2 to states 3, 5 and 12 and there is a transition to state 2 from state 10. So we must sign the fields 10 of columns 3, 5 and 12.

Continuing the procedure in turn for all states we would obtain as a result a figure like Fig. 4, but with a simpler denotation.

According to the Theorem 2 we have now to examine the states to be examined. According to the definition these are the states the matrix rows corresponding to which are void right of the principal diagonal (indicated by shade lines in Fig. 4).

Accordingly, states to be examined are 7, 8, 11 and 12. The first state to be examined is 7. This surely is an indicator of an ergodic class. According to the theorem, the states of the first ergodic class are (1, 4, 5, 7) as these fields are signed in row 7, and these states are accessible from the state 7 in a signed way. The next state to be examined is 8. From this there are signed transitions to a smaller ergodic class, therefore state 8 is a transient one. Signed transitions from the state to be examined 11 are possible to states 6, 9 and 11 none of them is an element of the ergodic class (1, 4, 5, 7), therefore (6, 9, 11) make again an ergodic class, indicated by 11. The last state is always to be examined. In the present case there are transitions from 12 to the ergodic class (6, 9, 11), so state 12 is a transient one. According to the Theorem 1 all other states not figuring up to now are also transient, so the class of transient states consists of states (2, 3, 8, 10, 12).

Next to the full matrix we have indicated in Fig. 4 in all rows the indicator of that ergodic class to which the state belongs, resp. we have written 0 for the transient states. This makes the marking vector of the Markov chain, which is therefore

$$(7, 0, 0, 7, 7, 11, 7, 0, 11, 0, 11, 0)\,.$$

Remark. It may be important to remark that the recognition of the states to be examined is possible during the algorithm and the algorithm allows to apply Theorem 2 to the state to be examined immediately after it had been recognized.

If the Markov chain has a greater number of states or it is built in a model, computer program is also required.

In case of machine procedure, we start from the real one-step transition matrix. The algorithm differs from the "manual" algorithm described above in that the signing of matrix-fields is replaced by the addition of the field contents. The signed field will mean now such a field which contains a positive quantity, and the non-signed field such which contains 0. We have made the program of the algorithm for the CDC 3300 computer of the Hungarian Academy of Sciences in both FORTRAN and ALGOL languages. We have the programs by applying them to matrices generated by a stochastic matrix generator. The procedure works quickly in FORTRAN language. The analysis of a 100×100 matrix lasted for instance 25 seconds. The FORTRAN program is presented in the form of a subroutine and the ALGOL program as a procedure.

```
      SUBRUTINE ERGOD (P, M, N)
      INTEGER N,M(N)
      REAL P(N,N)
C     THIS IS AN ALGORITHM TO PROVIDE THE ERGODIC
C     CLASSES OF A HOMOGENEOUS FINITE-STATE MARKOV
C     CHAIN CHARACTERISED BY THE TRANSITION MATRIX
C     P(N,N)
      DO 20 I=1,N
   20 M(I)=0
      N1=N-1
      DO 21 I=1,N1
      I1=I+1
      DO 22 J=1,N
      IF (P(I,J).EQ.0.) GO TO 22
      DO 23 K=I1,N
   23 P(K,J)=P(K,J)+P(K,I)
   22 CONTINUE
   21 CONTINUE
      DO 24 I=1,N
      IF (I.EQ.N) GO TO 29
      J1=I+1
      DO 25 J=J1,N
      IF (P(I,J).GT.0.) GO TO 24
   25 CONTINUE
```

```
29  DO 26 J=1,I
    IF (P(I,J)).EQ.0.) GO TO 26
    IF (M(J).NE.0) GO TO 27
    M(J)=I
    GO TO 26
27  DO 28 K=1,J
28  IF (M(K).EQ.I) M(K)=0
    GO TO 24
26  CONTINUE
24  CONTINUE
    RETURN
    END
```

```
procedure ERGOD (P,M,N);
  value N; integer N; integer array M; real array P;
  comment THIS IS AN ALGORITHM TO PROVIDE THE ERGODIC
        CLASSES OF A HOMOGENEOUS FINITE-STATE MARKOV-
        CHAIN CHARACTERISED BY THE TRANSITION MATRIX
        P[N,N];
  begin integer I,J,K;
        for I:=1 step 1 until N do M[I]:=0;
        for I:=1 step 1 until N do
        for J:=1 step 1 until N do
        if P[I,J] > 0 then for K:=I+1 step 1 until N do
        P[K,J]:=P[K,J]+P[K,I];
        for I:=1 step 1 until N do begin
        for J:=I+1 step 1 until N do
        if P[I,J] > 0 then go to LINE;
        for J:=1 step 1 until I do
        if P[I,J] > 0 ∧ M[J] > 0 then begin
        for K:=1 step 1 until J do
        if M[K]=I then M[K]:=0;
        go to LINE end else
        if P[I,J] > 0 then M[J]:=I;
LINE:   end I; end ERGOD
```

COLLOQUIA MATHEMATICA SOCIETATIS JÁNOS BOLYAI
9. EUROPEAN MEETING OF STATISTICIANS, BUDAPEST (HUNGARY), 1972.

CONCAVE UTILITIES ARE DISTINGUISHED BY THEIR OPTIMAL STRATEGIES

E. THORP — R. WHITLEY

1. INTRODUCTION

Mossin [5], Thorp [7], and Samuelson [6] showed for specific pairs of utility functions that different utilities can lead to different optimal strategies. In particular the optimal investment strategy for the utility $\log x$ is not necessarily the optimal strategy for the utility $\frac{1}{\gamma} x^\gamma$ $(\gamma \neq 0)$.

These examples suggest the following generalization, of obvious importance to general utility theory.

Consider a T stage investment process. At each stage allocate resources among the available investments. Each chosen sequence A of allocations ("strategy") yields a corresponding terminal probability distribution F_T^A of assets at the completion of stage T. For each utility function $U(\cdot)$, consider those strategies $A^*(U)$ which maximize the expected value $\int U(x) dF_T^A(x)$ of terminal utility. Assume sufficient hypotheses on U and the set of F_T^A so that the integral is defined and that furthermore the maximizing strategy $A^*(U)$ exists. Then is it true in general that $A^*(U_1)$ is not $A^*(U_2)$ for "distinct" utilities U_1 and U_2?

As we now show, the answer is yes: the Mossin – Thorp – Samuelson results for specific utility pairs generalizes to the principal class of interest in modern utility theory.

2. THE MAIN THEOREM

We prove this for the class of "interesting" concave utilities. We begin with more special hypotheses.

Theorem 1. *Let U and V be utilities defined and differentiable on $(0, \infty)$ with $U'(x)$ and $V'(x)$ positive and strictly decreasing as x increases. Then if U and V are inequivalent, there is a one period investment setting such that U and V have distinct sets of optimal strategies. Furthermore, the investment setting may be chosen to consist only of cash and a two-valued random investment, in which case the optimal strategies are unique.*

Corollary 2. *If the utilities U and V have the same (sets of) optimal strategies for each finite sequence of investment settings, then U and V are equivalent.*

Two utilities U_1 and U_2 are equivalent if and only if there are constants a and b such that $U_2(x) = aU_1(x) + b$ $(a > 0)$, otherwise U_1 and U_2 are inequivalent.

Let X_i $(1 \leqslant i \leqslant k)$ be the (random) outcome per unit invested in the ith "security". We call $(X_1, \ldots, X_k)$ the investment setting. We assume X_i is independent of the amount invested. Let the initial capital be Z_0 and let the final capital be Z_1. A strategy is an allocation $W = (w_1, \ldots, w_k)$ where w_i is the fraction of Z_0 allocated to security i. We assume $w_i \geqslant 0$ for all i, that $\sum_i w_i = 1$, and that wealth is infinitely divisible. Thus the w_i may assume any real values consistent with the constraints and with the requirement that $\sum_i w_i X_i$ is in the domain of the utility function U.

Given a particular U satisfying the hypotheses of the theorem, suppose $\mathsf{E}U(Z_1(W))$ is maximized by some strategy W^*. Then W^* is an *optimal* (or *best*) *strategy* for U relative to the given investment setting.

Proof of Theorem. Suppose that U and V have the same optimal strategies for every one period investment setting consisting of cash and a two-valued random investment. It will be shown that U and V are equivalent, which will establish the logical contrapositive to the theorem and hence the theorem itself.

In the proof of theorems we shall assume for technical simplicity that the initial capital $Z_0 = 1$. When theorems have been established for this case, consideration of the transformation $U_0(s) = U(Z_0 s) = U(t)$ gives the theorems for arbitrary $Z_0 > 0$. We shall therefore state the general results without further comment after proving the $Z_0 = 1$ case.

Let the only investment (besides cash) be X where $\mathsf{P}(X = 1 - b) = q = 1 - p$ and $\mathsf{P}(X = 1 + a) = p$, where $a > 0$ and $0 < p, b < 1$. The choice $0 < b < 1$, rather than simply $b = 1$, has been made because for $b = 1$ and $w = 1$, the expression $U(0)$ would arise and 0 is not necessarily in the domain of U (e.g., $U(x) = \log x$). The available strategies are to allocate the fraction w of recources to X and $1 - w$ to cash, with $0 \leqslant w \leqslant 1$.

At the end of the period, we have

$$\text{(2.1)} \qquad \mathsf{E}U(Z_1(w)) = pU(1 + aw) + qU(1 - bw) = f(w)\,.$$

To find the maximum, consider $f'(w) = apU'(1 + aw) - bqU'(1 - bw)$. Since $U'(t)$ strictly decreases as t increases, we have $f'(w)$ decreasing strictly as w increases. Thus there is a unique maximum. If $f'(w^*) = 0$ for some w^* with $0 \leqslant w^* \leqslant 1$, then the maximum is at this unique w^*. If $f'(w) > 0$ for all w with $0 \leqslant w \leqslant 1$, then the unique maximum is at $w = 1$. If instead $f'(w) < 0$ for $0 \leqslant w \leqslant 1$, then the unique maximum is at $w = 0$.

If $f'(w) = 0$ we have $\frac{U'(1+aw)}{U'(1-bw)} = \frac{bq}{ap}$. Suppose $a > 0$ and $\frac{1}{2} < b < 1$ are given and we wish $f'\left(\frac{1}{2b}\right) = 0$. Letting $\lambda = \frac{U'\left(1 + \frac{a}{2b}\right)}{U'\left(\frac{1}{2}\right)}$, we can solve $\lambda = \frac{bq}{ap}$ for p, with $0 < p < 1$. Thus for each $a > 0$ there is a choice of p, hence an X, such that $w^* = \frac{1}{2b}$ is optimal for U.

Now suppose that U and V have the same optimal strategies for all such investment settings. Then $w^* = \frac{1}{2b}$ for V also and we have $\frac{U'\left(1+\frac{a}{2b}\right)}{U'\left(\frac{1}{2}\right)} = \frac{V'\left(1+\frac{a}{2b}\right)}{V'\left(\frac{1}{2}\right)}$ for all $a > 0$. Letting $V'\left(\frac{1}{2}\right) = \alpha U'\left(\frac{1}{2}\right)$ we find $V'(t) = \alpha U'(t)$ $(t > 1)$ whence $V(t) = \alpha U(t) + \beta$ $(t > 1)$.

When $t < 1$, we proceed similarly. Choose X so that $\mathsf{P}(X = 2) = p$ and $\mathsf{P}(X = 1 - b) = q$, where $0 < b < 1$. Then

$$\mathsf{E}U(Z_1(w)) = pU(1+w) + qU(1-bw) = f(w)$$

$$f'(w) = pU'(1+w) - bqU'(1-bw)$$

and the maximum is unique and located as before.

If $f'(w) = 0$ we have $\lambda = \frac{U'(1-aw)}{U'(1+w)} = \frac{p}{aq}$ and given $w = b$, $0 < b < 1$, we can choose p with $0 < p < 1$ such that $\lambda = \frac{p}{aq}$. Then as before we find $V'(1-ab) = \gamma U'(1-ab)$ and since a and b can be any numbers such that $0 < a, b < 1$, then $V'(t) = \gamma U'(t)$ $(0 < t < 1)$ where $\gamma = \frac{V'(1+b)}{U'(1+b)}$. But γ was shown to be α.

Thus $V(t) = \alpha U(t) + \delta$ $(0 < t < 1)$. Also $V(1) = \alpha U(1) + \epsilon$. Hence $V(t) - \alpha U(t) = \beta$ if $t > 1$, δ if $t < 1$ and ϵ if $t = 1$. But $V(t) - \alpha U(t)$ is continuous so $\beta = \delta = \epsilon$ so $V(t) = \alpha U(t) + \beta$. Thus U and

V are equivalent under the assumption that they have the same optimal strategies for all one period investment settings containing only (cash and) a two-valued random investment. The logical contrapositive assertion is the Theorem. This completes the proof. The Corollary follows a fortiori.

Note that a single investment setting of the type in the proof will not in general distinguish inequivalent utility functions. For instance, if $\mathsf{E}(X) \leqslant \leqslant 0$ then $w = 0$ is the unique optimal strategy for all the utilities of Theorem 1 (more generally, for all strictly concave utilities, as defined below) so such X distinguish between none of these utilities. It may be of interest to characterize each investment setting by the pairs of utility functions it distinguishes between or "separates", and to similarly characterize collections of investment settings.

For a security X, let $m(X)$ and $M(X)$ be the greatest and least numbers, respectively, such that $\mathsf{P}(m(X) \leqslant X \leqslant M(X)) = 1$. Then for a collection C of investment settings whose securities are $\{X_\alpha : \alpha \in A\}$, where A is some index set, let $m_A = \inf\{m(X_\alpha) : \alpha \in A\}$ and $M_A = \sup\{M(X_\alpha) : \alpha \in A\}$. Evidently, if $U(t) = V(t)$ for $m_A \leqslant t \leqslant M_A$, the collection C will not separate U and V. Thus a collection with $m_A = 0$ and $M_A = \infty$ will be needed in general to prove the conclusion of Theorem 1.

Next we generalize Theorem 1 to concave non-decreasing utilities defined on $(0, \infty)$. We do not make the common assumption that first or even second derivatives exist. A function f is *concave* on an interval I if for each pair of points $x_1 \neq x_2$ in I and each number s with $0 < s < 1$, then $f(sx_1 + (1-s)x_2) \geqslant sf(x_1) + (1-s)f(x_2)$. If $f(sx_1 + (1-s)x_2) > sf(x_1) + (1-s)f(x_2)$ always, then f is *strictly concave.* (We use "concave" to mean "concave from below".)

The more general definition includes such computationally and empirically natural functions as the "polygonal" utilities. In these, the utility is a sequence of linear segments. The vertices are such that the function lies on or below each segment extended, and the ordinates of the vertices increase as the abscissas increase.

First, recall some facts from the elementary theory of concave functions. (Most texts give results for convex functions. But f is concave exactly when $-f$ is convex so the theories of concave and convex functions are equivalent.) A concave function is either continuous in the interior of its domain or non-measurable. An increasing function is always measurable so our utilities are continuous. A continuous concave function f defined on an open interval has a left derivative f'_- and a right derivative f'_+ defined everywhere. (If the left endpoint a is included in the interval of definition, then $f'_-(a)$ is not defined and $f'_+(a)$ may or may not be defined. Similarly, if the right endpoint b is included in the interval of definition, then $f'_+(b)$ is not defined and $f'_-(b)$ may or may not be defined.) Furthermore, $f'_-(t) \geqslant f'_+(t)$ for all t except the endpoints in the domain of f and whenever $t_1 < t_2$ then $f'_-(t_1) \geqslant f'_-(t_2)$ and $f'_+(t_1) \geqslant f'_+(t_2)$. There are at most countably many points where $f'_-(t) > f'_+(t)$; otherwise $f'_-(t) = f'_+(t) = f'(t)$ and f is differentiable. Proofs of these assertions and further theorems on concave functions are given for instance in Hardy, Littlewood, Polya [3].

Theorem 3. *Let U and V be concave utilities defined on $(0, \infty)$, one of which is strictly increasing on $(0, 1 + e)$ for some $e > 0$. If U and V are inequivalent then there is a one period investment setting such that the sets of optimal strategies for U and for V are distinct. The investment setting may be chosen to consist only of cash and a two-valued random investment. If U and V are each strictly concave on the same one of the sets $(0, Z_0]$ or $[Z_0, \infty)$, then the optimal strategies are unique and U and V therefore have distinct optimal strategies.*

Proof. We proceed as in the proof of Theorem 1 until we obtain equiation (2.1).

Note that f is concave and that if U is strictly concave on either $(0, 1]$ or $[1, \infty)$ then f is strictly concave. Now $f(w)$ is a continuous function defined on the closed bounded set $\{w\colon 0 \leqslant w \leqslant 1\}$ hence f has an absolute maximum. Let w^* be a point where f attains its maximum. It follows from the continuity of f that the set of all such w^* is closed.

From the concavity of f, the set of points w^* where f attains its maximum is also convex, hence it is a closed interval in $[0, 1]$. If f is strictly concave the maximum is unique.

For any w^* with $0 < w^* < 1$, f is a maximum if and only if $f'_-(w^*) \geqslant 0 \geqslant f'_+(w^*)$. A maximum occurs at $w^* = 0$ if and only if $f'_+(0) \leqslant 0$. A maximum occurs at $w^* = 1$ if and only if $f'_-(1) \geqslant 0$. If the maxima occur on an interval $[a, b]$ with $0 \leqslant a < b \leqslant 1$, then $f'_-(a) \geqslant 0$ and $f'_+(a) = 0$, $f'_-(b) = 0$ and $f'_+(b) \leqslant 0$, and $f'(w^*)$ exists and is zero for $a < w^* < b$.

Equation (2.1) yields

$$f'_-(w) = apU'_-(1 + aw) - bqU'_-(1 - bw) \geqslant \tag{2.2}$$
$$\geqslant apU'_+(1 + aw) - bqU'_+(1 - bw) = f'_+(w) .$$

Since $U'_-(t)$ and $U'_+(t)$ are non-increasing as t increases, it follows from equation (2.2) that $f'_-(w)$ and $f'_+(w)$ are non-increasing as w increases.

Let c be such that $0 < c < b$ and $U'(1 - c)$ and $V'(1 - c)$ are defined. This is possible because U' and V' are both defined except at countably many points hence there are uncountably many points in $(0, 1)$ where both U' and V' exist. With a and b already given, choose $w = \frac{c}{b}$. Consider now the case where $U'_-\left(1 + \frac{ac}{b}\right) > 0$. Then we may choose p with $0 < p < 1$ in equation (2.2) so that $f'_-\left(\frac{c}{b}\right) = 0$. This means $f'_+\left(\frac{c}{b}\right) \leqslant 0$ and since $w = \frac{c}{b}$ is not an endpoint of $[0, 1]$ this means f attains its maximum at $\frac{c}{b}$, thus $\frac{c}{b}$ is optimal for U in the given investment setting.

Since U and V have the same optimal strategies, $w = \frac{c}{b}$ is optimal for V hence V attains its maximum there so for $w = \frac{c}{b}$, $g'_-(w) = apV'_-(1 + aw) - bqV'_-(1 - bw) \geqslant 0$ and $apV'_+(1 + aw) -$

$-bqV'_+(1-bw) = g'_+(w) \leqslant 0$. Note that $g'_-\left(\frac{c}{b}\right) \geqslant 0$ and the fact $V'_-(1-c) > 0$ implies that $V'_-\left(1+\frac{ac}{b}\right) > 0$. We may show similarly that if $V'_-\left(1+\frac{ac}{b}\right) > 0$ then $U'_-\left(1+\frac{ac}{b}\right) > 0$. Since a is chosen independently of b and c this means that for each $t > 1$, $U'_-(t) > 0$ if and only if $V'_-(t) > 0$. But this is readily shown to be equivalent to the statement that $\{t\colon U(t) = \sup U(t)\} = \{t\colon V(t) = \sup V(t)\}$, i.e. that if either U or V become horizontal for $t \geqslant e > 1$ then they both become horizontal for $t \geqslant e > 1$. For $t > e$, we have of course $U'(t) = V'(t) = 0$. For $t < e$, the argument continues as follows.

From $f'_-(w) = 0$, $apU'_-\left(1+\frac{ac}{b}\right) = bqU'(1-c)$, noting that $U'_-(1-c) = U'(1-c)$. Thus $\dfrac{U'_-\left(1+\frac{ac}{b}\right)}{U'(1-c)} = \dfrac{bq}{ap}$. From $g'_-(w) \geqslant 0$, it follows similarly that $\dfrac{V'_-\left(1+\frac{ac}{b}\right)}{V'(1-c)} \geqslant \dfrac{bq}{ap}$. Letting $\alpha = \dfrac{V'(1-c)}{U'(1-c)}$ yields

$V'_-\left(1+\frac{ac}{b}\right) \geqslant \alpha U'_-\left(1+\frac{ac}{b}\right)$. Since the choices of b and c were independent of that a, the result holds for all $a > 0$, therefore $V'_-(t) \geqslant \alpha U'_-(t)$ for all $t > 1$.

A similar argument shows that $V'_+(t) \leqslant \alpha U'_+(t)$ for all $t > 1$. Thus, except for at most countably many points, $V'(t) = \alpha U'(t)$ for $t > 1$. Now U and V are readily shown to be absolutely continuous on any closed subinterval of $(1, \infty)$, as a consequence of the fact they are continuous, concave, and non-decreasing, thus $V - \alpha U$ is absolutely continuous. The absolute continuity of $U - \alpha V$ and the fact that $(V - \alpha U)' = 0$ almost everywhere implies that $V - \alpha U = \beta$, a constant (Goffman [2], p. 242, Prop. 12).

A similar argument shows that $V(t) = \alpha U(t) + \gamma$ for $t < 1$. The role of 2 in the proof of Theorem 1 is played by any number c such that $1 < c < e$ and $U'(c)$ and $V'(c)$ are both defined. One then shows as in the proof of Theorem 1 that $V(t) = \alpha U(t) + \beta$ for $0 < t < \infty$. We have established the contrapositive assertion as in the proof of Theorem 1. This completes the proof.

The hypothesis that either U or V (hence both, from the proof) is strictly increasing for a positive distance to the right of 1 is required. If instead U and V are merely concave and non-decreasing, the conclusion of Theorem 3 need not hold. For instance, let $U(t) = V(t) = 0$ if $t \geqslant d$, where $0 < d \leqslant 1$. Let $U(t)$ and $V(t)$ each be extended to $(0, d)$ so that they are continuous, concave, and strictly increasing on $(0, d)$. Then all such utilities have the same optimal strategies, yet many pairs are inequivalent.

To obtain an inequivalent pair, let $U(t) = t - d$ if $0 < t < d$ and let $V(t) = -(t-d)^2$. If for some constants α and β, $V(t) = \alpha U(t) + \beta$ then $V'(t) = \alpha U'(t)$. But $V'(t) = -2(t-d) \not\equiv \alpha = \alpha U'(t)$.

To see that all such utilities U have the same optimal strategies, note that $W = (w_1, \ldots, w_k)$ is optimal for the investment setting $(X_1, \ldots$ $\ldots, X_k)$ if and only if $\mathrm{P}\left(\sum_i w_i X_i \geqslant d\right) = 1$, in which case $\mathrm{E}U(Z_1(W)) =$ $= 0$. If instead $\mathrm{P}\left(\sum_i w_i X_i < d\right) > 0$ then for some $\epsilon > 0$, $\mathrm{P}\left(\sum_i w_i X_i \leqslant\right.$ $\left.\leqslant d - \epsilon\right) = \delta > 0$. Then $\mathrm{E}U(Z_1(W)) \leqslant \delta U(d-\epsilon) < 0$ so W is not optimal.

3. OTHER SEPARATING FAMILIES

We next establish the conclusion of Theorem 1 using investment settings with n points in their range. We determine the effect of varying the payoffs $(x_1, \ldots, x_n)$ and their probabilities $(p_1, \ldots, p_n)$ separately. One surprising conclusion (part (b)) can be stated in terms of an example. Suppose X consists of betting on a wheel of fortune divided into red, white and blue sectors, with payoffs of $\frac{1}{2}, \frac{3}{2}$, and $\frac{3}{4}$ respectively. Then if U and V are inequivalent on $\left[\frac{1}{2}, \frac{3}{2}\right]$ the areas of the sectors may be chosen so U and V have distinct optimal strategies. But if the wheel is divided into just red and blue sectors, with payoffs of $\frac{1}{2}$ and $\frac{3}{2}$, then there are two inequivalent utilities on $\left[\frac{1}{2}, \frac{3}{2}\right]$ which have the same optimal strategies for every choice of areas for the two sectors.

Theorem 4. *Suppose* U *and* V *are increasing strictly concave utilities on* $(0, \infty)$. *Let* X *be a random variable with outcomes* $0 \leqslant x_1 \leqslant \leqslant x_2 \leqslant \ldots \leqslant x_n$ *with* $x_1 < 1$ *and* $x_n > 1$. *Suppose* $\mathsf{P}(X = x_i) = p_i > 0$, $\sum_{i=1}^{n} p_i = 1$.

(a) *Let* n *and the* p_i *be given. If* U *and* V *have the same optimal strategies for each* X *(i.e.* $x_1, \ldots, x_n$ *vary), then* U *and* V *are equivalent.*

(b) *Let* n *and the* x_i *be given. Suppose* U' *and* V' *exist and are continuous at* 1. *If* U *and* V *have the same optimal strategies for each* X *(i.e.* $p_1, \ldots, p_n$ *vary) and at least three* x_i *are unequal to* 1, *then* U *and* V *are equivalent on* $[Z_0 x_1, Z_0 x_n]$. *If exactly two of the* x_i*'s are unequal to one, there are utilities* U *and* V *which are not equivalent on* $[Z_0 x_1, Z_0 x_n]$, *but which have the same optimal strategy for each* X.

Proof. Assume $Z_0 = 1$. Let $R = X - 1$ and $r_i = x_i - 1$. Then investing w in X gives an expected return (with respect to U) of $\mathsf{E}(U(wR + 1)) = \sum_{i=1}^{n} p_i U(wr_i + 1)$. Each function $U(wr_i + 1)$ is differentiable except at a countable set C_i of points, so except for w in the countable set $C_1 \cup \ldots \cup C_n$ the expectation $\mathsf{E}(U(wR + 1))$ is differentiable at w with $\frac{d\mathsf{E}(U(wR+1)}{dw} = \sum_{i=1}^{n} p_i r_i U'(wr_i + 1)$. Similarly each function $V(wr_i + 1)$ is differentiable except at a countable set. Thus, except at a countable set D of points in $[0, \infty]$ both $\mathsf{E}(U(wR + 1))$ and $\mathsf{E}(V(wR + 1))$ are differentiable functions of w. They are also strictly concave functions of w.

For part (a) let $p_1, \ldots, p_n$ be given and choose w_0 in $(0, 1) - D$. Consider the vectors $\alpha = (U'(w_0 r_1 + 1), \ldots, U'(w_0 r_n + 1))$ and $\beta = (V'(w_0 r_1 + 1), \ldots, V'(w_0 r_n + 1))$. Suppose that the non-zero vector

$\gamma = (c_1, c_2, \ldots, c_n)$ is perpendicular to α, i.e., the inner product $(\alpha, \gamma) = 0$. Choose $r_i = \dfrac{c_i}{p_i\left\{\epsilon \sum_{j=1}^{n} |c_j|\right\}}$ with $\epsilon = \max \dfrac{1}{p_i}$, $1 \leqslant i \leqslant n$.

Since each component of α is positive, some $c_i > 0$ and some $c_j < 0$, hence some $x_i > 1$ and some $x_j < 1$. Also $r_i + 1 = x_i > 0$. Then $\dfrac{d\mathsf{E}(U(wR+1))}{dw}\Big|_{w=w_0} = \sum_{i=1}^{n} p_i r_i U'(w_0 r_i + 1) = 0$ and $\mathsf{E}(U(wR+1))$ has a maximum at w_0. By hypothesis $\mathsf{E}(V(wR+1))$ has a maximum at w_0 and, since it is differentiable there, $\dfrac{d\mathsf{E}(V(wR+1))}{dw}\Big|_{w=w_0} =$ $= \sum_{i=1}^{n} p_i r_i V'(w_0 r_i + 1) = 0$ i.e., $(\beta, \gamma) = 0$. Hence the set of vectors perpendicular to α is also perpendicular to β which implies that $\beta = a\alpha$. Since the components of α and β are non-negative, $a \geqslant 0$. Equating components

$$U'(w_0 r_i + 1) = aV'(w_0 r_i + 1) \tag{3.1}$$

where a is a non-negative function of $r_1, \ldots, r_n$ and w_0. Since U and V are strictly concave there is a point t_0 not in D, $w_0 < t_0 < 1$, with $V'(t_0) > 0$ and $U'(t_0) > 0$. Choose r_1 so that $w_0 r_1 + 1 = t_0$, choose $r_2 > 0$ with $t = w_0 r_2 + 1$ not in D, and choose $r_3 < \ldots < r_n$ so they are not in D. Then $U'(t_0) = a(r_1, \ldots, r_n, w_0)V'(t_0)$ and $U'(w_2 r_2 + 1) = a(r_1, \ldots, r_n, w_0)V'(w_0 r_2 + 1)$. Thus $a = \dfrac{U'(t_0)}{V'(t_0)} > 0$ is constant. So $V'(t) = aU'(t)$ for any $t > 1$ not in D. Since V and U are absolutely continuous on any closed subinterval, $V(t) = aU(t) + b$ for all $t > 1$. A similar argument shows that $V(t) = cU(t) + d$ for $t < 1$ with $c = \dfrac{V'(t_0)}{U'(t_0)} = a$. The equivalence of U and V now follows (as in the proof of Theorem 1) from their continuity.

For part (b) suppose that the x_i are given, with $0 < x_1 < x_2 < \ldots$ $\ldots < x_p \leqslant 1 \leqslant x_{p+1} < \ldots < x_n$. We proceed as before, but now consider, for $0 < w_0 < 1$ and U, V differentiable at $w_0 r_j + 1$, $1 \leqslant j \leqslant n$, the vectors $\tilde{\alpha} = (r_1 U'(w_0 r_1 + 1), \ldots, r_n U'(w_0 r_n + 1))$ and $\tilde{\beta} = (r_1 V'(w_0 r_1 + 1), \ldots, r_n V'(w_0 r_n + 1))$. Since $\tilde{\alpha}$ has both positive and negative components there is a vector $(d_1, d_2, \ldots, d_n)$ perpendicular to $\tilde{\alpha}$ with each $d_i > 1$. Choose $p_i = \dfrac{d_i}{\sum_{i=1}^{n} d_j}$, thus $p_i > 0$ and $\sum_i p_i = 1$, and define X by $\mathsf{P}(X = x_i) = p_i$. Thus

$$0 = (\tilde{\alpha}, (d_1, d_2, \ldots, d_n)) = \sum_{i=1}^{n} \frac{d_i}{\sum_{i=1}^{n} d_j} r_i U'(w_0 r_i + 1) =$$

$$= \frac{d\mathsf{E}(U(wR + 1))}{dw}\bigg|_{w=w_0}$$

By hypothesis $\dfrac{d\mathsf{E}(U(wR + 1))}{dw}\bigg|_{w=w_0} = \dfrac{(\tilde{\beta}, (d_1, \ldots, d_n))}{\sum_i d_j} = 0$ so

$(\tilde{\beta}, (d_1, \ldots, d_n)) = 0$. Suppose that $\tilde{\gamma} = (e_1, e_2, \ldots, e_n)$ is perpendicular to $\tilde{\alpha}$. Let $d_0 > \max|e_i|$ and choose $p_i = \dfrac{e_i + d_0 d_i}{\sum_{i=1}^{n} e_j + d_0 d_j}$. Note that

$p_i > 0$ and $\sum_{i=1}^{n} p_i = 1$. Thus

$$\frac{d\mathsf{E}(U(wR + 1))}{dw}\bigg|_{w=w_0} = \sum_{i=1}^{n} p_i[r_i U'(w_0 r_i + 1)]$$

and letting $D = \sum_{j=1}^{n} (e_j + d_0 d_j)$ gives

$$\frac{1}{D} \sum_{j=1}^{n} e_i r_i U'(w_0 r_i + 1) + \frac{1}{D} d_0 \sum_i d_i r_i U'(w_0 r_i + 1) =$$

$$= \frac{1}{D} (\tilde{\alpha}, \tilde{\gamma}) + \frac{1}{D} d_0 (\tilde{\alpha}, (d_1, \ldots, d_n)) = 0\,.$$

Hence, $\left.\frac{d\mathrm{E}(V(wR+1))}{dw}\right|_{w=w_0} = 0$. This yields $(\widetilde{\beta}|\widetilde{\gamma}) + d_0(\widetilde{\beta},(d_1,\ldots$ $\ldots,d_n)) = (\widetilde{\beta},\widetilde{\gamma}) = 0$. Thus

$$U'(w_0 r_i + 1) = a(p_1,p_2,\ldots,p_n,w_0)V'(w_0 r_i + 1) \quad (1 \leqslant i \leqslant n)\,. \tag{3.2}$$

For w in $(0, 1)$ with U, V differentiable at $wr_i + 1$ $(1 \leqslant i \leqslant n)$ we have (3.2) with $w_0 = w$. Consider the quotient

$$h(w) = a(p_1,\ldots,p_n,w) = \frac{U'(wr_i+1)}{V'(wr_i+1)} \qquad (1 \leqslant i \leqslant n)\,. \tag{3.3}$$

First look at the case where at least three x_i's are unequal to one. Suppose that $x_1 < 1 < x_{n-1} < x_n$; the proof where two or more points fall to the left of 1 is similar.

Let $\varphi_i(w) = wr_i + 1$. The countable collection of functions $\{U, V, U\circ\varphi_n^{-1}, V\circ\varphi_n^{-1}, U\circ\varphi_n^{-1}\circ\varphi_{n-1}\circ\varphi_n^{-1}, V\circ\varphi_n^{-1}\circ\varphi_{n-1}\circ\varphi_n^{-1},$ $U\circ\varphi_n^{-1}\circ\varphi_{n-1}\circ\varphi_n^{-1}\circ\varphi_{n-1}\circ\varphi_n^{-1}, V\circ\varphi_n^{-1}\circ\varphi_{n-1}\circ\varphi_n^{-1}\circ\varphi_{n-1}\circ$ $\circ\varphi_n^{-1},\ldots\}$ is simultaneously differentiable except at a countable set of points D_0 in $(0, 1)$.

Choose t in $(1, x_n) - D_0$ and write $t = w_1 r_n + 1$, so $w_1 =$ $= \varphi_n^{-1}(t)$, and set $t_1 = w_1 r_{n-1} + 1 = \varphi_{n-1}[\varphi_n^{-1}(t)]$. We can also write $t_1 = w_2 r_n + 1$; $w_2 = \varphi_n^{-1}(t_1)$. By (3.3) we have $h(w_1) = h(w_2)$, since U and V are differentiable at w_1 and w_2. Note that $w_2 < w_1$, in fact, $w_2 = \lambda w_1$ with $\lambda = \frac{r_{n-1}}{r_n}$. Setting $t_2 = w_2 r_{n-1} + 1 = \varphi_{n-1}(w_2)$, $t_2 = w_3 r_n + 1$ and $w_3 = \varphi_n^{-1}(t_2)$. Then $h(w_2) = h(w_3)$ since U and V are differentiable at $w_2 = \varphi_n^{-1}\circ\varphi_{n-1}\circ\varphi_n^{-1}(t)$ and at $w_3 = \varphi_n^{-1}\circ$ $\circ\varphi_{n-1}\circ\varphi_n^{-1}\circ\varphi_{n-1}\circ\varphi_n^{-1}(t)$. Continuing inductively $t_j = w_j r_{n-1} + 1 =$ $= w_{j+1} r_n + 1$ and $w_{j+1} = \lambda w_j$. Iterating this equation $w_{j+1} = \lambda^j w_1 \to 0$ as $j \to \infty$, thus $h(w_1) = \ldots = h(w_n) \to h(1)$ since U' and V' are continuous at 1. Hence the equation $\frac{U'(t)}{V'(t)} = h(1)$ holds except for countably many t in $(1, x_n)$ and thus, since U and V are absolutely continuous on any closed subinterval, $U(t) = h(1)V(t) + c$ for all t in $[1, x_n)$.

Let t belong to $(x_1, 1)$ with U and V differentiable at $wr_j + 1$, $1 \leqslant j \leqslant n$. Then $t = wr_1 + 1$ and from equation (3.3) $\frac{U'(t)}{V'(t)} = \frac{U'(wr_n + 1)}{V'(wr_n + 1)} = h(1)$. Since U and V are absolutely continuous on closed subintervals of $(x_1, 1)$, $U(t) = h(1)V(t) + d$. The continuity of U and V at 1 implies that $c = d$ and thus U and V are equivalent on $[x_1, x_n]$.

To complete the proof we must consider the case where there are only two x_i's distinct from one, say, $0 < x_1 < 1 < x_2$. Let g_0 be any non-constant positive function on $[1, x_2]$ with a continuous derivative which is zero at 1. Define g on $[x_1, 1]$ by $g(wr_1 + 1) = g(wr_2 + 1)$ for $0 \leqslant w \leqslant 1$. Choose a so that $\max_{x_1 \leqslant t \leqslant x_2} |g'(t)| - a \cdot \min_{x_1 \leqslant t \leqslant x_2} |g(t)| < 0$ and define $U(t) = \int_0^t e^{-at} dt = \frac{1 - e^{-at}}{a}$ and $V(t) = \int_0^t e^{-at} g(t) dt$. Because $U''(t) = -ae^{-at} < 0$ and $V''(t) = e^{-at}(g'(t) - ag(t)) < 0$, U and V are strictly concave. Also $U'(t) = e^{-at}$ and $V'(t) = e^{-at}g(t)$ are positive so U and V are strictly increasing. Clearly U and V are not equivalent on $[x_1, x_2]$.

For these two functions U and V and $0 < w < 1$,

$$\frac{d\mathrm{E}(V(wR + 1))}{dw} = r_1 p_1 V'(wr_1 + 1) + r_2 p_2 V'(wr_2 + 1) =$$

$$= r_1 p_1 g(wr_1 + 1) U'(wr_1 + 1) + r_2 p_2 g(wr_2 + 1) U'(wr_2 + 1) =$$

$$= g(wr_1 + 1) \frac{d\mathrm{E}(U(wR + 1))}{dw} .$$

Hence $\frac{d\mathrm{E}(U(wR + 1))}{dw} = 0$ if and only if $\frac{d\mathrm{E}(V(wR + 1))}{dw} = 0$, and so w_0, $(0 < w_0 < 1)$, is an optimal strategy for U (with respect to X) if and only if it is an optimal strategy for V. If the derivative $\frac{d\mathrm{E}(U(wR + 1))}{dw}$ is never 0, the equation above shows that it has the same sign as $\frac{d\mathrm{E}(V(wR + 1))}{dw}$; so 0 (or 1) is an optimal strategy for U if and only if it is an optimal strategy for V.

We have seen that U and V are two utilities on $[x_1, x_2]$ which are not equivalent, but which have the same optimal strategies for all random variables with outcomes x_1 and x_2.

Remark. Our proofs may be modified readily to prove the theorems when U and V are defined on the *closed* interval $[0, \infty)$ and also when the interval is (c, ∞) or $[c, \infty)$, with $c < Z_0$. Presumably $c > 0$. (Alternately, the $[c, \infty)$ result implies the (c, ∞) result: if $U(x) = V(x)$ on every interval $[c + \epsilon, \infty)$ $(0 < \epsilon < Z_0 - c)$ then $U(x) = V(x)$ on (c, ∞).)

4. QUESTIONS FOR FURTHER INVESTIGATION

Friedman – Savage [1] and Markowitz [4] have shown that utilities which are not everywhere concave are of interest. This leads us to a question which we have not been able to answer yet:

Is the class of utilities which are continuous and strictly increasing (and differentiable everywhere, bounded, and even strictly positive derivative, if you like) distinguished by their optimal strategies?

In the real world factors such as human error, the discreteness of assets and monetary units, etc. make it in general not possible to choose the optimal allocation $W^* = (w_1^*, \ldots, w_k^*)$. The continuity of the utility in conjunction with boundedness of the *attainable* utilities implies that "sufficiently small" deviations from W^* will ensure that the realized utility is "close" to the optimum.

One feels as well that in the real world, the exact values of the utility function should not be critical. In other words, if two utility functions are somehow "close," the consequences of choosing one rather than the other should be "close."

What should it mean for two utility functions to be "close?" First, observe that we must define closeness not for functions, but for equivalence classes of functions. Let U be a utility. The equivalence class of U, written $[U]$, is the set $\{V\colon V = \alpha U + \beta,\ \alpha > 0\}$. For the class β of bounded utilities, i.e., $M(U) \equiv \sup U(t) < \infty$, $m(U) \equiv \inf U(t) > -\infty$, we suggest that each $[U]$ equivalence class be represented by $\widetilde{U} = \frac{U - M}{M - m} + 1$. Note that $M(\widetilde{U}) = 1$ and $m(\widetilde{U}) = 0$. Then the "closeness" of U and V, i.e., of $[U]$ and $[V]$, is defined to be $\sup[\widetilde{U}(t) - \widetilde{V}(t)]$ and written either $d(U, V)$ or $d([U],[V])$ or $d(\widetilde{U}, \widetilde{V})$.

We now show that U and V can be "close" yet the optimal strategies for U and V need not be. For $n \geq 2$, let $\widetilde{U}_n$ and $\widetilde{V}_n$ be defined as follows:

$$\widetilde{U}_n(t) = \frac{2nt}{n+1} - 1 \text{ if } 0 \leq t \leq 1 + \frac{1}{n} \text{ and } 1 \text{ if } t > 1 + \frac{1}{n};$$

$$\widetilde{V}_n(t) = \frac{2n-1}{n+1} t - 1 \text{ if } 0 \leq t \leq 1 + \frac{1}{n},$$

$$\frac{t + n - 3}{n - 1} \text{ if } 1 + \frac{1}{n} < t \leq 2, \text{ and } 1 \text{ if } t > 2.$$

Then $d(\widetilde{U}_n, \widetilde{V}_n) = \frac{1}{n}$. Now choose an investment setting consisting only of cash and the security X, where $\mathsf{P}(X = 1 - \epsilon) = q$, $\mathsf{P}(X = 1 + a) = p$, $\frac{1}{n} < a < 1$, and $0 < \epsilon, p, q < 1$. Assume $Z_0 = 1$. A calculation shows that if $ap > q\epsilon \frac{(2n-1)(n-1)}{n+1}$, then the unique optimal strategy for U_n is $w^* = \frac{1}{an}$ and for V_n the unique optimal strategy is $w^* = 1$.

Thus for any $\delta > 0$ we can construct sequences $\widetilde{U}_n$ and $\widetilde{V}_n$ such that $d(\widetilde{U}_n, \widetilde{V}_n) \to 0$ as $n \to \infty$ and $|w^*(\widetilde{V}_n) - w^*(\widetilde{U}_n)| \geq 1 - \delta$, where $w^*(\widetilde{U})$ means an optimal strategy for $\widetilde{U}$.

Even though a small "error" in the utility function can lead to a large change in optimal strategy, it can only lead to a small change in consequences, in the following sense. (We use the abbreviation $U(W)$ for $\mathsf{E}U\left(Z_0 \sum_i w_i X_i\right)$. Thus for each W, $U(W)$ is a number and $U\left(Z_0 \sum_i w_i X_i\right)$ is a random variable.)

Lemma. *If* $d(\tilde{U}, \tilde{V})$ *is "small," then* $\tilde{U}(W^*(\tilde{V})) \doteq \tilde{U}(W^*(\tilde{U}))$ *and* $\tilde{V}(W^*(\tilde{U})) \doteq \tilde{V}(W^*(\tilde{V}))$, *i.e., if* U *and* V *are "close," an optimal strategy for one is "nearly optimal" for the other.*

Proof. Let $d(\tilde{U}, \tilde{V}) \leqslant \epsilon$ so $\tilde{V}(t) + \epsilon \geqslant \tilde{U}(t)$. Then for any allocation W, $\tilde{V}\left(Z_0 \sum_i w_i X_i\right) + \epsilon \geqslant \tilde{U}\left(Z_0 \sum_i w_i X_i\right)$ and $\mathrm{E}\left(\tilde{V}\left(Z_0 \sum_i w_i X_i\right) + \right.$ $\left. + \epsilon\right) = \mathrm{E}\left(\tilde{V}\left(Z_0 \sum_i w_i X_i\right)\right) + \epsilon \geqslant \mathrm{E}\tilde{U}\left(Z_0 \sum_i w_i X_i\right)$, or $\tilde{V}(W) + \epsilon \geqslant \tilde{U}(W)$. Interchanging $\tilde{U}$ and $\tilde{V}$ in the argument yields $\tilde{U}(W) + \epsilon \geqslant \tilde{V}(W)$ so $|\tilde{U}(W) - \tilde{V}(W)| \leqslant \epsilon$. The choices for W of $W^*(\tilde{U})$ and $W^*(\tilde{V})$ yield the conclusion of the lemma.

The lemma and the example show us what may happen if we replace a U by a nearby V which may have more desirable properties, such as differentiability (of various orders), strictly increasing, etc.: The optimal strategies may change drastically but the maximum utility over all strategies changes only slightly.

Note added in proof: The authors have since extended the central result of the paper, Theorem 3, as follows.

Theorem. *Let* U *and* V *be continuous non-decreasing functions defined on an arbitrary interval* I *of the real line. Then if* U *and* V *are inequivalent, there is a one-period two security investment setting such that* U *and* V *have distinct optimal strategies if either* (a) U *and* V *are in the class of all functions which are either concave or convex, or* (b) U *and* V *are in the class of all functions with a second derivative which exists and is continuous, except perhaps for a set of isolated points.*

Thus the Theorem includes the utility functions generally encountered.

REFERENCES

[1] M. Friedman – L.J. Savage, The Utility Analysis of Choices Involving Risk, *Journal of Political Economy,* 56 (1948), 279-304.

[2] C. Goffman, *Real Functions,* Holt, Rinehart and Winston Inc., New York, 1953.

[3] G. Hardy – J. Littlewood – G. Polya, *Inequalities,* Cambridge University Press, 1959.

[4] H. Markowitz, The Utility of Wealth, *Journal of Political Economy,* (1952), 151-158.

[5] J. Mossin, Optimal Multiperiod Portfolio Policies, *Journal of Business,* (April 1968).

[6] P.A. Samuelson, The 'Fallacy' of Maximizing the Geometric Mean in Long Sequences of Investing or Gambling, unpublished preliminary preprint, 1971.

[7] E.O. Thorp, Optimal Gambling Systems for Favorable Games, *Review of the International Statistical Institute,* 37 (1969), 273-293.

COLLOQUIA MATHEMATICA SOCIETATIS JÁNOS BOLYAI
9. EUROPEAN MEETING OF STATISTICIANS, BUDAPEST (HUNGARY), 1972.

BAYESIAN SINGLE SAMPLING ACCEPTANCE PLANS FOR LIFE-TESTING

P. THYREGOD

0. INTRODUCTION

It is a characteristic feature of many life-testing experiments that testing time is rather costly and therefore tests are often stopped before all items under test have failed.

The statistical properties of truncated life-testing plans for exponentially distributed lifetimes have been studied by Epstein and Sobel [1] and Epstein [2], but not very much is known about the optimal design of life-testing plans, i.e., the economical choice of sample size and truncation number in a given testing situation.

The determination of the optimal sample size has been discussed for a wide class of acceptance sampling problems by Guthrie and Johns [3], Hald [4] and Thyregod [7], but none of these approaches considers the special problems arising when testing time is costly. In the present paper we shall give some results useful for the construction of Bayesian sampling schemes for exponentially distributed random variables when the cost of testing is proportional to the time under test.

We shall consider a situation where lots of size N are presented for inspection. The inspector takes out a sample of size n and puts all n items under test. Testing is continued until r items have failed, at which time the inspector decides whether to accept or reject the remainder of the lot. We derive an algorithm which, for given cost and prior distribution, produces a table of the optimal sample size, truncation number and acceptance number as a function of the lot size. Finally we sketch the asymptotic solution, valid for large lot sizes.

A detailed proof and a discussion of the results will be published elsewhere.

1. THE PROBABILITY MODEL

We shall assume that lots consisting of N items are presented for sampling inspection. Moreover we shall assume that the item characteristics $X_1, X_2, \ldots, X_N$ in the lot for all lot sizes $N = 1, 2, \ldots$ may be described as exchangeable random variables.

It is then well known that the joint density of the item characteristics, say lifetimes, in a lot of size N may be represented by

$$(1) \qquad f_N(x_1, \ldots, x_N) = \int \prod_{i=1}^{N} f(x_i \mid \theta) dW(\theta) ,$$

where $f(x \mid \theta)$ is a probability density and $W(\theta)$ is a probability measure.

Throughout the paper we shall assume that f is an exponential density

$$(2) \qquad f(x \mid \theta) = \frac{1}{\theta} e^{-\frac{x}{\theta}} .$$

A simple interpretation of the model defined by (1) and (2) is as follows: All items in each lot are produced under identical conditions according to the density (2). The parameter θ, the process quality, varies from lot to lot according to the known distribution $W(\theta)$.

2. THE COST FUNCTIONS

We shall assume that the costs of accepting, rejecting or sampling an item with lifetime X are given by $k_a^*(X)$, $k_r^*(X)$ and $k_s^*(X)$ respectively. The notation for the conditional expected cost given θ, and the overall expected cost then follows from Table 1.

Table 1.

The cost functions

Action	Cost per item	Cond. exp. cost	Marg. cost
Accept	$k_a^*(X)$	$k_a(\theta) = \int k_a^*(x) f(x \mid \theta) dx$	$\kappa_a = \int k_a(\theta) dW(\theta)$
Reject	$k_r^*(X)$	$k_r(\theta) = \int k_r^*(x) f(x \mid \theta) dx$	$\kappa_r = \int k_r(\theta) dW(\theta)$
Sample	$k_s^*(X)$	$k_s(\theta) = \int k_s^*(x) f(x \mid \theta) dx$	$\kappa_s = \int k_s(\theta) dW(\theta)$

Furthermore we shall assume that the cost of testing the sample is proportional to the time the sample is under test. We shall assume that the testing cost is k_t per time-unit as long as at least one item is under test. The marginal expected testing cost is then $\kappa_t = k_t \int \theta dW(\theta)$.

It will be assumed that the cost difference $k_r^*(X) - k_a^*(X)$ has only one change of sign, from negative to positive values of $k_r^* - k_a^*$. It then follows that the conditional expected cost difference, $l(\theta)$, defined by

$$l(\theta) = k_r(\theta) - k_a(\theta)$$

only has one change of sign, from negative to positive values of l. The break-even value θ_0 is then defined as the solution to $l(\theta) = 0$.

Example 1. As an example we consider the case where the distribution of lot-quality is a gamma-distribution,

$$dW(\theta) = \frac{1}{\theta T(\nu)}\left(\frac{\tau}{\theta}\right)^{\nu} e^{-\frac{\tau}{\theta}} d\theta$$

and the cost-unit has been chosen such that the cost of rejecting an item is 1. Suppose furthermore that the cost of accepting an item is $k > 1$ if the lifetime is less than 1 time-unit and 0 otherwise. Assume finally that the sampling cost equals the cost of rejection, i.e., that $k_s^*(X) = 1$ and that the testing cost is k_t per time-unit as long as an item is under test. The conditional expected costs and the marginal costs may then be found from Table 2.

Table 2.

An example of the cost-functions

Action	Cost per item	Cond. exp. cost	Marg. cost
Accept	$\begin{cases} k & \text{if } X < 1 \\ 0 & \text{if } X > 1 \end{cases}$	$k\left(1 - e^{-\frac{1}{\theta}}\right)$	$k\left[1 - \left(\frac{\tau}{\tau + 1}\right)^{\nu}\right]$
Reject	1	1	1
Sample	1	1	1

We obtain

$$l(\theta) = 1 - k\left(1 - e^{-\frac{1}{\theta}}\right)$$

and therefore the break-even quality is

$$\theta_0 = \frac{1}{\ln \frac{k}{k-1}} .$$

Finally we find the expected cost of testing an item until failure,

$$\kappa_t = \left(\frac{\tau}{\nu - 1}\right) k_t .$$

3. THE SAMPLING PLAN AND THE TOTAL COST

We shall assume that a sample of size n is taken from the lot and that all n items simultaneously are put under test and observed until the r'th failure occurs, at which time testing is terminated and a final decision is made regarding the remaining $N - n$ items in the lot.

The marginal density of the lifetimes in the sample is

$$f_n(x_1, \ldots, x_n) = \int \prod_{i=1}^{n} f(x_i \mid \theta) dW(\theta) ,$$

but we do not observe all the lifetimes, only the r smallest lifetimes $X_{(1)}, X_{(2)}, \ldots, X_{(r)}$ are recorded.

It can be shown that the optimal decision rule is of the form: Accept if $S_r > c$, where

$$S_r = X_{(1)} + X_{(2)} + \ldots + X_{(r)} + (n - r)X_{(r)}$$

denotes the total time under test.

The overall expected cost under this decision rule is

$$K(n, r, c, N) = n\kappa_s + \kappa_t \{n^{-1} + (n-1)^{-1} + \ldots$$
$$\ldots + (n - r + 1)^{-1}\} + (N - n)k(r, c)$$

with $k(r, c) = \kappa_r - \int l(\theta) P(r, c, \theta) dW(\theta)$, and our aim is thus to determine values of n, r and c such that K is minimized. Furthermore we want to tabulate the optimal values of (n, r, c) as functions of the lot size N.

We find

Theorem.

(i) $\min_c K(n, r, c, N)$ *is obtained for* $c = c_0(r)$, *where* $c_0(r)$ *is the solution to*

$$\int l(\theta)\theta^{-r} e^{-\frac{c}{\theta}} dW(\theta) = 0 .$$

$c_0(r)$ *is increasing and* $\min_c k(r, c)$ *is decreasing functions of* r.

Furthermore

(ii) $\min_n K(n, r, c, N)$ *is obtained for* $n = n_0(r, c)$, *where* $n_0(r, c) =$
$= \frac{1}{2}r + \sqrt{\frac{r^2}{4} + \frac{r\kappa_t}{\kappa_s - k(r, c)}}$. $n_0(r, c_0(r))$ *is increasing.*

(iii) *The optimal value of the truncation number,* $r_0(N)$ *is an increasing function of the lot-size.*

(iv) *The graph of* $K_0(N) = \min_{n,r,c} K(n, r, c, N)$ *is a concave polygonal line with slopes* $k(r, c_0(r))$ *for* $r = r_0(N)$.

If follows that it is rather easy to determine $K_0(N)$ by successively intersecting the linear functions $K(n_0(r), r, c_0(r), N)$ for $r = 1, 2, \ldots$.

Example 2. Assume that the situation is as in Example 1 with $k_t =$ $= 12$, $k = 1.2$ and $(\tau, \nu) = (1.8, 4)$. We then find $\kappa_a = 0.9950$ such that the cost of accepting without sampling is $K^0(N) = 0.9950\ N$.

For $r = 1$ the optimal sampling plan is given by $c_0(1) = 0.5203$, $n_0(1) = 15$ with $K(15, 1, 0.5203, N) = 0.948 + 0.9648\ N$.

We thus have the first iterate of the optimal cost

$$K^1(N) = \begin{cases} 0.9950\,N & \text{for} \quad N \leqslant 36.1 \\ 0.948 + 0.9648\,N & \text{for} \quad 36.1 < N . \end{cases}$$

For $r = 2$ we find $c_0(2) = 1.074$, $n_0(2) = 19$; $K(19, 2, 1.074, N) =$ $= 1.56 + 0.9588\ N$.

Hence the second iterate of the optimal cost is

$$K^2(N) = \begin{cases} 0.9950\,N & \text{for} \quad N \leqslant 36.1 \\ 0.9482 + 0.9648\,N & \text{for} \quad 36.1 < N \leqslant 61.5 \\ 1.56 \quad + 0.9588\,N & \text{for} \quad 61.5 < N . \end{cases}$$

Continuing this procedure for $r = 3, 4, 5, \ldots$ we find the table of optimum plans given in Table 3.

4. THE ASYMPTOTIC SOLUTION

The asymptotic properties of the optimal plans for large lot sizes may be derived by the method given by Hald [5].

Define

$$\kappa_m = \int \min \{k_a(\theta),\ k_r(\theta)\} dW(\theta)$$

$$\delta_a = \kappa_a - \kappa_m, \quad \delta_r = \kappa_r - \kappa_m, \quad \delta_s = \kappa_s - \kappa_m;$$

$$d(r, c) = k(r, c) - \kappa_m .$$

We shall assume that the prior distribution has a positive density which may be expanded around θ_0.

$$\frac{dW(\theta)}{d\theta} = w_0 \{1 + O(\theta - \theta_0)\}$$

with $w_0 > 0$, and suppose further that

$$l(\theta) = l_0 \left(\frac{\theta - \theta_0}{\theta_0} \right)^{\mu} \{1 + O(\theta - \theta_0)\} .$$

For $r \to \infty$ we have

(3) $$c_0(r) = r\theta_0 + O(1) ,$$

$$d(r, c_0(r)) = \frac{\gamma}{r^{\frac{\mu+1}{2}}} \left(1 + O\left(\frac{1}{r}\right) \right)$$

with

$$\gamma = \frac{2^{\frac{\mu+1}{2}} \Gamma\left(\frac{\mu}{2} + 1\right)}{(\mu + 1)\sqrt{\pi}} l_0 w_0 \theta_0 .$$

TABLE 3.

BAYESIAN SINGLE SAMPLING ACCEPTANCE PLANS FOR LIFE-TESTING, (EPSTEIN-SOBEL).

SAMPLING DISTRIBUTION: EXPONENTIAL WITH MEAN THETA.
PRIOR DISTRIBUTION OF 1/THETA: GAMMA WITH SCALE PARAMETER TAU=1.80, SHAPE PARAMETER NU=4.00.
MEAN OF THETA=0.6000

COST OF SAMPLING PER ITEM=1.0
COST OF OBSERVATION PER TIME UNIT=12.
COST OF ACCEPTING AN ITEM=1.2 IF THE LIFETIME IS LESS THAN 1.0 AND 0 OTHERWISE.
COST OF REJECTING AN ITEM=1.00

BREAK-EVEN QUALITY, THETA=0.5581, ACCEPTABLE PART OF PRIOR DISTRIBUTION: 0.4031
MINIMUM COST PER ITEM: 0.9295
ACCEPTANCE LOSS PER ITEM: 0.0655537, REJECTANCE LOSS PER ITEM: 0.0705, SAMPLING LOSS PER ITEM: 0.0705.

THE OPTIMAL DECISION RULE IS OF THE FORM:
TAKE A SAMPLE OF SIZE N AND OBSERVE THE LIFETIMES UNTIL THE R'TH FAILURE.
ACCEPT THE REMAINDER OF THE LOT IF X1+X2+...+XR+(N−R)XR EXCEEDS C.

LOT SIZE		N	R	C	REGRET	OC-CURVE: VALUE OF THET A GIVING RISE TO ACCEPTANCE PROB. P				
						P=0.99	P=0.95	P=0.50	P=0.10	P=0.05
1–	36	SINGULAR SOLUTION			2.3599					
37–	61	15	1	0.52035	3.3446	51.7746	10.1446	0.7507	0.2260	0.1737
62–	87	19	2	1.0735	4.1124	7.2264	3.0209	0.6396	0.2760	0.2263
88–	114	23	3	1.6281	4.7548	3.7337	1.9911	0.6088	0.3059	0.2586
115–	143	25	4	2.1835	5.3333	2.6523	1.5981	0.5946	0.3268	0.2816
144–	173	28	5	2.7396	5.8481	2.1418	1.3905	0.5865	0.3427	0.2993
174–	205	31	6	3.2960	6.3329	1.8462	1.2614	0.5813	0.3554	0.3135
206–	239	33	7	3.8528	6.7924	1.6534	1.1727	0.5777	0.3658	0.3253
240–	275	35	8	4.4098	7.2324	1.5174	1.1078	0.5750	0.3746	0.3354
276–	312	37	9	4.9669	7.6447	1.4161	1.0579	0.5730	0.3822	0.3441
313–	351	39	10	5.5242	8.0447	1.3375	1.0182	0.5713	0.3889	0.3517
352–	392	41	11	6.0816	8.4339	1.2746	0.9858	0.5701	0.3947	0.3585
393–	434	43	12	6.6391	8.8049	1.2231	0.9588	0.5690	0.4000	0.3646
435–	479	45	13	7.1967	9.1768	1.1800	0.9359	0.5681	0.4047	0.3701
480–	524	47	14	7.7543	9.5257	1.1433	0.9162	0.5673	0.4090	0.3752
525–	572	49	15	8.3120	9.8770	1.1117	0.8989	0.5667	0.4130	0.3798
573–	621	51	16	8.8697	10.2160	1.0842	0.8838	0.5661	0.4166	0.3840
622–	671	53	17	9.4274	10.5440	1.0599	0.8703	0.5656	0.4199	0.3879
672–	723	54	18	9.9852	10.8690	1.0384	0.8583	0.5652	0.4230	0.3916
724–	777	56	19	10.5430	11.1910	1.0191	0.8474	0.5648	0.4259	0.3950

LOT SIZE	N	R	C	REGRET	OC-CURVE: VALUE OF THETA GIVING RISE TO ACCEPTANCE PROB. P				
					P=0.99	P=0.95	P=0.50	P=0.10	P=0.05
	SINGULAR SOLUTION								
778– 832	58	20	11.1010	11.5040	1.0017	0.8375	0.5644	0.4286	0.3982
833– 888	59	21	11.6590	11.8090	0.9859	0.8285	0.5641	0.4311	0.4012
889– 947	61	22	12.2170	12.1170	0.9716	0.8203	0.5638	0.4335	0.4040
948– 1005	63	23	12.7750	12.4080	0.9584	0.8127	0.5636	0.4357	0.4066
1006– 1067	64	24	13.3320	12.7080	0.9463	0.8056	0.5633	0.4378	0.4092
1068– 1130	66	25	13.8900	13.0010	0.9352	0.7991	0.5631	0.4398	0.4115
1131– 1192	67	26	14.4480	13.2800	0.9248	0.7931	0.5629	0.4417	0.4138
1193– 1259	69	27	15.0060	13.5710	0.9152	0.7874	0.5627	0.4435	0.4160
1260– 1324	70	28	15.5640	13.8440	0.9062	0.7821	0.5625	0.4452	0.4180
1325– 1395	72	29	16.1220	14.1320	0.8978	0.7771	0.5624	0.4468	0.4200
1396– 1461	73	30	16.6800	14.3930	0.8900	0.7724	0.5622	0.4484	0.4218
1462– 1535	75	31	17.2380	14.6760	0.8826	0.7680	0.5621	0.4499	0.4236
1536– 1605	76	32	17.7960	14.9350	0.8756	0.7639	0.5620	0.4513	0.4254
1606– 1680	78	33	18.3540	15.2060	0.8690	0.7599	0.5619	0.4527	0.4270

Furthermore

$$(4) \qquad n_0(r) = r + \frac{\kappa_t}{\delta_s} + O\left(\frac{1}{r}\right).$$

For $N \to \infty$ we have for the optimal truncation number

$$(5) \qquad r_0(N) = (\rho N)^{\frac{2}{\mu + 3}} + O(1)$$

with

$$\rho = \frac{(\mu + 1)\gamma}{2\delta_s}.$$

The optimal cost has the expansion

$$(6) \qquad K_0(N) = N\kappa_m + \frac{\mu + 3}{\mu + 1}\delta_s r + \kappa_t \ln r + O(1)$$

with

$$r = r_0(N).$$

If the test is not truncated the optimal sample size is given by (5) and the corresponding cost of the optimal sampling plan is

$$(7) \qquad K^* = N\kappa_m + \frac{\mu + 3}{\mu + 1}\delta_s n + \kappa_t \ln n + O(1)$$

with n given by (5).

It may be shown by taking further terms into consideration that the asymptotic gain of truncation is (7) – (6),

$$-\delta_s\left[\frac{\kappa_t}{\delta_s}\right] + \kappa_t\left(1 + \frac{1}{2} + \ldots + \frac{1}{\left[\frac{\kappa_t}{\delta_s}\right]}\right),$$

corresponding to a gain from shorter testing time but an extra expense due to the greater number of items tested.

REFERENCES

[1] B. Epstein – M. Sobel, Life testing, *J.A.S.A.*, 48 (1953), 486-502.

[2] B. Epstein, Truncated life tests in the exponential case, *Ann. Math. Statist.*, 25 (1954), 555-564.

[3] D. Guthrie – M.V. Johns, Bayes acceptance sampling procedures for large lots, *Ann. Math. Statist.*, 30 (1959), 896-925.

[4] A. Hald, The compound hypergeometric distribution and a system of single sampling inspection plans based on prior distributions and costs, *Technometrics,* 2 (1960), 275-352 and 370-372.

[5] A. Hald, Asymptotic properties of Bayesian single sampling plans, *J. Roy. Statist. Soc.*, 29 (1967), 162-173. Corrigenda on p. 586.

[6] R.M. Soland, Bayesian analysis of the Weibull process with unknown scale parameter and its application to acceptance sampling, *IEEE Trans. Reliability,* 17 (1968), 84-90.

[7] P. Thyregod, Bayesian single sampling acceptance plans for finite lot sizes, Mimeographed report, The Technical University of Denmark, 1972.

COLLOQUIA MATHEMATICA SOCIETATIS JÁNOS BOLYAI
9. EUROPEAN MEETING OF STATISTICIANS, BUDAPEST (HUNGARY), 1972.

ON THE RAREFACTION OF MULTIVARIATE POINT PROCESSES

J. TOMKÓ

1. INTRODUCTION

A series of limit theorems has been obtained for a variety of operations on point processes. The most frequently considered operations having much importance in applications are the superposition of independent streams of event, the random translation of points and the deletion of events from a process. For a short review on these results the references [1], [5], [6] might be mentioned. In these results the Poisson process plays a central role. Namely, it turns out that under rather broad conditions the operations mentioned lead, in the limit, to a Poisson point process. Mogyoródi [9] has recently considered a general thinning procedure for renewal processes an iteration of which results in a renewal process with recurrence time distribution $G(x)$ appearing in limit theorems for Galton – Watson processes.

The results mentioned briefly above refer to point processes the events of which are of the same type. Many fields of applications justify a generalization considering a stream of events of two or more types. Cox and

Lewis, in [4], call the latter stream a multivariate point process while they call the former a univariate one. Then the question of extending the above limit theorems to multivariate case arises naturally. The first step in doing this was made by Činlar [2], who considered the superposing operation and discussed various conditions under which the resulting process is a multivariate Poissonian one. Grigelionis [7] treats the same problem for more general situations.

This paper is devoted to a discussion of the thinning operation. In Sections 2 and 3 a multivariate point process somewhat more general than a Markov renewal one will be considered. Then a "sub"-process is introduced which, usually, is thinner than the original one. Let $\{\mathsf{P}_\epsilon,\ \epsilon > 0\}$ be a continuous family of probability measures governing the multivariate point process such that the intensity of the introduced sub-process approaches 0 as $\epsilon \to 0$. A basic result of this paper (Section 4) is that on scaling the time suitably the sub-process becomes, in the limit, a Poisson one with independent components. Section 5 deals with a way of deleting events from a Markov renewal process. Two examples from queuing theory are discussed in Section 6.

2. THE ORIGINAL PROCESS AND A WAY OF DELETING ITS EVENTS

Throughout the paper we deal with a stream of events of a finite number of types. We assume that the types are designated by integers $1, 2, \dots, n$. It is convenient for us to describe a multi-variate point process as a sequence of pairs of r.v.'s

(1) $\{t_k, \nu_k;\ k \geqslant 0\}$

satisfying the relation $0 = t_0 \leqslant t_1 \leqslant \dots \leqslant t_k \leqslant \dots$ where t_k and ν_k indicate the occurring time and the type of the kth event, respectively. We specify the dependence structure of the sequence (1) as follows.

Let

$$\{T_k = (J_k, r_k),\ X_k;\ k \geqslant 0\}$$

be a process with state space charecterized by $J_k = 1, 2, \ldots, n$; $r_k = 0, 1$; $0 \leqslant X_k < \infty$ and satisfying the conditions:

$$\mathsf{P}\{(J_0, r_0) = (j, 1)\} = a_j \geqslant 0\,, \qquad \sum_1^n a_j = 1$$

and

$$\mathsf{P}\{J_k = j,\ r_k = u,\ X_k \leqslant x \mid T_0, T_1, X_1, \ldots, T_{k-1}, X_{k-1}\} \overset{\text{a.s.}}{=}$$
$$\overset{\text{a.s.}}{=} G^{r_{k-1}, u}_{J_{k-1} j}(x)$$

for all $0 \leqslant x < \infty$ and $0 < j < n + 1$, $u = 0, 1$.* Set $t_k = \sum_0^k x_i$ and ν_k = first component of T_k.

The sequence

(2) $\{t_k, \nu_k;\ k \geqslant 0\}$

represents an n-variate point process that we call an original one. The probability measure governing this process is described by four transition matrices

(3) $(G^{u,v}_{ij}(x)) \qquad (i, j = 1, 2, \ldots, n;\ u, v = 0, 1)$

satisfying

(i) $G^{u,v}_{ij}(x) = 0$ for $x < 0$ and in non-decreasing in x,

(ii) $\displaystyle\sum_{j=1}^n \{G^{u,0}_{ij}(\infty) + G^{u,1}_{ij}(\infty)\} = 1 \qquad (u = 0, 1)\,.$

We shall be concerned with a family of processes or probability measures, so instead of (3) a family of transition matrices

(4) $(G^{u,v}_{ij}(\epsilon, x)) \qquad (u, v = 0, 1;\ \epsilon > 0)$

is considered. Without the lower indices, $G^{u,v}(x)$ denotes an $n \times n$ matrix. The corresponding Laplace – Stieltjes (L-S) transforms are denoted

*The process introduced is similar to that of [10] under Definition 3.3.

by $\Gamma_{ij}^{u,v}(\epsilon, s)$ and $\Gamma^{u,v}(\epsilon, s)$, respectively.

We introduce now a "sub-process" of the original one by taking into account only events occurring with value $r_k = 1$. More precisely, set

$$\delta_0 = 1 ,$$

- -

$$\delta_k = \min\{l > \delta_{k-1}, r_l = 1\} ,$$

- ;

$$X_0^* = X_0 \quad X_k^* = \sum_{l=\delta_{k-1}+1}^{\delta_k} X_l \quad (k > 0);$$

$$t_k^* = \sum_{l=0}^{k} X_l^* , \quad \nu_k^* = \text{first component of } T_{\delta_k} .$$

Then the sub-process to be introduced is

$$\{t_k^*, \nu_k^*;\ k \geqslant 0\} . \tag{5}$$

Put now

$$N(t) = \max\{k > 0,\ t_k^* < t\} ,$$

$N_j(t)$ = the counting process of the events of j-type $(j = 1, 2, \ldots, n)$. The n-dimensional process $\{N_1(t), N_2(t), \ldots, N_n(t)\}$ is a Markov renewal process, and the process $\{Z_t;\ t \geqslant 0\}$ defined by $Z_t = \nu^*_{N(t)}$ is a semi-Markov process (see for details [10]). Because of these remarks one can describe the probability law governing the sub-process (5) by the transition matrix of functions

$$G_{ij}(\epsilon, x) = \mathrm{P}\{\nu_k^* = j,\ \ t_k^* - t_{k-1}^* \leqslant x \mid \nu_{k-1}^* = i\} ,$$
$$(i, j = 1, 2, \ldots, n) .$$

An obvious consideration yields a relation between the G and $G^{u,v}$ matrices. We give this relation in terms of (L-S) transform:*

*This relation was pointed out to the author by R. Pyke.

(6) $$\Gamma(\epsilon, s) = \Gamma^{1,1}(\epsilon, s) + \Gamma^{1,0}(\epsilon, s)[I - \Gamma^{0,0}(\epsilon, s)]^{-1}\Gamma^{0,1}(\epsilon, s) .$$

3. THE ASSUMPTIONS AND TWO LEMMAS

It is evident that the subprocess is thinner than the original one. The less is the probability of an event with $r_k = 1$ the thinner is the resulting process. We turn to formulate the conditions under which the limiting behaviour of the subprocess can be examined.

To make the probability of an event with $r_k = 1$ decrease assume that for any $x > 0$

$$G^{1,1}(\epsilon, x) + G^{0,1}(\epsilon, x) \to 0$$

as $\epsilon \to 0$. For the other two matrices suppose that for $x > 0$

$$G^{0,0}(\epsilon, x) \to G(x) ,$$

$$G^{1,0}(\epsilon, x) \to \hat{G}(x)$$

as $\epsilon \to 0$ where $G(\cdot)$, $\hat{G}(\cdot)$ denote transition matrices with the property

$$\sum_{j=1}^{n} G_{ij}(\infty) = \sum_{j=1}^{n} \hat{G}_{ij}(\infty) = 1$$

for $1 \leqslant i \leqslant n$. In terms of (L-S) transform these conditions are as follows: for any s ($\mathrm{Re}\, s > 0$)

$$\Gamma^{1,1}(\epsilon, s) + \Gamma^{0,1}(\epsilon, s) \to 0 ,$$

(I) $$\Gamma^{0,0}(\epsilon, s) \to \Gamma(s) ,$$

$$\Gamma^{1,0}(\epsilon, s) \to \hat{\Gamma}(s)$$

as $\epsilon \to 0$, where $\Gamma(0) = G(\infty)$, $\hat{\Gamma}(0) = \hat{G}(\infty)$ are proper transition matrices.

Note that the property (ii) involves the validity of (I) for $s = 0$.

Furthermore we assume that the limits

$$\rho_{ij} = \lim_{\epsilon \to 0} \frac{\Gamma_{ij}(0) - \Gamma_{ij}^{0,0}(\epsilon, 0)}{\epsilon},$$

(II)

$$r_{ij} = \lim_{\epsilon \to 0} \frac{1}{\epsilon} \Gamma_{ij}^{0,1}(\epsilon, 0)$$

and the integrals

$$\mu_{ij}(\epsilon) = \int_0^\infty x dG_{ij}^{0,0}(\epsilon, x),$$

(III)

$$\mu_{ij} = \int_0^\infty x dG_{ij}(x)$$

exist and are finite, and finally

$$\mu_{ij}(\epsilon) \to \mu_{ij},$$

(IV)

$$\int_0^\infty x dG^{0,1}(\epsilon, x) \to 0$$

as $\epsilon \to 0$.

Note that because of the conditions imposed the intensity of the sub-process decreases to zero as $\epsilon \to 0$. In such cases, in general, we change the time scale so that the expected number of events occurring in a unit time interval remains near a positive quantity, as $\epsilon \to 0$. The time scaling considered has the form

(7) $\tau = \gamma_\epsilon t$

where γ_ϵ is a function of ϵ such that $\gamma_\epsilon \to 0$ as $\epsilon \to 0$ and it has derivative $0 < \gamma' < \infty$ at $\epsilon = 0$. For the (L-S) transform the time scaling is equivalent to a replacement of s with $\gamma_\epsilon s$. Thus $\Gamma(\epsilon, \gamma_\epsilon s)$ is the (L-S) transform of the transition matrix governing the normalized sub-process (obtained after the time scale has been changed).

In what follows it will be convenient to make use of some elementary results that we now give.

Let us introduce the notation A_* for an arbitrary element of the matrix $A = (A_{ij})$. The requirement (II) ensures the smoothness of $\Gamma^{0,0}(\epsilon, 0)$ and $\Gamma^{0,1}(\epsilon, 0)$ at $\epsilon = 0$. Since $\Gamma_*^{0,0}(\epsilon, s)$ does not decrease as $|s| \to 0$, $\lim_{\epsilon \to 0} \Gamma^{0,0}(\epsilon, \gamma_\epsilon s) = \Gamma(0)$. Then the whole complex of conditions imposed assures its smoothness at $\epsilon = 0$.

Lemma 1. *Any element of the matrices* $\Gamma^{0,0}(\epsilon, \gamma_\epsilon s)$, $\Gamma^{0,1}(\epsilon, \gamma_\epsilon s)$ *has finite derivative at* $\epsilon = 0$.

Proof. Write $\dfrac{\Gamma_*(0) - \Gamma_*^{0,0}(\epsilon, \gamma_\epsilon s)}{\epsilon}$ in the form

$$\frac{\Gamma_*(0) - \Gamma_*^{0,0}(\epsilon, 0)}{\epsilon} + \frac{\Gamma_*^{0,0}(\epsilon, 0) + \Gamma_*^{0,0}(\epsilon, \gamma_\epsilon s)}{\epsilon} .$$

Because of (II) it suffices to examine the second term. It may be written as follows

$$\frac{\gamma_\epsilon s}{\epsilon} \int_0^{T_\epsilon} \frac{1 - e^{-\gamma_\epsilon s x}}{\gamma_\epsilon s} \, dG_*^{0,0}(\epsilon, x) + \frac{1}{\epsilon} \int_{T_\epsilon}^{\infty} (1 - e^{-\gamma_\epsilon s x}) dG_*^{0,0}(\epsilon, x)$$

where $T_\epsilon = \dfrac{E}{\gamma_\epsilon |s|}$ with arbitrary small $E > 0$. It is a simple matter to show that the first term has a limit equal to $\gamma' s \mu_*$. The limit of the second term is zero as it can easily be seen from

$$\frac{1}{\epsilon} \int_{T_\epsilon}^{\infty} (1 - e^{-\gamma_\epsilon s x}) dG_*^{0,0}(\epsilon, x) \leqslant \frac{1}{\epsilon T_\epsilon} \int_{T_\epsilon}^{\infty} x dG_*^{0,0}(\epsilon, x) ,$$

observing that because of (I), (III) and (IV) the last integral converges to zero for any $T_\epsilon \to \infty$ as $\epsilon \to 0$. Thus we have shown that

$$\lim_{\epsilon \to 0} \frac{\Gamma_*(0) - \Gamma_*^{0,0}(\epsilon, \gamma_\epsilon s)}{\epsilon} = \rho_* + \gamma' s \mu_* . \tag{8}$$

By repetition of this argument one can prove the assertion for $\Gamma^{0,1}(\epsilon, \gamma_\epsilon s)$ and obtain

$$\lim_{\epsilon \to 0} \frac{1}{\epsilon} \Gamma_*^{0,1}(\epsilon, \gamma_\epsilon s) = r_* . \tag{9}$$

Introduce the notation $\langle B \rangle$ for $(\bar{b}_{ij})'$ where $B = (b_{ij})$ and $\bar{b}_{ij}$ is the cofactor of its (i, j)th element. Thus if the determinant $|B| \neq 0$

then $\langle B\rangle = |B|B^{-1}$. We shall be concerned with matrices the rows of which are identical. Such matrices will be called i.r. matrices.

Lemma 2. *Let* $\Gamma(s)$ *be the (L-S) transform of a transition matrix such that* $\Gamma(0) = P = (p_{ij})$ *is irreducible.*

(i) $\langle I - P\rangle$ *is an i.r. matrix and if* $\pi = (\pi_1, \ldots, \pi_n)$ *is the ergodic distribution of* P *then the rows of* $\langle I - P\rangle$ *differ from* π *only by a constant factor.*

(ii) *If* $\Gamma'(0) = (\mu_{ij})$ *exists then*

$$\lim_{s\to 0} s[I - \Gamma(s)]^{-1} = \frac{P_0}{\sum_i \pi_i \sum_j \mu_{ij}} \tag{10}$$

where P_0 *is an i.r. matrix with rows identical to* π.

Proof. To prove the first part of (i) we have to show for the cofactors $\bar{q}_{ij}$ of $I - P$ that $\bar{q}_{il-1} = \bar{q}_{il}$ for any i and $2 \leqslant l \leqslant n$. One can see this upon replacement of the $(l-1)$st column of $\bar{q}_{il}$ by the sum of all its columns. The positiveness of $\bar{q}_{ii}$ may be shown by a quite elementary argument e.g. by induction. The second part of (i) follows from the remark that the ergodic distribution π and the vector

$$v_i = (\bar{q}_{1i}, \bar{q}_{2i}, \ldots, \bar{q}_{ni})$$

for any $1 \leqslant i \leqslant n$ satisfy the same equation $(I - P)v' = 0$ the solutions of which may differ from each other only by a multiplier constant. To give reasoning for the statement (ii) write $s[I - \Gamma(s)]^{-1}$ in the form

$$\frac{\langle I - \Gamma(s)\rangle}{\frac{1}{s}|I - \Gamma(s)|}\,.$$

Let $s \to 0$ and apply L'Hospital's rule for examining the limit of the denominator. One obtains formula (10) after simplifying by the multiplier constant.

Remark. Some deeper results on the behaviour of $[I - \Gamma(s)]^{-1}$ in the neighborhood of $s = 0$ are presented in [8]. Naturally they require much stronger tools of treatment than that employed above.

4. THE LIMIT PROCESS

In this section we prove our main result on the limiting behaviour of the sub-process as $\epsilon \to 0$. First we take note of the following fact.

Let $\{\tau_k, \beta_k;\ k \geqslant 0\}$ be an n-variate Poisson process the components of which are independent and have intensities $\lambda_i > 0$ $(1 \leqslant i \leqslant n)$. Then the (L-S) transform of the governing transition matrix has the form

$$\frac{\lambda}{\lambda + s} P_0 \tag{11}$$

where $\lambda = \sum_1^n \lambda_i$, P_0 is an i.r. matrix rows of which are equal to $\left(\frac{\lambda_1}{\lambda}, \ldots, \frac{\lambda_n}{\lambda}\right)$.

Theorem. *Consider the sub-process under* (5) *and change the time in accordance with* (7). *Suppose that* $\Gamma(0)$ *appearing in assumption* (I) *is irreducible. Then under conditions* (I), (II), (III) *and* (IV) *the sub-process becomes in the limit, as* $\epsilon \to 0$, *a Poissonian one with independent components.*

Proof. It suffices to show that the limit of $\Gamma(\epsilon, \gamma_\epsilon s)$ under (6) has the form (11) with some P_0 and $\lambda > 0$. The first term $\Gamma^{1,1}(\epsilon, \gamma_\epsilon s)$ converges to zero as $\epsilon \to 0$; this follows from (I) and its validity for $s = 0$ The same reasoning is applicable to show that

$$\lim_{\epsilon \to 0} \Gamma^{1,0}(\epsilon, \gamma_\epsilon s) = \hat{\Gamma}(s) .$$

Examine now the product left in (6). Write it in the form

$$\frac{\langle I - \Gamma^{0,0}(\epsilon, \gamma_\epsilon s)\rangle}{\frac{1}{\epsilon} |I - \Gamma^{0,0}(\epsilon, \gamma_\epsilon s)|} \cdot \frac{\Gamma^{0,1}(\epsilon, \gamma_\epsilon s)}{\epsilon} .$$

When $\epsilon \to 0$, $\langle I - \Gamma^{0,0}(\epsilon, \gamma_\epsilon s)\rangle$ approaches to $\langle I - \Gamma(0)\rangle$, which, by (i) of Lemma 2, is an i.r. matrix. To inspect the denominator observe that for any $n \times n$ matrices A and B

$$|A+B| = \sum_{i=1}^{n+1} |C_i| ,$$

where C_i has its first $i-1$ rows from B, its ith row from A while its other rows are that of $A+B$. Apply this remark to

$$|I - \Gamma^{0,0}(\epsilon, \gamma_\epsilon s)| = |\Gamma(0) - \Gamma^{0,0}(\epsilon, \gamma_\epsilon s) + I - \Gamma(0)|$$

with an obvious choice of the notation. Then divide the ith row of C_i by ϵ. The last term, $|C_{n+1}|$ in our case, equals to zero. Letting $\epsilon \to 0$, we obtain by Lemma 1 and formula (8) that

$$\lim_{\epsilon \to 0} \frac{1}{\epsilon} |I - \Gamma^{0,0}(\epsilon, \gamma_\epsilon s)| =$$

$$= \sum_{i=1}^{n} \begin{vmatrix} (1-p_{11}) - p_{12} & \cdots & -p_{1n} \\ \cdots & \cdots & \cdots \\ \rho_{i1} + \gamma' s \mu_{i1} & \cdots & \rho_{in} + \gamma' s \mu_{in} \\ \cdots & \cdots & \cdots \\ -p_{n1} & \cdots & (1-p_{nn}) \end{vmatrix} =$$

$$= \sum_{i=1}^{n} \sum_{j=1}^{n} \rho_{ij} \bar{q}_{ij} + \gamma' s \sum_{i=1}^{n} \sum_{j=1}^{n} \mu_{ij} \bar{q}_{ij} = a + bs .$$

Similarly, by Lemma 1 and formula (9) we have

$$\lim_{\epsilon \to 0} \frac{\Gamma^{0,1}(\epsilon, \gamma_\epsilon s)}{\epsilon} = R = (r_{ij}) .$$

Observe now that $\hat{\Gamma}(0)\langle I - \Gamma(0)\rangle = \langle I - \Gamma(0)\rangle$ and that $\langle I - \Gamma(0)\rangle R$ is again an i.r. matrix. We norm the rows of the latter matrix by the sum of the elements of a row. For this sum we have

$$\sum_j \sum_i \bar{q}_{ij} r_{ij} = \sum_i \bar{q}_{ii} \sum_j r_{ij} .$$

One can very easily show that the latter sum coincides with a. Namely, $a = \sum_i \bar{q}_{ii} \sum_j \rho_{ij}$ and $\sum_j \rho_{ij} = \sum_j r_{ij}$ for any $1 \leqslant i \leqslant n$; this follows from

$$\frac{1}{\epsilon}\sum_{j=1}^{n}{}' \{\Gamma_{ij}^{0,1}(\epsilon, 0) - [\Gamma_{ij}(0) - \Gamma_{ij}^{0,0}(\epsilon, 0)]\} = 0 ,$$

which is a straightforward consequence of (ii) of Section 2. Consequently we have that

$$\lim_{\epsilon \to 0} \Gamma(\epsilon, \gamma_\epsilon s) = \frac{\lambda}{\lambda + s} P_0$$

with $\lambda = \frac{a}{b}$ and an i.r. matrix P_0 the row sum in which is equal to 1. Thus the proof of the Theorem is complete.

5. DELETING EVENTS FROM A MARKOV RENEWAL PROCESS

Let $\{t_k, \nu_k;\ k \geqslant 0\}$ be an n-variate Markov Renewal Process with transition matrix $G(x)$ such that $P = G(\infty)$ is irreducible and $\int_0^\infty x dG(x) < \infty$.

Consider every possible transition of the process and assign to each of them a random variable r taking only two values 0 and 1. When a transition from any state to state j terminates an event of type j occurs. If the value of r assigned to this transition is 0 then the occuring event is deleted. We suppose that the probability $\mathsf{P}(r = 1)$ depends on the initial and the final state and on the elapsing time of the transition. As to the dependence structure of r's assigned to the sequential transitions the Markov property is imposed.

Thus denote by $r_{ij}^0(x)$ $(r_{ij}^1(x))$ the probability of $r = 0$ for a transition from state i to state j with elapsing time x if for the previous transition r took value 0 (1). The matrices

$$R^0(x) = (r_{ij}^0(x)) , \quad R^1(x) = (r_{ij}^1(x))$$

are called thinner matrices.

Let now r_k be the r.v. assigned to the kth transition of the process and to be determined assume that whatever the initial value ν_0 is $\mathsf{P}(r_0 = 1) = 1$. The process $\{t_k, \nu_k, r_k;\ k \geqslant 0\}$ is similar to that of Section 2 and has transition matrices

$$G^{0,0}(x) = R^0(x) * G(x) ,$$

(12) $$G^{1,0}(x) = R^1(x) * G(x) ,$$

$$G^{u,1}(x) = (E - R^u(x)) * G(x) , \qquad (u = 0, 1)$$

where the following notations are introduced: for any $n \times n$ matrices A and B

$$A * B = (a_{ij} b_{ij})$$

and E is a matrix all elements of which are 1.

The process resulting by the described thinning procedure (which might be called the thinned process) corresponds to the sub-process under (5).

Consider now a family of thinner matrices

(13) $$\{R^0(\epsilon, x), R^1(\epsilon, x);\ \epsilon > 0\}$$

satisfying the requirements: if $\epsilon > 0$, then, for $u = 1, 2$, $R^u(\epsilon, x) \to E$ a.s. with respect to $\sum_{i,j} G_{ij}(x)$, $\frac{1}{\epsilon}(E - R^0(x)) \to \rho(x)$ a.s. with respect to $\sum_{i,j} G_{ij}(x)$, and for any i, j

$$\rho_{ij} = \int_0^\infty \rho_{ij}(x) dG_{ij}(x) < \infty .$$

It is a simple matter to verify that for the transition matrices $G^{u,v}(\epsilon, x)$ defined by (13) in accordance with (12) the conditions (I), (II), (III) and (IV) of Section 2 hold. Thus, considering a time scale transformation of type (7), the normalized thinned process approaches, as $\epsilon \to 0$, to a Poisson process with independent components.

6. TWO EXAMPLES FROM QUEUING THEORY

Consider a server dealing with customers of several types. Let the number of the types be n and suppose that the arrivals form an n-variate

Markow renewal Process with transition matrix $G(x)$. We assume that $G(\infty)$ is irreducible and $\int_0^\infty x dG(x) < \infty$. There is no place for waiting i.e. we consider a loss system. If a customer finds the server idle then its service starts and has an exponentially distributed length with parameter $\mu > 0$ common for all types.

Let, for $k \geqslant 0$, t_k and ν_k accordingly be the moment and the type of the kth arrival. Define r_k by setting $r_k = 0$ ($r_k = 1$) if the customer arriving at t_k is (is not) rejected. Assume that the server is idle at $t = 0$ so $\mathsf{P}(r_0 = 1) = 1$. Let now $\{t_l^*, \nu_l^*;\ l \geqslant 0\}$ be the subsequence selected from $\{t_k, \nu_k;\ k \geqslant 0\}$ in accordance with $r_k = 1$ i.e. the flow of the customers who were served. This flow is the thinned process resulting from the arrival process in a manner described in the previous section with thinner probabilities

$$r_{ij}^0(x) = r_{ij}^1(x) = e^{-\mu x} \qquad (i, j = 1, 2, \ldots, n) .$$

Let us now suppose that the expected service time increases indefinitely i.e., writing ϵ for μ, $\epsilon \to 0$. For the thinner probabilities we have

$$\lim_{\epsilon \to 0} e^{-\epsilon x} = 1$$

and

$$\lim_{\epsilon \to 0} \frac{1 - e^{-\epsilon x}}{\epsilon} = x .$$

We are now in a position to deduce that the flow of the served customers, after contracting the time scale accordingly, approaches a Poisson process with independent components. The rate of the components can be calculated and thus an approximation may be given for the loss ratio of the customers for different types.

Our second example is also concerned with a server dealing with customers of several types. There is now one place for waiting to be served. If the server busy then this place is occupied by the customer who came last. Thus any arriving customer spends some time at the waiting place;

zero time if at his arrival the server is idle. A waiting customer is rejected if a new customer arrives before he could engage the server.

Let the number of the types be n. To describe the manner of arrivals we introduce the process $\{\xi_t, t \geq 0\}$ where ξ_t is the type of the last last customer coming before t. Our assumption the input is that $\{\xi_t, t \geq 0\}$ is a time homogeneous Markov process. Let $A = (a_{ij})$, $(i, j = 1, 2, \ldots, n)$ be the infinitesimal matrix of $\{\xi_t, t \geq 0\}$ and suppose that in addition to

$$\sum_{j=1}^{n} a_{ij} = 0 \qquad (i = 1, 2, \ldots, n)$$

A satisfies the requirement $a_{ij} > 0$ for all $i \neq j$ and $a_{ii} < 0$ for all i. This ensures that $\{\xi_t, t \geq 0\}$ is ergodic. The process $\{\xi_t, t \geq 0\}$ might be considered a semi-Markov process with transition matrix $G(x)$ given by

$$G_{ij}(x) = \begin{cases} 0 & \text{for} \quad i = j \\ r_{ij}(1 - e^{-q_i x}) & \text{otherwise,} \end{cases}$$

where $q_i = -a_{ii}$, $r_{ij} = \dfrac{a_{ij}}{q_i}$, $(i \neq j)$ (see Example (b) on pp. 1252 of [11]). This semi-Markov process has the particular property that the events of the same type may not occur repeatedly. However we would like to allow the possibility that events of the same type occur after one another successively. For this one may consider the time spent in a state, say e.g. in the state i, as a randomly stopped sum of independent exponentially distributed r.v.'s with rate λ_i, $\lambda_i(1 - p_i) = q_i$, $0 \leq p_i < 1$ for the state i, i.e. as $\sum_{l=1}^{\nu} \eta_l^{(i)}$, where the random number ν is independent of the terms and has geometric distribution with parameter p_i. Thus one arrives at a semi-Markov process with transition matrix

$$G(x) = (p_{ij}(1 - e^{-\lambda_i x}))$$

with $p_{ij} = r_{ij}(1 - p_i)$, $(i \neq j)$ and $p_{ii} = p_i$. This latter process might be considered as stochastically equivalent to the process $\{\xi_t, t \geq 0\}$.

In the sequel we shall be concerned with a family of input processes $\{\xi_t^{(\lambda)},\ t \geqslant 0\}$, $(\lambda > 0)$ such that the intensity of every marginal process of the input increases infinitely as $\lambda \to \infty$. For the sake of simplicity we shall consider only the case characterized as follows:

$$(14) \qquad \lambda_i = \lambda r_i, \qquad (i = 1, 2, \ldots, n),$$

r_i, p_{ij} being fixed, i.e. the elements of the infinitesimal matrix A are

$$a_{ij} = \begin{cases} \lambda r_i p_{ij}, & i \neq j, \\ -\lambda r_i (1 - p_{ii}), & i = j. \end{cases}$$

Finally let $P_\lambda(t) = (P_{ij}^{(\lambda)}(t))$ be the matrix of the transition probabilities for the process $\{\xi_t^{(\lambda)},\ t \geqslant 0\}$ and $G_i(x)$ be the service time distribution function for a customer of type i $(1, 2, \ldots, n)$.

We shall examine the busy period of the server. Such a busy period starts with the service of a customer finding the server idle and terminates when a customer departs from the system leaving no customer behind him. For the classical M/G/1 system with finite queue capacity the busy period has been studied in [3]. An earlier paper of the author [12] is devoted to a discussion of limiting behaviour of the busy period when the input intensity increases indefinitely; we now deal with the extended problem. For an unrestricted queue and multivariate input, results on busy period are available in [13].

Consider now the output of the system given above. It is a multi-variate point process the structure of which is similar to that of the process under (2). More precisely denote by t_k the time point of the kth departure, put $\nu_k = \xi_{t_k - 0}$ and define the r.v. r_k by setting $r_k = 1$ or $r_k = 0$ according as a busy period terminates at t_k or not. Suppose that

$$\mathsf{P}\{t_0 = 0,\ \nu_0 = i,\ r_0 = 1\} = \mathsf{P}(\nu_0 = i) = a_j, \quad \sum_j a_j = 1 .$$

Then the process $\{t_k, \nu_k, r_k;\ k \geqslant 0\}$ has the same property as the original process in Section 2. The elements of the corresponding four transition matrices are as follows:

$$G_{ij}^{0,0}(x) = \int_0^x \{\delta_{ij}[P_{ii}(u) - e^{-\lambda_i u}] + (1 - \delta_{ij})P_{ij}(u)\}dG_i(u);$$

$$G_{ij}^{0,1}(x) = \delta_{ij}p_{ii}\int_0^x e^{-\lambda_i u}\, dG_i(u) \qquad (\delta_{ij} = 0,\ (i \neq j);\ \delta_{ii} = 1);$$

$$G_{ij}^{1,0}(x) = \int_{u=0}^{x} \sum_{k=1}^{n} \int_{v=0}^{u} \lambda_i e^{-\lambda_i v} p_{ik} P_{kj}(u - v)dG_k(u - v)dv;$$

$$G_{ij}^{1,1}(x) = \int_{u=0}^{x} \int_{v=0}^{u} \lambda_i e^{-\lambda_i v} p_{ij} e^{-\lambda_j(u-v)} dG_j(u - v)dv\,.$$

Consider now the case (14). Set

$$\Gamma_i(s) = \int_0^\infty e^{-sx}\, dG_i(x)$$

and

(15) $\qquad \epsilon = \sum_{j=1}^{n} \Gamma_j(\lambda r_j)\,.$

Thus instead of $\lambda \to \infty$ one may let $\epsilon \to 0$. To be compatible with the previous notations consider ϵ as an independent parameter and denote by λ^ϵ the root of (15). Furthermore, as to the d.f.'s

$$G_i(x) \qquad (i = 1, 2, \ldots, n)$$

we require that their first moments

(16) $\qquad \mu_i = \int_0^\infty x dG_i(x) \qquad (i = 1, 2, \ldots, n)$

exist, and the limits

(17) $\qquad \lim_{\epsilon \to 0} \dfrac{\Gamma_i(\lambda^\epsilon r_i)}{\epsilon} = \rho_i \qquad (i = 1, 2, \ldots, n)$

also exist and are positive.

Proposition. *The four transition matrices governing the output of the system described satsify the conditions* (I), (II), (III) *and* (IV) *of*

Section 3 *provided that the requirements* (16), (17) *hold.*

We shall not give detailed arguments to prove this statement. We note only two facts on which a rigorous proof might be based. For any fixed $t > 0$

$$(18) \qquad \lim_{\lambda \to \infty} P_{ij}^{(\lambda)}(t) = P_j \qquad (j = 1, 2, \ldots, n)$$

independently of i and $\sum_1^n P_j = 1$, $P_j > 0$. Our second remark is that the convergence rate in (18) is exponential.

Because of the proposition one deduces that the flow of the customers served and leaving the server idle after a time scale transformation of type (7) becomes a Poisson process with independent components as $\epsilon \to 0$. Since the idle period tends to zero in probability as $\epsilon \to 0$ $(\lambda \to \infty)$, this result implies that the distribution of the busy period of the server is asymptotically exponential.

Acknowledgement. I would like to thank Professor D. Cox for drawing my attention to the multivariate point processes.

REFERENCES

[1] Ju.B. Beljaev, Limit theorems for dissipative flows, *Prob. Appls.*, 8 (1963), 165-173.

[2] E. Činlar, On the superposition of *m*-dimensional point processes. *J. Appl. Prob.*, 5 (1968), 169-176.

[3] J.W. Cohen, On the busy periods for the M/G/1 queue with finite and with infinite waiting room, *J. Appl. Prob.*, 8 (1971), 821-827.

[4] D.R. Cox – P.A.W. Lewis, Multivariate point processes, *Proceedings of the Sixth Berkeley Symposium on Mathematical Statistics and Probability*, 1971.

[5] B.V. Gnedenko – I.N. Kovalenko, *Introduction to Queuing Theory,* Israel Program for Scientific Translation, Jerusalem, 1968.

[6] J.R. Goldman, Stochastic point processes: limit theorems, *Ann. Math. Statist.*, 38 (1967), 771-779.

[7] B. Grigelionis, Limit theorems for sums of multidimensional step-like stochastic processes, (in Russian), *Litovsk. Mat. Sb.*, 10 (1970), 29-49.

[8] J. Keilson, On the matrix renewal function for Markov Renewal Processes, *Ann. Math. Statist.*, 40 (1969), 1901-1907.

[9] J. Mogyoródi, On the rarefaction of Renewal Processes, *Studia Sci. Math. Hungar.*, (to appear).

[10] R. Pyke, Markov Renewal Processes: definitions and preliminary properties, *Ann. Math. Statist.*, 32 (1961), 1231-1242.

[11] R. Pyke, Markov Renewal Processes with finitely many states, *Ann. Math. Statist.*, 32 (1961), 1243-1259.

[12] J. Tomkó, A limit theorem for a finite queue with indefinitely increasing input intensity, *Studia Sci. Math. Hungar.*, 2 (1967), 447-454.

[13] P.D. Welch, On the busy period of a facility which serves customers of several types, *J. Roy Statist. Soc. Ser. B*, 27 (1965), 361-370.

COLLOQUIA MATHEMATICA SOCIETATIS JÁNOS BOLYAI
9. EUROPEAN MEETING OF STATISTICIANS, BUDAPEST (HUNGARY), 1972.

CONTRIBUTION TO GIRAULT'S LECTURE

V. TZONEV

I have only one general remark to make in connection with Professor Girault's very stimulating paper (in this volume pp. 227-240).

I fully agree with Professor Girault that Statistics has always felt the need of some unifying ideas concerning the task of the statistical methods. But I am somewhat more pessimistic than he is in the evaluation of the present state of Statistics. It seems to me that modern Statistics still lacks truly great unifying ideas inspite of some fine achievements and new developments in the field of factor and discriminant analysis.

Some may not accept this verdict. Together with R.A. Fisher, they may be inclined to point out that Statistics has the task to reduce bulks of initial data and, furthermore, to split their variance into components, to compare them, and so on. Personally, I am not convinced that this is a deep going formulation of the task of Statistics.

For the questions still remains why should we reduce the bulk of data altogether; why should we split their variance into components and compare them; what cognitive object is aimed at?

These questions are apparently very simple but an analytical answer to them can be given only by the philosophy of science, or more specifically, by Metastatistics as a source of unifying ideas for Statistics.

What can Metastatistics reveal?

This is a topic on which a lot of hard work is already carried out in my country. I personally spoke in some detail on this issue at the IVth International Congress for Logic, Methodology and Philosophy of Science held last year in Bucharest.

The following results appear to me most essential.

The cognitive process can be realized in two basic forms: a microform and a macroform.

The microform of the cognitive process deals with micronotions, micropropositions and deductions. The macroform, on the other hand, deals with macronotions, macropropositions, and so on. In order to draw a sharp difference between the two forms, it is of the utmost importance to emphasize the difference between a set and an aggregate, (the Germans would say between "Vielheit" and "Gesamtheit").

And now to the point. What we expect from a probe into the realm of Metastatistics is the following: an immensely deeper understanding of the cognitive functions of the statistical procedures and techniques. All these procedures and techniques will reveal themselves as instruments of the cognitive process in its more sophisticated form: the macroform. They help the realization of the macroform according to its cognitive tasks and inherent logic.

To make people understand this logic and to recognize its requirements is, in may opinion, the real test for the succesful teaching of Statistics.

Let me summarize: Statistics should be extended through the development and teaching of Metastatistics.

COLLOQUIA MATHEMATICA SOCIETATIS JÁNOS BOLYAI
9. EUROPEAN MEETING OF STATISTICIANS, BUDAPEST (HUNGARY) 1972.

ON THE CONCEPT OF INFORMATION

K. URBANIK

The theory of information has been developed in recent years and has found wide application in different fields. The first definition of the notion of information was given in its full generality by C. E. Shannon in 1948 in [9]. The formula for the information $H = -\sum_{k=1}^{n} p_k \log p_k$ adopted by Shannon from statistical physics (Boltzmann's formula for entropy) has been deduced from various systems of axioms. The most of axiomatic definitions of information are essentially based on the notion of probability. For the complete list of references the reader is refered to [1]. We mention also presentations of information independent from probability, given in [3] – [8], where really the notion of probability is based on that of information. This approach may be considered as theoretically interesting because it explains the relationship between such basic concepts as information and probability. As it was pointed out in [6] information seems intuitively a much simpler and more elementary notion than that of probability. It gives more a cruder and global description of some physical situations than probability does. Therefore, information represents a more primary step of knowledge than that of cognition of probabilities. It should be noted that the axioms proposed in [6] and [7], in particular that which gives a connection between information of rings and their subrings is very hard to grasp in their intuitive meaning. On the other hand the simplified axioms for information given in [3], [4] and [5] contain the weight factors which are nothing else as classical Laplacean probabilities for *H*-homogeneous rings. The aim of the present note is to propose a modification of the system of axioms from [7], which has rather simple intuitive meaning and the informational feature. Furthermore, we shall show that the information determines the probability uniquely except for some degenerate cases.

In probability theory one assumes that all the events concerning a statistical experiment constitute a Boolean algebra. From the point of view of information theory it is more convenient to consider these sets of events as Boolean rings. In the present note we confine ourselves to finite Boolean rings. The rings are considered with the meet and the joint operations. Further, we assume that the zero element 0_A of the ring A is contained in A. By 1_A we shall denote the unit element of A, i.e. the join of all elements of A. In the present note we shall consider non-trivial rings only, i.e. rings A with the property $0_A \neq 1_A$. Moreover, non-trivial subrings will be called briefly subrings. Let $a \in A$ and $a \neq 0_A$. By $a \cap A$ we shall denote the subring of A consisting of all elements of A contained in a. Further, let $a_1, a_2, \ldots, a_m$ and $A_1, A_2, \ldots, A_n$ be elements and subrings of the ring A respectively. By $[a_1, a_2, \ldots, a_m, A_1, A_2, \ldots \ldots, A_n]$ we shall denote the least subring of A containing all elements $a_1, a_2, \ldots, a_m$ and all subrings $A_1, A_2, \ldots, A_n$.

A set $\mathscr{L}$ of finite Boolean rings satisfying the conditions

(L1) if $A \in \mathscr{L}$ and B is a subring of A, then $B \in \mathscr{L}$,

(L2) for any $A \in \mathscr{L}$ there exists such a $B \in \mathscr{L}$ that A is a proper subring of B,

will be called a *Boolean ladder*. We assume that the ladder $\mathscr{L}$ is equipped with a convergence satisfying except the standard Fréchet conditions the following ones:

(L3) if $A_n \longrightarrow A$, then for every subring B of A there exist subrings B_n of A_n, respectively, such that $B_n \longrightarrow B$,

(L4) if $A_n \longrightarrow A$, then each sequence of subrings of A_n contains a subsequence convergent to a subring of A.

If on a Boolean ladder $\mathscr{L}$ a real-valued function F is defined, we get a decomposition of $\mathscr{L}$ into classes of equivalent rings. More precisely, we say that two rings A and B from $\mathscr{L}$ are F-equivalent, in symbols $A \underset{F}{\sim} B$, if there exists an isomorphism h of A onto B such that $F(C) = F(h(C))$ for any subring C of A. A ring A from $\mathscr{L}$ is said to be F-homogeneous if for every automorphism g of A and every subring B of A we have $F(B) = F(g(B))$. Let $\mathscr{L}_F$ denote the set of all F-homogeneus rings from $\mathscr{L}$ with at least three atoms and all their subrings. We say that the function F is regular on $\mathscr{L}$ if it is continuous and $\mathscr{L}_F$ is dense in $\mathscr{L}$.

Now we define information as a real-valued regular function H on a Boolean ladder $\mathscr{L}$ satisfying the following axioms:

(H1) The law of the broken choice. If $a_1, a_2, \ldots, a_n \in A$, $a_i \cap a_j = 0_A$ $(i \neq j)$, $a_1 \cup a_2 \cup \ldots \cup a_n = 1_A$ and $a_i \cap A \underset{H}{\sim} a_j \cap A$ $(i, j = 1, 2, \ldots \ldots, n)$, then

$$H(A) = H([a_1, a_2, \ldots, a_n]) + H(a_1 \cap A).$$

(H2) The local character of information. Let a, b and A be elements and a subring of a ring from $\mathscr{L}$ respectively. If a, b and 1_A are disjoint, then

$$H([a, b, A]) - H([a \cup b, A]) = H([a, b, 1_A]) - H([a \cup b, 1_A]).$$

(H3) Indistinguishability. Isomorphic H-homogeneous rings with at least three atoms are H-equivalent.

(H4) The principle of increase of information. If A is H-homogeneous and B an H-homogeneous proper subring of A, then

$H(B) < H(A)$.

The intuitive meaning of these axioms is rather simple and may be explained as follows:

(H1) Suppose that a statistical experiment A is broken down into two succesive choices. In the first step we select an event a_j from the set of informationally equivalent events $a_1, a_2, \ldots, a_n$. Afterwards we choose an event from $a_j \cap A$. The total information $H(A)$ obtained by selecting should be the sum of the partial informations $H([a_1, a_2, \ldots, a_n])$ and $H(a_j \cap A)$. This property of information was formulated already by the founder of information theory, C.E. Shannon, in [9].

(H2) Consider a complete system of disjoint events $a, b, c_1, c_2, \ldots, c_n$ and suppose that $a \cup b$, $c_1, c_2, \ldots, c_n$ are possible outcomes of a statistical experiment. It seems reasonable to assume that the loss of information due to the indistinguishability of the events a and b is independent upon some particular properties of the remaining events $c_1, c_2, \ldots, c_n$ and depends only upon the joint $c_1 \cup c_2 \cup \ldots \cup c_n$, i.e. the complement of $a \cup b$. Our axiom expresses the local character of the change of information when a statistical experiment does not distinguish between two events. It is a simplified version of axiom II from [7] and corresponds in the language of probability theory to axiom II* from the paper [2] by Z. Daróczy.

(H3) This is an obvious requirement that rings which are maximally informationally uniform cannot be distinguished from the point of view of information theory.

(H4) The axiom expresses the natural property of information that it increases when the number of elements of the respective H-homogeneous rings increases.

A connection between concepts of information and probability is given by the following theorem.

Theorem. *Let H be an information on a Boolean ladder $\mathscr{L}$. Then for every $A \in \mathscr{L}$ there exists a strictly positive probability measure p_A*

defined on A *such that for every subring* B *of* A *the formula*

$$H(B) = -\kappa \sum_{k=1}^{r} \frac{p_A(b_k)}{p_A(1_B)} \log \frac{p_A(b_k)}{p_A(1_B)}$$

holds, where κ *is a positive constant and* $b_1, b_2, \ldots, b_r$ *are atoms of* B. *In particular,*

$$H(A) = -\kappa \sum_{k=1}^{n} p_A(a_k) \log p_A(a_k) \tag{1}$$

where $a_1, a_2, \ldots, a_n$ *are atoms of* A. *Moreover, the measure* p_A *is uniquely determined for all rings* A *with at least three atoms. Finally,* p_A *is uniquely determined on two-atomic rings up to a rearrangement of atoms.*

We note that the converse is also true. Namely, suppose we have a Boolean ladder $\mathscr{L}$ consisting of rings A equipped with a strictly positive probability measure p_A such that for every subring B of A the corresponding measure p_B is induced by p_A on B. We identify two rings A and C whenever they are isomorphic and $p_A(a) = p_C(h(a))$ for an isomorphism h from A onto C and for all $a \in A$. We define a convergence on $\mathscr{L}$ by setting $A_n \longrightarrow A$ if for sufficiently large n the rings A_n and A are isomorphic and, for all $a \in A$,

$$\lim_{n \to \infty} p_{A_n}(h_n^{-1}(a)) = p_A(a)$$

where h_n are suitably chosen isomorphisms from A_n onto A. Further, suppose that the set consisting of all rings with uniform probability distribution and all their subrings is dense in $\mathscr{L}$. Then it is easy to verify that the function H defined by formula (1) is regular and satisfied all axioms (H1)-(H4).

REFERENCES

[1] J. Aczél, On different characterizations of entropies, *Lecture Notes in Mathematics, Probability and Information Theory,* 89 (1969), 1-11.

[2] Z. Daróczy, Über eine Charakterisierung der Shannonschen Entropie, *Statistica,* 27 (1967), 199-205.

[3] R.S. Ingarden, A simplified axiomatic definition of information, *Bull. Acad. Polon. Sci.,* 11 (1963), 209-212.

[4] R.S. Ingarden, Simplified axioms for information without probability, *Prace Mat.,* 9 (1965), 273-282.

[5] R.S. Ingarden, Information theory and thermodynamics of light, I., *Fortschritte der Physik,* 12 (1964), 567-594.

[6] R.S. Ingarden – K. Urbanik, Information as fundamental notion of statistical physics, *Bull. Acad. Polon. Sci.,* 9 (1961), 313-316.

[7] R.S. Ingarden – K. Urbanik, Information without probability, *Colloquium Math.,* 9 (1962), 131-150.

[8] J. Kampé de Feriet, Mesure de l'information fournie par un événement, *Les probabilités sur les structures algébriques,* Paris, 1970, 191-221.

[9] C.E. Shannon, A mathematical theory of communication, *The Bell System Technical Journal,* 27 (1948), 379-423, 623-656.

COLLOQUIA MATHEMATICA SOCIETATIS JÁNOS BOLYAI
9. EUROPEAN MEETING OF STATISTICIANS, BUDAPEST (HUNGARY), 1972.

ON THE MAXIMUM PROBABILITY PRINCIPLE IN STATISTICAL PHYSICS

I. VINCZE

0. INTRODUCTION

In the works of S. Boltzmann and M. Planck a general procedure is elaborated for determining the distribution of the total energy of a system among its single components, when the assumption is made that all the components can be considered as independent and of the same distribution. The method belongs completely to probability theory and is closely connected but not identical with the maximum likelihood method used in mathematical statistics. E.T. Jaynes [4] and independently G.W. Mackey [10] have given information-theoretical formulation of the Boltzmann – Planck method but for the discrete case only. Jaynes remarked that the continuous case is basically more complicated and not at all settled from the point of view of statistical physics.

In his lecture entitled "The theory of the energy distribution law of the normal spectrum" ("Zur Theorie des Gesetzes der Energieverteilung in Normalspectrum") [12] Planck refers to the circumstance that if there is no limit to the division of the total energy into parts then infinitely many different distributions may occur. ("Wenn E als unbeschränkt teilbare Grösse angesehen wird, ist die Verteilung auf unendlich viele Arten möglich" l.c. p. 58.) In the next sentence – considering this circumstance as the reason of the difficulties – he assumes that the total energy consists of a well defined, finite number of parts of the same type and he introduced the well-known quality h. ("Wir betrachten aber – und dies ist der wesentliche Punkt der ganzen Berechnung – E als zusammengesetzt aus einer ganz bestimmten Anzahl endlicher gleicher Teile und bedienen uns dazu der Naturkonstante $h = 6{,}55 \cdot 10^{-23}$ erg. sec.") Planck was confronted here, of course, by a combinatorial difficulty arising from the continuous nature of the distribution in question.

Our aim is to establish an exact and unified formulation of the principle used by Boltzmann and by Planck. The formulation of this maximum probability principle given below is valid for both continuous and discrete random quantities. It makes possible e.g. the derivation of the Planck-distribution (density formula) without the quantum hypothesis. We have to emphasize that we do not oppose the quantum theory or the necessity of the quantum hypothesis which is still now a key to the explanation of a series of physical phenomena. All we claim is that there is no need to introduce a quantum hypothesis to derive the (continuous) energy distribution in the case of black body radiation. – Planck's combinatorial difficulty is of the same nature as occurs when Shannon's formula for the discrete entropy is extended to continuous random variables. The two problems are not independent and so we begin with the treatment of this latter question which we consider from the point of view of the measure theoretical foundation of probability theory given by A.N. Kolmogorov. The problem of the extension of Shannon's entropy formula to continuous random variables was considered by the author [15], [16] and later by Jaynes [5], who came essentially to the same result using a slightly different argumentation. –

The maximum probability principle in the discrete case was formulated by Jaynes as an information theoretical foundation of statistical physics (see also A. Katz [6]). This is rather the reverse of the historical development: Shannon explicitly took his machinery from the concepts of the thermodynamics introduced by Boltzmann.

Planck's distribution will be discussed in §4. It turns out that the maximum probability principle does not yield precisely the Planck density formula, if we assume the Bose – Einstein statistics but instead another formula (not explicitly derivable). Our procedure leads to the Planck formula only if certain elements arising in our procedure are neglected. Consequently the maximum probability principle, if acceptable, implies that the Bose – Einstein statistics and the Planck density formula cannot both be exact, but rather one would be an approximation to the other. – We shall discuss furthermore Einstein's formulation concerning indistinguishable particles; here – at least from the point of view of probability theory – the independence or dependence of the particles (i.e. the corresponding random variables) are essential. Given e.g. two coins – indistinguishable ones – the *HH*, *HT*, *TT* outcomes with probability $\frac{1}{4}, \frac{2}{4}, \frac{1}{4}$ will occur – if they will be thrown independently, even if the coins are indistinguishable. But any other distribution may happen using different kinds of dependence.

As mentioned, the Boltzmann – Planck method has the same character as the maximum likelihood method used in mathematical statistics. We shall point out in §5 that this is an analogy only: the appropriateness (consistency, efficiency) of the maximum likelihood method can be proved under fairly general conditions. The Boltzman – Planck method cannot be justified by any mathematical method; it can be justified only by agreement between theoretical principles and empirical data. However, this question is related to certain problems of probability theory which will be treated as well.

We emphasize repeatedly that our aim is to give an exact formulation of a method which is unobjectionable from the mathematical point of view and to give an interpretation which is plausible, does not need any information theoretical argumentation, or principle unrelated to physical reasoning. As our results do not agree completely with certain formulas of the classical quantum theory it would be interesting to make the customary comparison between theory and experiment. Perhaps this may not require new experiments so much as new numerical studies (in the light of the new formulae) of existing data.

The author expresses his thanks to Professor Imre Fényes for many valuable discussions concerning this topic and to Professor Lee Lorch for the aid in checking the manuscript.

§1. EXTENSION OF SHANNON'S DISCRETE ENTROPY FOR CONTINUOUS RANDOM VARIABLES

Shannon introduced the concept of entropy for a discrete probability distibution $\left(p_1, p_2, \ldots, p_n;\ \sum_{i=1}^{n} p_i = 1\right)$ by means of the formula

$$E_n = \sum_{i=1}^{n} p_i \log \frac{1}{p_i}\ .$$

In the present chapter we accept the following interpretation for E_n: this measures our uncertainty – against the random phenomenon considered – before experimentation. Indeed E_n lies between the bounds 0 and $\log n$. Its smallest value occurs if and only if the distribution is degenerate, i.e. when for some index j, $p_j = 1$ and $p_i = 0$, $i \neq j$, while $E_n = \log n$ in the unique case when uniform distribution is present: $p_1 = p_2 = \ldots = p_n = \frac{1}{n}$. These two cases correspond really to the "smallest" and "largest" uncertainty. The question is now "What is the corresponding formula for a continuous distribution $F(x) = \int_{-\infty}^{x} f(t)dt$ $(-\infty < x < \infty)$?" As known, the quantities E_n belonging to partitions of the real line

$$(x_{0,n} = -\infty < x_{1,n} < x_{2,n} < \ldots < x_{n-1,n} < x_{n,n} = \infty)$$

in general tend to infinity with order of magnitude $\log n$. Indeed when our interest against a phenomenon is highly detailed (e.g. which point of the real line will be the outcome of an experiment) it is not surprising to get an infinite value for the measure of our uncertainty. But we have "infinite entropy" even in certain discrete cases with an infinite number of possible values. For a plausible extension of Shannon's formula to the continuous case we begin with some remarks.

1. We would like to consider which of the quantities

$$E_n = \sum_{i=1}^{n} p_i \log \frac{1}{p_i}$$

and

$$I_n = \log n - E_n = \sum_{i=1}^{n} p_i \log \frac{p_i}{\frac{1}{n}}$$

is a more appropriate measure, for both theoretical and practical purposes of the uncertainty in connection with a random phenomenon. The second quantity I_n can be and usually is taken as "information", showing the opposite character to E_n: it is zero when the uncertainty is maximal and is maximal, when we have no uncertainty concerning the outcome of the experiment (for n given). As

$$I_n + E_n = \log n$$

the two quantities are completely equivalent. However, I_n measures the opposite side of the same property. But as E_n tends to infinity in practically important cases we accept the quantity I_n measuring "uncertainty by means of information". We shall see that under mild conditions concerning the continuous distribution – although E_n tends to infinity – the remaining information I_n will have a finite limit. We shall see furthermore that an interpretation of I_n given by a formula of I.N. Sanov [14] justifies this choice as well.

2. The next remark is closely connected with the measure theoretical foundation of probability theory. According to this theory, a starting point of a stochastic model, i.e. of a random phenomenon, is a triple $\{\Omega, \mathscr{A}, \mathrm{P}\}$, where Ω is the set of the possible outcomes (called elementary events) of the random phenomenon or random experiment; a special element of Ω, i.e. an elementary event will be denoted by ω. $\mathscr{A}$ is a σ-algebra

formed by subsets of Ω, the elements of which are called events. P is a probability distribution (measure) on $\mathscr{A}$, i.e. $\mathsf{P}(A)$ is the probability that the outcome ω of the experiment belongs to the event A, if $A \in \in \mathscr{A}$.

Information theoretical authors referring to Kolmogorov's theory came soon after Shannon's concept to a slightly modified definition of entropy. According to them entropy belongs to a decomposition of the space of elementary event rather than to a probability distribution. From our point of view this circumstance is very essential.

Let $\mathfrak{A}_n = (A_1, A_2, \ldots, A_n)$ be a decomposition of Ω into mutually exclusive events, i.e. $\mathfrak{A}_n$ is a complete system of events:

$$\bigcup_1^n A_i = I; \quad A_i \cap A_j = \phi \qquad (i \neq j), \quad A_i \in \mathscr{A},$$
$$(i, j = 1, 2, \ldots, n,).$$

I denotes here the certain event which corresponds to the "subset" Ω. Let the corresponding probability distribution be

$$\mathscr{P}_n = \left\{ p_i = \mathsf{P}(A_i), \; i = 1, 2, \ldots, n; \; \sum_{i=1}^n p_i = 1 \right\}.$$

Our remarks suggests now the following notations for entropy and information:

$$E(\mathfrak{A}_n) = \sum_{i=1}^n p_i \log \frac{1}{p_i}$$

and

$$I(\mathfrak{A}_n) = \sum_{i=1}^n p_i \log \frac{p_i}{\frac{1}{n}}$$

expressing the fact that they belong to the decomposition $\mathfrak{A}_n$.

An important consequence of this notation is now that information and entropy are not characteristics of the random phenomenon, but of the actual decomposition in which we are interested. In case the same random phenomenon arises from another point of view we must consider then the decomposition

$$\mathfrak{B}_m = \left\{ B_1, B_2, \ldots, B_m ;\ B_i \cap B_j = \phi\ (i \neq j),\ B_i \in \mathscr{A},\ \bigcup_1^m B_i = I \right\}$$

with corresponding probabilities

$$q_i = \mathsf{P}(B_i) \quad \left(i = 1, 2, \ldots, m;\ \sum_{i=1}^m q_i = 1 \right)$$

and the information becomes

$$I(\mathfrak{B}_m) = \sum_{i=1}^m q_i \log \frac{q_i}{\frac{1}{m}} .$$

This quantity differs in general from $I(\mathfrak{A}_n)$. This "subjective" feature of the concept of entropy and information can be motivated simply by the following example:

Suppose two persons are sitting at a roulette table in the Grand Casino in Monte Carlo. Then, the physical phenomenon is the same for both of them. However, if they play different systems, they will have, or at least may have, very different uncertainty (information).

Hence entropy and information are "relative" concepts, they are measures relative to a specified decomposition of the space of elementary events. Consequently it is not surprising that the situation is the same for a continuous random variable.

3. Let $X = X(\omega)$ be a random variable, i.e. an $\mathscr{A}$-measurable function for which the events

$$A_x = \{\omega\colon X(\omega) < x\}$$

are elements of $\mathscr{A}$ for each real value x. Let $F(x)$ be the distribution function of X:

$$\mathsf{P}(A_x) = \mathsf{P}(X < x) = F(x) .$$

If we wish to trace the notion of entropy or of information back to the discrete formula, then decompositions of Ω are needed for $n = 2, 3, \ldots$. Let

$$\mathfrak{A}_n = \{A_1^{(n)}, A_2^{(n)}, \ldots, A_n^{(n)}\}, \qquad (n = 2, 3, \ldots)$$

be a sequence of decompositions of interest. We shall use now a heuristic argument to indicate the nature of a decomposition-system which can be considered as a "reasonable" system or "admissible" system.

Suppose for the sake of simplicity that the random variable $X = X(\omega)$ maps the events $A_i^{(n)}$ in consecutive intervals of the real line:

$$X(\omega) \in (x_{i-1}^{(n)}, x_i^{(n)}), \quad \text{if} \quad \omega \in A_i^{(n)}, \quad (i = 1, 2, \ldots, n).$$

This means that we have a partition of the real line for $n = 2, 3, \ldots$ with point of division:

$$\text{(1)} \qquad \{x_0^{(n)} = -\infty < x_1^{(n)} < x_2^{(n)} < \ldots < x_{n-1}^{(n)} < x_n^{(n)} = \infty\}.$$

Such a system gives a reasonable partition if it has the (usual) separation property: each point $x_i^{(n)}$ belongs to the interval $(x_i^{(n+1)}, x_{i+1}^{(n+1)})$, for $i = 1, 2, \ldots, n-1$; $n = 2, 3, \ldots$. This requirement can be justified in the simplest way if we consider the transition from $n = 2$ to $n = 3$. For $n = 2$ a single point $x_1^{(2)}$ is chosen which means that we are interested only in whether the outcome $X = x$ of the experiment is smaller or larger than $x_1^{(2)}$. But, if assuming this interest it would be meaningless to consider for $n = 3$, whether both $x_1^{(3)}$ and $x_2^{(3)}$ are larger (or smaller) than $x_1^{(2)}$; the plausible situation would be $x_1^{(3)} < x_1^{(2)} < x_2^{(3)}$. Similar reasoning holds for larger values of n. (A more convenient treatment of the question involves choosing the values $n = 2^s$, $s = 1, 2, \ldots$, which means that the next partition comes from the former one by division of the intervals. This helps also in the consideration of the general case, namely when the mapping $X = X(\omega)$ from Ω to the real line has not as simple structure as we here just assumed.)

It can be easily seen that a point system like (1) having the separation property determines uniquely a (left or right continuous) distribution function $\Phi(x)$ the quantiles of which agree with the points of the system (1):

$$\Phi(x_i^{(n)}) - \Phi(x_{i-1}^{(n)}) = \frac{1}{n}, \qquad (i = 1, 2, \ldots, n; \; n = 2, 3, \ldots).$$

The converse is trivial: a distribution function $\Phi(x)$ determines a system (1) having the separation property.

In possession of remarks 1, 2 and 3 we are in the position to derive the information for a continuous random variable belonging to a given decomposition. A partition of the space of elementary event will be considered being in accordance with our interest, and leading to the point-system (1) and to the distribution function $\Phi(x)$.

For finite n we have the complete system of events

$$\mathfrak{A}_n = \{A_i^{(n)} = \{\omega \colon x_{i-1}^{(n)} \leqslant X(\omega) < x_i^{(n)}\}, \; i = 1, 2, \ldots, n\},$$

where

$$\Phi(x_i^{(n)}) - \Phi(x_{i-1}^{(n)}) = \frac{1}{n} \qquad (i = 1, 2, \ldots, n; \; n = 2, 3, \ldots)$$

As $\mathsf{P}(A_i^{(n)}) = F(x_i^{(n)}) - F_{i-1}^{(n)})$, we may write by definition

$$I(\mathfrak{A}_n) = \sum_{i=1}^{n} \mathsf{P}(A_i^{(n)}) \log \frac{\mathsf{P}(A_i^{(n)})}{\frac{1}{n}} =$$

$$= \sum_{i=1}^{n} (F(x_i^{(n)}) - F(x_{i-1}^{(n)})) \log \frac{F(x_i^{(n)}) - F(x_{i-1}^{(n)})}{\Phi(x_i^{(n)}) - \Phi(x_{i-1}^{(n)})}.$$

We assume now that the two measures μ_F and μ_Φ defined on the Borel sets of the real line by the distribution functions $F(x)$ and $\Phi(x)$ respectively (in the usual way) are absolutely continuous with respect to each other: $\mu_\Phi \ll \mu_F$ and $\mu_F \ll \mu_\Phi$, i.e. $\mu_F \sim \mu_\Phi$. But under this assumption the expression $I(\mathfrak{A}_n)$ tends to a limit, which is usually called the I-divergence.

$$\lim_{n \to \infty} I(\mathfrak{A}_n) = \int_{-\infty}^{\infty} dF(x) \log \frac{dF(x)}{d\Phi(x)} = \int_{-\infty}^{\infty} f(x) \log \frac{f(x)}{\varphi(x)} dx,$$

where we have assumed also the existence of the density functions belonging to $F(x)$ and $\Phi(x)$. This expression vanishes if and only if $f(x) = \varphi(x)$ almost everywhere with respect to $F(x)$. The I-divergence was introduced in mathematical statistics by Kullback and Leibler [7], in information theory by Perez [11], used in the theory of stochastic processes by Hájek [3], in theory of central limit theorems by Linnik [9], it was generalized by Csiszár [1].

§2. A FURTHER INTERPRETATION OF I-DIVERGENCE, THE FORMULA OF SANOV

2.1. Let X be random variable defined on the probability field $\{\Omega, \mathscr{A}, \mathsf{P}\}$, let $\Phi(x)$ be its distribution function

$$\mathsf{P}(X < x) = \Phi(x) \qquad (-\infty < x < \infty) .$$

The measure on the Borel sets of the real line defined by the distribution function $F(x)$ will be denoted by μ_F. The set $\mathscr{F}_\Phi$ of distribution functions will be considered for the elements $F(x)$ of which the measures μ_F and μ_Φ are equivalent, i.e. they are absolutely continuous with respect to each other:

$$\mathscr{F}_\Phi = \{F(x)\colon F(x) \text{ distr. function, } \mu_F \sim \mu_\Phi\} .$$

Let $X_1, X_2, \ldots, X_n, \ldots$ be a sequence of independent random variables with the common distribution function $\Phi(x)$:

$$\mathsf{P}(X_i < x) = \Phi(x) \qquad (i = 1, 2, \ldots) .$$

We seek now the probability of the event that the empirical distribution function $\Phi_N(x)$ belonging to the first N elements of the above sequence of random variables lies close to $F(x)$ instead of $\Phi(x)$, where $F(x) \in \mathscr{F}_\Phi$. As known from Glivenko's theorem, $\Phi_N(x)$ tend to $\Phi(x)$ uniformly on the whole real line. Consequently the probability in question will tend to zero as $N \to \infty$. But the following holds:

Theorem (Sanov [14]). *Let $\Phi_N(x)$ be the empirical distribution function of the random variable X after N independent trials; let $\Phi(x)$*

be the distribution function of the random variable X *and let* $F(x)$ *be another distribution function such that* $\int_{-\infty}^{\infty} dF(x) \log \frac{dF(x)}{d\Phi(x)}$ *exists. Then for every* $\epsilon > 0$

$$P(\sup_x |\Phi_N(x) - F(x)| < \epsilon) =$$

$$= \exp\left\{N\left(-\int_{-\infty}^{\infty} \log \frac{dF(x)}{d\Phi(x)} dF(x) + \gamma_\epsilon + \delta_{N,\epsilon}\right)\right\},$$

where

$$\lim_{\epsilon\to 0} \gamma_\epsilon = 0, \qquad \lim_{\epsilon\to 0} \lim_{N\to\infty} \delta_{N,\epsilon} = 0$$

and hence

$$\lim_{\epsilon\to 0} \lim_{N\to\infty} \frac{1}{N} \log P(\sup_x |\Phi_N(x) - F(x)| < \epsilon) =$$

$$= -\int_{-\infty}^{\infty} \log \frac{dF(x)}{d\Phi(x)} dF(x).$$

This means that the probability of the mentioned event tends to zero as N tends to infinity but at the same time its N-th root tends to a certain limit

$$\lim_{\epsilon\to 0} \lim_{N\to\infty} \sqrt[N]{P(\sup_x |\Phi_N(x) - F(x)| < \epsilon)} =$$

(3)

$$= \exp\left\{-\int_{-\infty}^{\infty} \log \frac{dF(x)}{d\Phi(x)} dF(x)\right\}$$

or

$$P(\sup_x |\Phi_N(x) - F(x)| < \epsilon) \approx \exp\left\{-N \int_{-\infty}^{\infty} \log \frac{dF(x)}{d\Phi(x)} dF(x)\right\}.$$

This relation is true for continuous and for discrete distribution as well. Denote by $f(x)$ and $\varphi(x)$ the density functions corresponding to $F(x)$ and $\Phi(x)$ respectively. In the continuous case, the exponent will have the form

$$-\int_{-\infty}^{\infty} f(x)\log\frac{f(x)}{\varphi(x)}\,dx\ ,$$

in the discrete case

$$-\sum_i p_i \log\frac{p_i}{q_i}$$

where p_i and q_i denote the probabilities belonging to the common jump points of the distribution functions $F(x)$ and $\Phi(x)$ respectively.

2.2. Seeking to a slightly different formulation of the theorem of Sanov we make the following remark:

The value of the density function $f(x)$ of a continuous random variable X is a quantity which is proportional to the probability that the event $\{x \leqslant X < x + \Delta\}$ occurs. Analogously for an element $F(x)$ of $\mathscr{F}_\Phi$ we may consider the quantity on the right side of (3) as a value which is "proportional" to the probability that in a large sample the event $\{-\epsilon < \Phi_N(x) - F(x) < \epsilon\}$ occurs. For this reason we shall call the value $\exp\left\{-\int_{-\infty}^{\infty} dF\log\frac{dF}{d\Phi}\right\}$ the pseudo-density or pseudo-likelihood function on $\mathscr{F}_\Phi$ belonging to the element $F(x)$. Similarly the quantity

$$-\int_{-\infty}^{\infty} dF(x)\log\frac{dF(x)}{d\Phi(x)}$$

the negative of the I-divergence will be used as the logarithm of the pseudo-likelihood function.

As is known, the I-divergence has the very important properties – preserved from the corresponding discrete entropy – that a) is nonnegative, b) it is invariant under a monotonic transformation of the x-axis. These properties are in favour of the I-divergence as opposed to Boltzmann's H-function introduced by Shannon as entropy for the continuous case.

§3. PRINCIPLE FOR THE SELECTION OF THE TRUE DISTRIBUTION FUNCTION: THE BOLTZMANN – GIBBS DISTRIBUTION

In mathematical statistics one way to determine (estimate) the true value of a parameter of a distribution function is the maximum-likelihood method. Having a parametric set of distribution functions which can be taken into consideration one can prove that – under mild conditions – this procedure applied to a sample yields a good result, i.e. an asymptotically consistent, efficient estimator.

In statistical physics the problem of determining the distribution function of a random variable is somewhat different in several respects. In many cases no parametric set of available distribution functions is at our disposal. Furthermore no sample is present; perhaps the random variable is not immediately observable or even not observable at all; finally no mathematical proof can be given for the justification of the formula obtained or methods used but agreement with experience will justify the method; this justification may happen by means of direct observations or in a roundabout way.

We turn now to a reformulation of the maximum probability principle introduced into statistical mechanics by S. Boltzmann. As we shall point out, this principle is completely analogous to those used in mechanics. Our formulation is valid for random quantities both with discrete and with continuous distributions and for *independent* random quantities was formulated by Jaynes. In the next chapter we shall generalize this procedure for quantities which cannot be considered independent.

Let us consider a physical system $\mathcal{E}$ (e.g. a particle of an ideal gas) and let X be a random quantity in connection with this system (e.g. energy, velocity of the particle etc.). Our starting point in now that in a certain initial state of the system (e.g. at temperature T_0) the distribution law of X is given by the function $\Phi(x)$:

$$\mathsf{P}(X < x \mid T_0) = \Phi(x) \qquad (-\infty < x < \infty) .$$

The question is now, "If a certain external effect occurs (e.g. change of the temperature of the system) what will be the new distribution of X?"

The following assumptions will be made:

a) The basic behaviour of the random variable X even after the assumed effect will be determined essentially by the law $\Phi(x)$ (or more precisely by the probability measure P mentioned in § 1 Remark 2), although the actual distribution will deviate from it in consequence of the constraint.

b) The set of possible values of the random variable X will remain unchanged even if the external factor has a durable effect. The precise meaning of this requirement is that the measures μ_Φ and μ_F are equivalent where $F(x)$ denotes the distribution after and consequent upon this effect.

c) The random variables $X_1, X_2, \ldots, X_N, \ldots$ belonging to system $\mathscr{E}_1, \mathscr{E}_2, \ldots, \mathscr{E}_N, \ldots$ can be considered as independent in each equilibrium state of the complete gas system.

d) The number N of system (of particles) is large.

The effect (constraint) will be taken into account in the simplest way: for the first moment after the effect on X, the value m will be prescribed different from the value $m_0 = \int_{-\infty}^{\infty} x d\Phi(x)$.

The procedure is now the following: Take the maximum of the pseudo-density, i.e. the "probability" that for a large sample the empirical distribution function lies close to $F(x)$ instead of to the initial $\Phi(x)$:

$$\exp\left\{-\int_{-\infty}^{\infty} \log \frac{dF(x)}{d\Phi(x)}\, dF(x)\right\} = \max$$

under the conditions

$$\int_{-\infty}^{\infty} x dF(x) = m \neq m_0$$

and

$$\int_{-\infty}^{\infty} dF(x) = 1 .$$

The solution of this problem is quite simple and may be obtained by means of the Lagrange-multiplier. We come to the well-known distribution called the Gibbs distribution (or Boltzmann-distribution in the discrete case) having the form

$$dF(x) = c_\mu e^{-\mu x} d\Phi(x) ,$$

where

$$c_\mu = \left(\int_{-\infty}^{\infty} e^{-\mu x} d\Phi(x) \right)^{-1} .$$

To specify $F(x)$, knowledge of $\Phi(x)$ is necessary. In many cases, however, $\Phi(x)$ is not known; certain assumptions can be made on it only. Such an assumption – hypothesis – either will be justified by experience, or not. In the latter case modification of the theory is needed. In any case $\Phi(x)$ *can be assumed either as discrete or as continuous according to the physical background.*

Finally we call attention to the complete analogy between statistical physics and mechanics in this respect. Having e.g. a field, the path of a particle will be determined by the law of this field (which is $\Phi(x)$ in the above treatment) and this will be essentially valid even when a constraint is present which must be taken into consideration and which influences the actual path in a noticeable fashion.

§4. A GENERAL DEFINITION OF THE ENTROPY: THE FUNDAMENTAL DISTRIBUTIONS OF THE QUANTUM MECHANICS

A system containing a large number, say N, of particles will be considered. Let $X_1, X_2, \ldots, X_N$ be the energies as random variables belonging to the units of the system. Dividing the energy-axis into the intervals

$$(x_{i-1}^{(n)}, x_i^{(n)}) \qquad (i = 1, 2, \ldots, n)$$

for each $n = 2, 3, \ldots$ the distribution of the particles into these intervals will be considered. If these intervals are of the same "probability" (e.g. the intervals correspond to the sets of equal Lebesgue measure in the phase space), then the expression

$$\frac{N!}{N_1!\,N_2!\ldots N_n!}, \qquad \sum_{i=1}^{n} N_i = N$$

– with N_i the number of particles in the ith interval – is called the "thermodynamic probability" of the system. The principal term for large N of its logarithm is used to define the Gibbs function

$$G = -k \sum_{i=1}^{n} \frac{N_i}{N} \log \frac{N_i}{N}.$$

Under the condition that the total energy of the system is given, this yields the entropy formula, which was identified by Boltzmann with the "thermodynamic entropy".

From now on we shall use a slightly more general and at the same time more exact formula for the thermodynamic probability:

$$(1) \qquad \frac{N!}{N_1!\,N_2!\ldots N_n!}\, q_1^{N_1} q_2^{N_2} \ldots q_n^{N_n},$$

where q_i denotes the probability that the value of the particle lies in the interval $(x_{i-1}^{(n)}, x_i^{(n)})$ in a basic (initial) state of the system. $\frac{1}{N}$ times the logarithm of this quantity tends to the pseudodensity – introduced in §2 – when we assume that in a large ensemble the energy distribution $F(x)$ will be realized instead of the original $\Phi(x)$ belonging to an initial state of the system. Roughly speaking: as $N \to \infty$, the relative frequencies $\frac{N_i}{N}$ will tend to $F(x_i^{(n)}) - F(x_{i-1}^{(n)})$ instead of tending to $\Phi(x_i^{(n)}) - \Phi(x_{i-1}^{(n)})$. Thus we come to the expression for a generalized Gibbs function

$$-k \int_{-\infty}^{\infty} \log \frac{dF(x)}{d\Phi(x)} dF(x) .$$

Our starting formula (1) was constructed under the assumption that the energies belonging to the different particles as random variables are independent. This is the reason why the multinomial distribution was used. If the assumption of independence does not hold, then the determination of the corresponding probability will be done under the new assumption and accordingly will change the Gibbs function for which we give now the following more general definition.

Having a system containing N particles let us denote by $P_N(\Phi(x), F(x); \epsilon)$ the probability of the event that the random quantity X belonging to the units of the system will be distributed, not as in the initial distribution $\Phi(x)$, but according to the law $F(x)$ in the sense that for the empirical distribution function the event (relation) $\{|\Phi_N(x) - F(x)| < \epsilon\}$ holds.

Then the generalized Gibbs function will be defined by the expression

(3) $$\lim_{\epsilon \to 0} \lim_{N \to \infty} \frac{1}{N} \log P_N(\Phi(x), F(x); \epsilon) .$$

Thus, to determine the entropy, the joint distribution of the random variables $X_1, X_2, \ldots, X_N$ is to be taken into consideration, since independence is not assumed.

We turn now to the determination of the entropy in the case when the interaction of the particles (i.e., the correlation of the random variables) results in the so-called Bose – Einstein statistics.

For the calculation of the expression (3) – referring to the cited article of Sanov – we shall determine the following limit:

$$\lim_{n \to \infty} \lim_{N \to \infty} \frac{1}{N} \log \mathrm{P}\left(\frac{\nu_i}{N} = \frac{N_i}{N} \qquad (i = 1, 2, \ldots, n)\right)$$

$$\left(\sum_{i=1}^{n} \nu_i = \sum_{i=1}^{n} N_i = N\right),$$

where the random variables ν_i $(i = 1, 2, \ldots, n)$ denote the frequencies of the intervals $(x_{i-1}^{(n)}, x_i^{(n)})$ respectively. As $N \to \infty$ we let $\frac{N_i}{N}$ tend instead of $\Phi(x_i^{(n)}) - \Phi(x_{i-1}^{(n)})$ to $F(x_i^{(n)}) - F(x_{i-1}^{(n)})$.

For the sake of simplicity the energy will be divided into n parts having equal (a priori) probabilities

$$\Phi(x_i^{(n)}) - \Phi(x_{i-1}^{(n)}) = \Delta\Phi(x_i^{(n)}) = \frac{1}{n} \qquad (i = 1, 2, \ldots, n).$$

The same number z of cells will be chosen in each interval $(x_{i-1}^{(n)}, x_i^{(n)})$ with the common probability (size) $h = \frac{1}{nz}$. With the usual notation the known relation

$$z_i = \frac{1}{nh} = \frac{\Phi(x_i^{(n)}) - \Phi(x_{i-1}^{(n)})}{h} = \frac{d\Phi(x_i^{(n)})}{h}$$

is valid.

We replace now the assumption c) introduced in §3 by the following one:

c*) The underlying random variables are not independent, they have a joint distribution, which results for the probabilty $P_{n,N}$ of the event $\left\{\frac{\nu_1}{N}, \frac{\nu_2}{N}, \ldots, \frac{\nu_n}{N}\right\}$ in the expression:

$$P_{n,N} = \frac{\prod_{i=1}^{n} \binom{N_i + z - 1}{N_i}}{\binom{N + nz - 1}{N}}$$

This is the case of the Bose – Einstein statistic.

We assume now besides a), b), c*), d) the additional assumption.

e) The number of cells is the same order of magnitude as the number of particles in an interval, (i.e. in z cells); for the sake of simplicity the choice $N = nz$ will be applied.

Using the asymptotic relation $N! \sim \left(\frac{N}{e}\right)^N \sqrt{2\pi N}$ we obtain for our joint probability the asymptotic relation

$$P_{n,N} \sim \frac{N^N (nz-1)^{nz-1}}{(N+nz-1)^{N+nz-1}} \prod_{i=1}^{n} \frac{\left(\frac{N_i}{N}+\frac{1}{n}\right)^{N_i+z}}{\left(\frac{N_i}{N}\right)^{N_i}\left(\frac{1}{n}\right)^z}.$$

For the Nth root of the term before the sign Π we have

$$\frac{\left(1+\frac{n}{N}\right)^{1+\frac{n}{N}}}{\left(2+\frac{n}{N}\right)^{2+\frac{n}{N}}} \to \frac{1}{4}, \qquad \text{as} \qquad \frac{n}{N} \to 0.$$

In the product term we have to take into consideration our assumption 2 in §3 according to which the measure μ_F and μ_Φ are absolutely continuous with respect to each other. This will be satisfied if we assure the existence of the measurable function (the Randon – Nikodym derivative)

$$\psi(x) = \frac{dF(x)}{d\Phi(x)} = \lim_{n\to\infty} \frac{F(x_i^{(n)}) - F(x_{i-1}^{(n)})}{\Phi(x_i^{(n)}) - \Phi(x_{i-1}^{(n)})},$$

where $x \in (x_{i-1}^{(n)}, x_i^{(n)})$ for each n. But we can write our probability in the form

$$P_{n,N} \sim \frac{1}{4^N} \prod_{i=1}^{n} \left(1+\frac{\frac{1}{n}}{\frac{N_i}{N}}\right)^{N_i+z} \left(\frac{\frac{N_i}{N}}{\frac{1}{n}}\right)^z.$$

As $\frac{N_i}{N} = \Delta F(x_i^{(n)})$, $\frac{1}{n} = \Delta\Phi(x_i^{(n)})$ the following relation is valid:

$$\frac{1}{N}\log P_{n,N} \sim \log\frac{1}{4} +$$

$$+ \sum_{i=1}^{n}\left[(\Delta F(x_i^{(n)}) + \Delta\Phi(x_i^{(n)})\log\left(1+\frac{\Delta\Phi(x_i^{(n)})}{\Delta F(x_i^{(n)})}\right) + \right.$$

$$\left. + \Delta\Phi(x_i^{(n)})\log\frac{\Delta F(x_i^{(n)})}{\Delta\Phi(x_i^{(n)})}\right].$$

Tending with N and n to infinity we obtain for the thermodynamic probability

$$H_{B\text{-}E} = -k\int dF \log \frac{dF(x)}{(1+\psi(x))^{1+\frac{1}{\psi(x)}}\, d\Phi(x)} =$$

$$= -k\int \psi(x) \log \frac{\psi(x)}{(1+\psi(x))^{1+\frac{1}{\psi(x)}}}\, d\Phi(x)\,.$$

By means of a similar procedure we obtain for the thermodynamic probability in the case of the Fermi – Dirac statistic the following expression:

$$H_{F\text{-}D} = -k\int dF \log \frac{dF}{\left(1-\frac{1}{\alpha}\psi(x)\right)^{\frac{\alpha}{\psi(x)}-1} d\Phi(x)} \qquad (\alpha > 1)\,.$$

The solution of the conditional extreme-value problem

$$H = \text{maximum}$$

$$\int x dF(x) = m \neq \int x d\Phi(x)\,, \qquad \int dF(x) = 1$$

for the different kind of entropies will be treated in the next paragraph.

§5. ON THE MAXIMUM PROBABILITY PRINCIPLE

As mentioned earlier, principles (methods) used in the mathematical statistics (maximum likelihood, least squares, momentum) are justified by proving theorems on their advantageous statistical properties (unbiasedness, consistency, efficiency etc.). But the legitimisation of principles in physics can only be done by comparing its theoretical consequences with experience, using observations or experiments. This is the situation with the maximum probability principle described above.

However the following investigation can be considered in a certain respect as a justification of our argumentation also.

As pointed out earlier, a probabilistic model starts with a triple

$\{\Omega, \mathscr{A}, \mathsf{P}\}$ called the probability field, where the measure P is the valid quantitative rule for any event in this frame. This distributions of the very different random variables occurring in the model are subject to this measure: denoting by Y a random variable, its distribution function $G(y)$ can be expressed in the form:

$$G(y) = \mathsf{P}(\{Y < y\})$$

where the event $\{Y < y\}$ is a subset of Ω and belongs to $\mathscr{A}$.

Returning to the question of our principle we can formulate the basic problem in the following way: knowing that the common distribution of the random variables $X_1, X_2, \ldots$ is $\Phi(x)$ what can we state on the distribution of – say – X_1 when we observed that in a large sample the arithmetic mean $\bar{X}$ lies close to a value m, which is different from the original $m_0 = \int_{-\infty}^{\infty} x d\Phi(x)$. In language of the probability theory, the limit of the conditional probability

$$\mathsf{P}(X_1 < x \mid Nm < X_1 + X_2 + \ldots + X_N < N(m + \Delta m))$$

is to be determined when N tends to infinity.

If this conditional probability converges with N to the distribution obtained by the maximum probability method then this fact may be considered as a support of the procedure.

In fact as was proved by P. Bártfai [17] the following statement is true:

Let $X_1, X_2, \ldots$ be independent random variables with the common distribution function $\Phi(x)$ for which the function $R(t) = \int_{-\infty}^{\infty} e^{yt} d\Phi(y)$ in an open interval around $t = 0$ exists, further the equation $\frac{R'(t)}{R(t)} = a$ has a solution in t for the values $|a - m| < \epsilon$ (ϵ small, but $\epsilon > 0$). Under these conditions the limiting relation

$$\lim_{N \to \infty} \mathrm{P}(X_1 < x \mid Nm < X_1 + X_2 + \ldots + X_N < N(m + \Delta m)) =$$

$$= \frac{\int_{-\infty}^{x} e^{-yt(m)} d\Phi(y)}{\int_{-\infty}^{\infty} e^{-yt(m)} d\Phi(y)}$$

holds, where $t(m)$ is the solution of the equation $\frac{R'(t)}{R(t)} = m.$

This theorem can be considered as a basis of a correspondence principle in the probability theory covering e.g. the following procedure:

Let $X_1, X_2, \ldots$ be independent, identically distributed random variables with the common distribution function $\Phi(x)$, the first moment of which exists. Consider the (not very easy) problem of the determination of the conditional probability (distribution) in the limit:

$$F(x) = \lim_{N \to \infty} \mathrm{P}(X_1 < x \mid Nm < X_1 + X_2 + \ldots + X_N < < N(m + \Delta m)) .$$

The following simpler problem has the same solution: Determine $F(x)$ so that

$$\exp\left(- \int_{-\infty}^{\infty} dF(x) \log \frac{dF(x)}{d\Phi(x)} \right)$$

should be maximal under the conditions

$$\int_{-\infty}^{\infty} x dF(x) = m \neq \int_{-\infty}^{\infty} x d\Phi(x) ,$$

$$\int_{-\infty}^{\infty} dF(x) = 1 .$$

This means that for solving a problem concerning conditional probability we turn to the maximum probability principle separating the probability of the event which is to be maximized and the condition which is added independently.

The solution of the latter problem is based on the simple non-negativity property of the I-divergence and is the following:

Introducing the Lagrangean multipliers λ and μ the unconditional extreme value problem

$$\int_{-\infty}^{\infty} dF(x)\log\frac{F(x)}{\Phi(x)} + \lambda\int_{-\infty}^{\infty} dF(x) + \mu\int_{-\infty}^{\infty} x\,dF(x) = \min .$$

is to be solved.

A slight modification of this expression results in the form

$$\int_{-\infty}^{\infty} dF(x)\log\frac{dF(x)}{e^{-\lambda-\mu x}\,d\Phi(x)} = \min .$$

Introducing the constant $c(\lambda, \mu)$ with the choice

$$c(\lambda, \mu) = \left(\int_{-\infty}^{\infty} e^{-\lambda-\mu x}\,d\Phi(x)\right)^{-1}$$

we have for the expression to be minimized

$$\int_{-\infty}^{\infty} dF(x)\log\frac{dF(x)}{c(\lambda, \mu)e^{-\lambda-\mu x}\,d\Phi(x)} + c(\lambda, \mu) .$$

For λ and μ fixed the minimum value of this expression coincides with the minimum of the first term which being nonnegative and having its smallest value (i.e. is zero) if and only if

$$dF(x) = c(\lambda, \mu)e^{-\lambda-\mu x}\,d\Phi(x)$$

i.e. the stated result is obtained.

The determination of the extremal distribution for $H_{B\text{-}E}$ and $H_{F\text{-}D}$ is not as simple as in the above case the problem being non-linear. An exact solution can be obtained in concrete cases by means of numerical methods only. If we neglect in the exponent the term $\frac{1}{\psi(x)}\left(\frac{\alpha}{\psi(x)}\right)$, further if we disregard the fact that the constants $c(\lambda, \mu)$ in these case depend on the unknown $\psi(x)$ too, then we come to the classical distributions:

$$dF(x) = \frac{c(\lambda, \mu) d\Phi(x)}{e^{\lambda + \mu x} - 1}$$

in the Bose – Einstein case,

$$dF(x) = \frac{c(\lambda, \mu) d\Phi(x)}{e^{\lambda + \mu x} + 1}$$

in the Fermi – Dirac case.

REFERENCES

[1] I. Csiszár, On generalized entropy, *Studia Sci. Math. Hungar.*, 4 (1969), 401-419.

[2] J. Fritz, Information Theory and Thermodynamics of Gas Systems, *Proc. Coll. on Information Theory,* Debrecen (Hungary), 1967.

[3] J. Hájek, A property of *I*-divergence of marginal probability distributions, *Czechoslovac Journal of Mathematics,* 8 (1958), 460-463.

[4] E.T. Jaynes, Information Theory and Statistical Mechanics, *Physical Reviews,* 106 (1957), 620-630.

[5] E.T. Jaynes – K.W. Ford, Ed.: *Statistical Physics,* ch. N, W.A. Benjamin Inc., 1963.

[6] A. Katz, *Principles of Statistical Mechanics,* W.H. Freeman and Company, Sanfrancisco – London, 1967.

[7] S. Kullback – R.A. Leibler, On information and sufficiency, *Ann. Math. Statist.*, 22 (1951), 79-86.

[8] S. Kullback, *Information and statistics,* John Wiley and Sons, New York, 1959.

[9] Ju.W. Linnik, Information-theoretical proof of the central limit theorem under Lindeberg's condition (in Russian), *Teorija Verojatn. i Primenen.*, 4 (1959), 311-321.

[10] G.W. Mackey, Quantum mechanics and Hilbert space, *Amer. Math. Monthly,* 64 (1957), 45-57.

[11] A. Pérez, Notions generalisées d'incertitude, d'entropie et d'information du point de vue de la théorie de martingales, *Transactions of the first Prague Conference on Information Theory,* (1957), 183-208.

[12] M. Planck, Zur Theorie des Gesetzes der Energieverteilung in Normalspektrum, *Verhandl. der Deutsch. Phys. Ges.,* (1900).

[13] A. Rényi, On the dimension and entropy of probability distributions, *Acta Math. Acad. Sci. Hungar.,* 10 (1959), 193-215.

[14] I.N. Sanov, On the probability of large deviations of random variables, *IMS and AMS Selected translations in Mathematical Statistics and Probability,* 1 (1961), 213-244.

[15] I. Vincze, An interpretation of the *I*-divergence of information theory, *Transactions of the second Prague Conference on Information Theory,* (1959), 681-684.

[16] I. Vincze, Some questions concerning the probabilistic concept of information, *IMS and AMS Selected Translations in Mathematical Statistics and Probability,* 5 (1965).

[17] P. Bártfai, On a conditional limit theorem, *Colloquia Math. Soc. J. Bolyai,* No. 9, *European Meeting of Statisticians,* Budapest, 1972, 81-91.

COLLOQUIA MATHEMATICA SOCIETATIS JÁNOS BOLYAI
9. EUROPEAN MEETING OF STATISTICIANS, BUDAPEST (HUNGARY), 1972.

EXCHANGEABILITY AND ASYMPTOTIC RANDOM MATRIX SPECTRA

K.W. WACHTER

The asymptotic behaviour of singular values and eigenvalues of random matrices is of interest to the statistical interpretation of principal component analyses, discriminant analyses, and canonical correlations for large samples of many variates as well as to physical theories of nuclear energy levels.* This behaviour is conveniently studied through the empirical measure associated with the set of singular values or eigenvalues of a matrix, the pure point measure assigning equal mass to each of the singular values or eigenvalues. Once loosely called *eigenvalue density*, this measure has lately been called the *stochastic spectrum***, *random spectrum*, or *raspectrum****. In asymptotic theory one considers the limiting distributions of raspectra for a sequence of probability measures Q_n on spaces

*A.P. Dempster [4], Sections 7.6, 9.2, 10.4, and especially Sections 13.8, 13.9; M.L. Mehta [8].

**U. Grenander [5], Section 7.5.3.

***K.W. Wachter [10].

of matrices of increasing dimensions. A great deal of work has been devoted to finding the limit distribution for certain special cases of Q_n*. In all of the cases cited, this distribution has turned out to be degenerate, and under strong conditions almost-sure sample raspectral convergence to the degenerate limit can usually be proved.

Previous work on raspectra has relied almost entirely on special properties of special distributions Q_n. The present paper is one part of a larger attempt to build a more general theory of raspectra examining in an abstract context the symmetry properties responsible for regularities of raspectral convergence. It takes a different approach from the published works with which the author is acquainted and investigates directly the structure of infinite singular value distributions for spaces of infinite matrices, for their own sake as well as for asymptotic theorems, calling to its aid the topological machinery of weak convergence. As an illustration of the power of this approach, the present paper will exploit the exchangeability of singular value distributions to prove that a limiting distribution of the raspectrum is degenerate if and only if the singular values are asymptotically independent.

The singular values of a $p \times m$ dimensional real rectangular matrix Z are the eigenvalues of the $(p + m) \times (p + m)$ symmetric matrix with Z in its upper right corner and Z transpose in its lower left, and zeros in the remaining, diagonal blocks. The singular values of a symmetric matrix are its eigenvalues up to signs, so eigenvalues of symmetric matrices are a special case of singular values. Like eigenvalues, singular values of course are produced by a matrix in no particular order. They can be regarded as exchangeable random variables, and this is the property with which we shall be concerned in the present paper. We shall assume that we are given a sequence of probability measures P_n which are effectively measures on R^n but will all be taken to be defined on R^∞ and to satisfy two conditions:

$$(1.1) \qquad P_n\{\theta \in R^\infty : \theta_{n+1} = \theta_{n+2} = \ldots = 0\} = 1$$

*Mehta [8] with references; Wigner [12]; Grenander [5]; Olson and Uppuluri [9]; Charles Stein (personal communication); Mallows and Wachter [7]; Arnold [1], [2].

(1.2) P_n is invariant under permutation of the first n coordinate indices of θ.

We envisage P_n as a joint distribution of singular values for random $p(n) \times m(n)$ matrices with $p + m = n$, scaled by $\frac{1}{\sqrt{n}}$, but we shall not use special properties of such distributions beyond exchangeability.

To define convergence, we place the product topology on R^∞, the topology of coordinate-wise convergence, and we require the P_n's to be Borel measures. The product topology is imposed by the metric

$$M(\theta, \gamma) = \sum_{i=1}^{\infty} \frac{|\theta_i - \gamma_i|}{2^i(1 + |\theta_i - \gamma_i|)}$$

under which R^∞ is a complete separable metric space. We let $\mathscr{L}R$ and $\mathscr{L}R^\infty$ denote the spaces of Borel probability measures on R and R^∞ with the topology of weak convergence. For properties of $\mathscr{L}R$ and $\mathscr{L}R^\infty$ we shall rely continually on P. Billingsley's lucid comprehensive account *Convergence of Probability Measures* [3]. We write $P_n \Longrightarrow P$ for "P_n converges weakly to P" and use the squiggle $\rightsquigarrow$ for convergence of random elements in distribution.

Definition 1.3. The map $S_k: R^\infty \longrightarrow \mathscr{L}R$ takes θ into the pure point measure which assigns mass $\frac{1}{k}$ to each of the real numbers $\theta_1, \theta_2, \ldots, \theta_k$. $\mathscr{F}_{nk}$ is the random element of $\mathscr{L}R$ which is distributed like the image $S_k(\theta)$ when θ has distribution P_n. Stated differently, $P\{\mathscr{F}_{nk} \in B\} = P_n\{\theta: S_k(\theta) \in B\}$ for Borel subsets B of $\mathscr{L}R$: $\mathscr{F}_{nk}$ has distribution given by the pull-back measure $P_n S_k^{-1}$. For any other measure P_∞ in $\mathscr{L}R^\infty$, $\mathscr{F}_{\infty k}$ is defined by the same token to be the random element in $\mathscr{L}R$ with distribution given by $P_\infty S_k^{-1}$.

We shall prove the following results:

Theorem 2.4. P_n *being invariant under permutations of the first* n *coordinates of* $\theta \in R^\infty$, *if* $P_n \Longrightarrow P_\infty$ *then* $\mathscr{F}_{nk} \underset{n}{\rightsquigarrow} \mathscr{F}_{\infty k}$ *and there exists a random element* $\mathscr{F}_{\infty\infty}$ *in* $\mathscr{L}R$ *such that both* $\mathscr{F}_{\infty k} \underset{k}{\rightsquigarrow} \mathscr{F}_{\infty\infty}$

and $\mathscr{F}_{jj} \underset{j}{\rightsquigarrow} \mathscr{F}_{\infty\infty}$.

Corollary 2.5. P_n *converges weakly iff (if and only if)* $\mathscr{F}_{nn}$ *converges in distribution.*

Theorem 3.1. *If* $\mathscr{F}_{nn} \underset{n}{\rightsquigarrow} \mathscr{R}$ *then* P_n *converges weakly to a measure* P_∞ *on* R^∞ *which equals almost* P_∞*-surely a conditional infinite product-power measure conditional on a function from* R^∞ *to* $\mathscr{L}R$ *whose image value is distributed like* $\mathscr{R}$.

Corollary 3.2. $\mathscr{F}_{nn}$ *converges in probability to a fixed element* F *in* $\mathscr{L}R$ *if* P_n *converges weakly to an infinite unconditional product-power measure on* R^∞*; iff, in other words,* $\theta_1, \theta_2, \ldots$ *are asymptotically independent.*

Two kinds of changes affect the raspectrum $\mathscr{F}_{jj}$ as j grows to infinity. First, the distribution of singular values changes, because the probability space of matrices and hence the measure P_n changes. Secondly, the random distribution function corresponding to $\mathscr{F}_{jj}$ acquires more and smaller steps, because the number of singular values reckoned in $\mathscr{F}_{jj}$ increases as the dimensions increase. In applications, one describes P_n by a distribution function on R^n and the marginal distribution functions for any fixed k out of the whole set of singular values may well converge to a limit as $\mathsf{P}_n \Longrightarrow \mathsf{P}_\infty$ in $\mathscr{L}R^\infty$. But a limiting distribution function for all n out of n singular values as n goes to infinity is an absurdity. Therefore it is of great practical use to be able to separate n and k and to study $\mathscr{F}_{nk}$ as $n \to \infty$ before k. But it is far from obvious that the double limit in distribution with $n \to \infty$, $k \to \infty$ will equal the single limit of $\mathscr{F}_{jj}$ as $j \to \infty$, for there is no guarantee that convergence be uniform. The proof of the equality of the limits in Theorem 2.4 is somewhat technical, and we first require three lemmas.

Lemma 2.1. *The map* $\mathsf{S}_k : R^\infty \longrightarrow \mathscr{L}R$ *defined in* 1.3 *is continuous between the product topology on* R^∞ *and the topology of weak convergence on* $\mathscr{L}R$.

Proof. Although the proof of continuity is simple, we write it out explicitly in order to emphasize two points, first, that continuity of the map S_k has nothing to do with continuity or discontinuity of the distribution functions of the image measures $S_k(\theta)$ for any θ, and, second, that the continuity argument would fail if k grew to infinity.

The map S_k is continuous as long as expectations of any continuous function g with respect to $S_k(\theta^{(t)})$, written $\int g dS_k(\theta^{(t)})$, converge to $\int g dS_k(\gamma)$ whenever all the coordinates $\theta_i^{(t)}$ of $\theta^{(t)}$ converge to the coordinates γ_i of γ. When the coordinates converge, they converge uniformly for $i = 1, 2, \ldots, k$ but not for $i = 1, 2, \ldots, k, \ldots$. The continuity of g then implies $g(\theta_i^{(t)}) \longrightarrow g(\gamma_i)$ uniformly for $i \leqslant k$, so the average given by $\int g dS_k(\theta^{(t)})$, specifically $\frac{1}{k}\sum_{i=1}^{k} g(\theta_i^{(t)})$ converges to the average $\frac{1}{k}\sum_{i=1}^{k} g(\gamma_i)$, which is $\int g dS_k(\gamma)$.

Q.E.D.

Lemma 2.2. *A separable locally compact space* Y *carries a countable collection* $g_1, g_2, \ldots$ *of bounded continuous real functions which form a convergence-determining class for* $\mathscr{L}Y$.

Proof. Y is sigma-compact; the class of continuous functions with compact support is separable in the supremum norm. As sketched in [3], p. 41, problem 7, the class is a convergence-determining class. A countable subclass exists by separability dense in this class; being dense in a convergence-determining class, it is itself a convergence-detemining class.

Q.E.D.

Lemma 2.3. *If* $P_n \Longrightarrow P_\infty$ *in* $\mathscr{L}R^\infty$, *then* P_∞ *is invariant under all finite permutations of coordinate indices and* $S_k(\theta)$ *in* $\mathscr{L}R$ *converges as* $k \to \infty$ *for* θ *in a set of* P_∞*-measure one.*

Proof. If α is a permutation of $1, 2, \ldots, k$, then the map $\underline{\alpha}$: $\langle \theta_1, \theta_2, \ldots \rangle \longmapsto \langle \theta_{\alpha^{-1}(1)}, \theta_{\alpha^{-1}(2)}, \ldots, \theta_{\alpha^{-1}(k)}, \theta_{k+1}, \ldots \rangle$ is continuous. $P_n \Longrightarrow P_\infty$ implies $P_n\underline{\alpha}^{-1} \Longrightarrow P_\infty\underline{\alpha}^{-1}$ for the pull-back measures by the continuous map $\underline{\alpha}$. By 1.2, for all $n \geqslant k$, $P_n\underline{\alpha}^{-1} = P_n$, so $P_n \Longrightarrow P_\infty\underline{\alpha}^{-1}$. By the uniqueness of weak limits we conclude that $P_\infty\underline{\alpha}^{-1} = P_\infty$.

By the uniqueness of weak limits we conclude that $\mathsf{P}_\infty \underline{\alpha}^{-1} = \mathsf{P}_\infty$.

Instead of adapting to $\mathscr{L}R$ the argument on p. 365 of [6] for distribution functions, we take a more abstract approach*. Let Σ_k be the sigmafield of Borel subsets of R^∞ invariant under the group of permutations of the first k coordinate indices of θ. Put $\Sigma_\infty = \bigcap_k \Sigma_k$. Because R is separable and locally compact, there exists by Lemma 2.2 a countable convergence-determining class of bounded continuous functions, $g_1, g_2, \ldots$ from R to R. $\mathsf{S}_k(\theta) \Rightarrow \mathsf{F}$ in $\mathscr{L}R$ iff $\int g_i d\mathsf{S}_k(\theta) \underset{k}{\longrightarrow} \int g_i d\mathsf{F}$, for all i. Let us write E_∞ for the expectation operator for P_∞. Now $\int g_i d\mathsf{S}_k(\theta)$ happens to be a conditional expectation. It equals $\frac{1}{k}\sum_{j=1}^{k} g_i(\theta_j)$ which we may write as $\mathsf{E}_\infty(g_i(\theta_1(\theta)) \mid \Sigma_k)$ almost P_∞-surely. The function g_i is bounded by assumption. The reversed martingale convergence theorem, as in [6] p. 396, then shows that

$$\mathsf{E}_\infty(g_i(\theta_1(\theta)) \mid \Sigma_k) \longrightarrow \mathsf{E}_\infty(g_i(\theta_1(\theta)) \mid \Sigma_k)$$

for θ in a P_∞-sure set depending on i. The intersection of these sets for all the countably many i is a P_∞-sure subset of R^∞ on which $\mathsf{S}_k(\theta)$ converges in $\mathscr{L}R$.

Q.E.D.

We are now in a position to prove Theorem 2.4.

Theorem 2.4. *If* $\mathsf{P}_n \Longrightarrow \mathsf{P}_\infty$, *then* $\mathscr{F}_{nk} \underset{n}{\rightsquigarrow} \mathscr{F}_{\infty k}$, $\mathscr{F}_{\infty k} \underset{k}{\rightsquigarrow} \mathscr{F}_{\infty\infty}$ *and also* $\mathscr{F}_{jj} \underset{j}{\rightsquigarrow} \mathscr{F}_{\infty\infty}$.

Proof of Theorem 2.4. The crux of the proof is the computation in 2.4.3 to establish the equality of certain double limits in n and k with diagonal limits $n = k = j \to \infty$. This computation would remain valid if instead of the real random variables $X_{nk}(i)$ which appear there we substituted the values at points x_i of the random distribution functions of the measures $\mathscr{F}_{nk}$. We might operate with distribution functions entirely, at the price of a discussion of which x_i would be enough and which too

*This approach is suited for generalizations of De Finetti's Theorem to groups more general than permutation groups, work on which the present author is now engagaed.

many to confirm the Theorem. Instead, we handle such topological preliminaries in 2.4.2 in the more general context of random elements of $\mathscr{X}R$. In 2.4.1 we begin the proof by verifying the easy limits $n \to \infty$ and $k \to \infty$.

2.4.1. Proof that $\mathscr{F}_{nk} \underset{n}{\rightsquigarrow} \mathscr{F}_{\infty k}$ and $\mathscr{F}_{\infty k} \underset{k}{\rightsquigarrow} \mathscr{F}_{\infty\infty}$.

We define the random element $\mathscr{F}_{\infty\infty}$ of $\mathscr{X}R$ by the requirement for every Borel subset B of $\mathscr{X}R$ that

$$\mathsf{P}\{\mathscr{F}_{\infty\infty} \in B\} = \mathsf{P}_\infty\{\theta\colon \lim_k \mathsf{S}_k(\theta) \text{ exists in } B\}.$$

By Lemma 2.3 there is a set of θ in R^∞ with P_∞-measure one such that the sequence of measures $\mathsf{S}_k(\theta)$ converges. Therefore the definition of $\mathsf{P}\{\mathscr{F}_{\infty\infty} \in B\}$ does yield a proper probability measure. Since almost-sure convergence entails convergence in distribution, the random elements $\mathscr{F}_{\infty k}$ with distributions $\mathsf{P}_\infty \mathsf{S}_k^{-1}$ converge in distribution to $\mathscr{F}_{\infty\infty}$. By Lemma 2.1 the map S_k is a continuous map and therefore $\mathsf{P}_n \mathsf{S}_k^{-1} \Longrightarrow \mathsf{P}_\infty \mathsf{S}_k^{-1}$ follows directly from $\mathsf{P}_n \Longrightarrow \mathsf{P}_\infty$, i.e. $\mathscr{F}_{nk} \underset{n}{\rightsquigarrow} \mathscr{F}_{\infty k}$.

2.4.2. The sufficiency of $X_{jj} \underset{j}{\rightsquigarrow} X_{\infty\infty}$.

If $g_1, g_2, \ldots$ is any countable collection of bounded continuous functions, put

$$X_{nk}(i) = \int g_i(t)d\mathscr{F}_{nk}(t)$$

$$X_{\infty k}(i) = \int g_i(t)d\mathscr{F}_{\infty k}(t)$$

$$X_{\infty\infty}(i) = \int g_i(t)d\mathscr{F}_{\infty\infty}(t)$$

X_{nk}, $X_{\infty k}$, and $X_{\infty\infty}$ are random elements of R^∞. Their distributions are pull-back measures of the distributions of the corresponding $\mathscr{F}$'s by a continuous map, namely the map α from R to R^∞ mapping G into $\langle \int g_1 d\mathrm{G}, \int g_2 d\mathrm{G}, \ldots \rangle$. This map is continuous in the topology of coordinatewise convergence on R^∞ and weak convergence in $\mathscr{X}R$ by the very definition of weak convergence. Therefore $\mathscr{F}_{nk} \underset{n}{\rightsquigarrow} \mathscr{F}_{\infty k}$ implies $X_{nk} \underset{n}{\rightsquigarrow} X_{\infty k}$ and $\mathscr{F}_{\infty k} \underset{k}{\rightsquigarrow} \mathscr{F}_{\infty\infty}$ implies $X_{\infty k} \underset{k}{\rightsquigarrow} X_{\infty\infty}$.

In a similar manner, if it is indeed true, as we intend to prove, that $\mathscr{F}_{jj} \underset{j}{\rightsquigarrow} \mathscr{F}_{\infty\infty}$, then we may expect to find that $X_{jj} \underset{j}{\rightsquigarrow} X_{\infty\infty}$. Therefore, we try to prove that $X_{jj} \rightsquigarrow X_{\infty\infty}$ and look for a converse to deduce $\mathscr{F}_{jj} \rightsquigarrow \mathscr{F}_{\infty\infty}$ back from $X_{jj} \rightsquigarrow X_{\infty\infty}$, for appropriate collections of functions g. We can deduce $\mathscr{F}_{jj} \rightsquigarrow \mathscr{F}_{\infty\infty}$ if we know $\chi(\mathscr{F}_{jj}) \rightsquigarrow \chi(\mathscr{F}_{\infty\infty})$ for all bounded continuous function χ: $\mathscr{L}R \longrightarrow R$; χ: $\mathrm{G} \longmapsto \chi(\mathrm{G})$. But only a few of these functions are functions depending on G only through integrals $\int g_i(t)d\mathrm{G}(t)$. We must prove that these few special χ are, under the circumstances, enough.

Let $g_1, g_2, \ldots$ be a countable convergence-determining class of bounded continuous functions for the probability measures on the separable locally compact real line R. Such a class exists by Lemma 2.2. Then $\mathrm{G}_n \Rightarrow \mathrm{G}$ in $\mathscr{L}R$ not merely implies but is implied by convergence $\int g_i d\mathrm{G}_n \underset{n}{\longrightarrow} \int g_i d\mathrm{G}$ for all i. Regarding $\int g_i d\mathrm{G}$ as the ith coordinate of a point in R^∞, this statement means that the continuous map α: $\mathrm{G} \longmapsto$ $\longmapsto \langle \int g_1 d\mathrm{G}, \int g_2 d\mathrm{G}, \ldots \rangle$ is the inverse of a continuous map β from a subset of R^∞, the range of α, onto $\mathscr{L}R$, a map β continuous, that is, in the relative topology induced on this subset by the product topology on R^∞.

There is no guarantee that the range of α will be a measurable subset of R^∞. However, we already know what the limit of X_{jj} in distribution is supposed to be. It is supposed to be $X_{\infty\infty}$, the limit of $X_{\infty k}$. This knowledge let us construct beforehand a Borel subset Δ of R^∞ which contains both $X_{\infty\infty}$ and each X_{jj} with probability one.

We may find in [3] pp. 239, 240 and 40, that, because R is separable and complete, $\mathscr{L}R$ is separable and topologically complete, that therefore the one-element subset of $\mathscr{L}\mathscr{L}R$ consisting of the distribution of $\mathscr{F}_{\infty\infty}$ is a tight subset of $\mathscr{L}\mathscr{L}R$ and so there is a countable union of compact subsets of $\mathscr{L}R$ which contains $\mathscr{F}_{\infty\infty}$ with probability one. The same is true for any $\mathscr{F}_{jj}$. The countable union of all these sets for all j and for ∞ is still a countable union of compact sets and contains all $\mathscr{F}_{jj}$ and $\mathscr{F}_{\infty\infty}$ with probability one. The continuous map α maps compact subsets of $\mathscr{X}R$ into compact subsets of R^∞, and the countable union of these compact image sets is a Borel set Δ of R^∞ such that $\mathsf{P}\{\alpha(\mathscr{F}_{jj}) \in \Delta\} = \mathsf{P}\{X_{jj} \in \Delta\} = 1 = ' \{\alpha(\mathscr{F}_{\infty\infty}) \in \Delta\} = \mathsf{P}\{X_{\infty\infty} \in \Delta\}$. The Borel subsets of Δ for the topology of R^∞ and for the relative topology induced on the range of α coincide, so $X_{jj} \rightsquigarrow X_{\infty\infty}$ in R^∞ implies $X_{jj} \rightsquigarrow X_{\infty\infty}$ in the relative topology for the range of α, on which β, the inverse of α, is defined and continuous. Thus $X_{jj} \rightsquigarrow X_{\infty\infty}$ in R^∞ implies $\beta(X_{jj}) \rightsquigarrow \beta(X_{\infty\infty})$ in $\mathscr{X}R$ or $\mathscr{F}_{jj} \rightsquigarrow \mathscr{F}_{\infty\infty}$. Thus we may be satisfied that, in the end, the topological technicalities do not sabotage our convergence-determining functions $g_1, g_2, \ldots$ in determining convergence.

2.4.3. Proof that $X_{jj} \underset{j}{\rightsquigarrow} X_{\infty\infty}$.

It remains to show that the limit in distribution of X_{jj} as $j \to \infty$ exists and equals the double limit in distribution $\lim\limits_{n\to\infty} \lim\limits_{k\to\infty} X_{nk} = X_{\infty\infty}$. In R^∞, we know $X_{jj} \rightsquigarrow X_{\infty\infty}$ iff for each q the finitely many coordinates $X_{jj}(1), X_{jj}(2), \ldots, X_{jj}(q)$ converge jointly in distribution to $X_{\infty\infty}(1), X_{\infty\infty}(2), \ldots, X_{\infty\infty}(q)$. Coordinatewise convergence in R^q is equivalent to convergence in the Euclidean root-sum-of-squares norm $\|\cdot\|$ which we shall employ. Fixing q, we keep the same symbols $X_{jj}, X_{\infty\infty}$, to denote the projections into R^q, which should cause no confusion, and we denote the q-tuple $g_1, g_2, \ldots, g_q$ simply by g. We know already that $X_{nk} \underset{n}{\rightsquigarrow} X_{\infty k}$ and $X_{\infty k} \underset{k}{\rightsquigarrow} X_{\infty\infty}$ in R^∞ and so in R^q. We shall establish $X_{jj} \underset{j}{\rightsquigarrow} X_{\infty\infty}$ in R^q for any q and so in R^∞ by appeal to Billingsley's Theorem 4.2 on page 25 of [3] which requires further that

$$\lim_{j\to\infty} \limsup_{n\to\infty} \mathsf{P}\{\|X_{nj} - X_{nn}\| \geqslant \delta\} = 0 \quad \text{for all } \delta.$$

Let E_n be the exceptation operator for P_n. By Čebyšev inequality

$$\mathsf{P}\{\|X_{nj} - X_{nn}\| > \delta\} = \mathsf{P}_n\{\theta \colon \|\textstyle\int g d\mathsf{S}_j(\theta) - \int g d\mathsf{S}_n(\theta)\| \geqslant \delta\} \leqslant$$

$$\leqslant \frac{1}{\delta^2} \mathsf{E}_n(\|\textstyle\int g d\mathsf{S}_j - \int g d\mathsf{S}_n\|^2) =$$

$$= \frac{1}{\delta^2} |\int g_1 d\mathsf{S}_j - \int g_1 d\mathsf{S}_n|^2 + \ldots + \frac{1}{\delta^2} |\int g_q d\mathsf{S}_j - \int g_q d\mathsf{S}_n|^2 .$$

Let us expand any one of the terms in the sum on the right in the same fashion as on p. 365 of [6], insisting that $n \geqslant j$. Our computation holds for any bounded measurable g and does not make use of continuity.

$$\mathsf{E}_n\left(\left|\int g d\mathsf{S}_j(\theta) - \int g d\mathsf{S}_n(\theta)\right|^2\right) =$$

$$= \mathsf{E}_n\left(\left|\frac{1}{n}\sum_{i=1}^{n} g(\theta_i) - \frac{1}{j}\sum_{i=1}^{j} g(\theta_i)\right|^2\right) =$$

$$= \left(\frac{1}{n} - \frac{1}{j}\right)^2 \mathsf{E}_n\left(\sum_{i=1}^{j} g(\theta_i) \sum_{k=1}^{j} g(\theta_k)\right) +$$

$$+ \frac{2}{n}\left(\frac{1}{n} - \frac{1}{j}\right) \mathsf{E}_n\left(\sum_{i=1}^{j} g(\theta_i) \sum_{i=j+1}^{n} g(\theta_i)\right) +$$

$$+ \frac{1}{n^2} \mathsf{E}_n\left(\sum_{i=j+1}^{n} g(\theta_i) \sum_{k=j+1}^{n} g(\theta_k)\right) .$$

We capitalize on the exchangeability of $\theta_1, \ldots, \theta_n$ under P_n to replace θ_i and θ_k by θ_1 and θ_2 inside the expectations.

$$\left(\frac{n-j}{nj}\right)^2 [j\mathsf{E}_n(g^2(\theta_1)) + (j^2 - j)\mathsf{E}_n(g(\theta_1)g(\theta_2))] -$$

$$- \frac{2}{n}\left[\frac{n-j}{nj} j(n-j)\mathsf{E}_n(g(\theta_1)g(\theta_2))\right] +$$

$$+ \frac{1}{n^2}[(n-j)\mathsf{E}_n(g^2(\theta_1)) + (n-j)(n-j-1)\mathsf{E}_n(g(\theta_1)g(\theta_2))] =$$

$$= \frac{n-j}{nj}[\mathsf{E}_n(g^2(\theta_1)) - \mathsf{E}_n(g(\theta_1)g(\theta_2))] \leqslant$$

$$\leqslant \frac{j}{n}\left(1 - \frac{j}{n}\right) 2\mathsf{E}_n(g^2(\theta_1)) .$$

Thus we may assert that

$$\mathsf{P}\{\| X_{nj} - X_{nn} \| > \delta\} \leqslant \frac{2q}{j\delta^2} \sup_{i,t} g_i^2(t)$$

and this expression has the limit zero for any δ as $j \to \infty$. By the Theorem 4.2 of Billingsley which we have cited we may conclude $X_{jj} \underset{j}{\rightsquigarrow} X_{\infty\infty}$ for any q, and therefore $\mathcal{F}_{jj} \underset{j}{\rightsquigarrow} \mathcal{F}_{\infty\infty}$.

Q.E.D.

Corollary 2.5. *P_n converges weakly iff $\mathcal{F}_{nn}$ converges in distribution.*

Proof of Corollary 2.5. Theorem 2.4 shows the forward implication. Conversely, suppose $\mathcal{F}_{nn}$ converges to some random element in $\mathcal{L}R$. Then for any bounded continuous functions g and γ, $\gamma\colon R^k \longrightarrow R^q$, $g\colon R^q \longrightarrow R$ the map $\mathsf{G} \longmapsto \int g(\gamma(t))d\mathsf{G}(t_1)\ldots d\mathsf{G}(t_k)$ from R to R is continuous. Therefore the expectation $\mathsf{E}(\int g(\gamma(t))d\mathcal{F}_{nn}(t_1)\ldots d\mathcal{F}_{nn}(t_k))$ always converges as $n \to \infty$. We recall $\int g(t)d\mathcal{F}_{nn}(t) = \frac{1}{n}\sum_{i=1}^{n} g(\theta_i)$. Now by 1.2, $\int g(\theta_1, \theta_2, \ldots, \theta_q)d\mathsf{P}_n(\theta) = \frac{(n-q)!}{n!}\int \sum_{\alpha} g(\theta_{\alpha(1)}, \ldots, \theta_{\alpha(q)})d\mathsf{P}_n(\theta)$ where the sum is taken over all permutations of $1, 2, \ldots, n$, for $q \leqslant n$. This integral is the sum

$$\sum_{k=1}^{q} \sum_{m_1, m_2, \ldots} \sum_{\gamma} \frac{n^k(n-q)!(-1)^{q+k}}{k!n!m_1!m_2!\ldots m_q!2^{m_2}3^{m_3}\ldots q^{m_q}} \times$$
$$\times \left(\mathsf{E}\int g\circ\gamma(t)d\mathcal{F}_{nn}(t_1)\ldots d\mathcal{F}_{nn}(t_k)\right)$$

where the sum is taken over $k = 1, 2, \ldots, q$, the q-tuple m over all tuples of integers such that $m_1 + 2m_2 + \ldots + qm_q = q$ and $m_1 + m_2 + \ldots + m_q = k$, and γ over maps $\gamma\colon R^k \longrightarrow R^q$ where $\gamma(t)$ equals a q-tuple in which exactly

m_1 of the t_i occur once

m_2 of the t_i occur twice

...

m_q of the t_i occur q times.

As $\frac{n^k(n-q)!}{n!}$ is bounded by $\left(1 - \frac{q}{n}\right)^{-q}$ which goes to 1 as n goes to infinity, we have proved that the expectations $\int g d\mathsf{P}_n$ always converge for bounded continuous $g\colon R^q \longrightarrow R$ and hence P_n converges in $\mathscr{X}R^\infty$.

Q.E.D.

Theorem 2.4 establishes a relationship between limiting raspectra and an asymptotic distribution of an infinite sequence of random variables which can be pictured as an infinite collection of singular values. This relationship has far-reaching consequences. The most immediate implication depends on De Finetti's Theorem and asserts that the limiting sample raspectrum is a sufficient statistic for the asymptotic singular value distribution.

Theorem 3.1. *If* $\mathscr{F}_{nn} \rightsquigarrow \mathscr{R}$ *in* $\mathscr{X}R$, *then* P_n *converges to a measure* P_∞ *on* R^∞ *which is almost* P_∞*-surely a conditional infinite product-power measure conditional on a function from* R^∞ *to* $\mathscr{X}R$ *whose image value is distributed like* $\mathscr{R}$.

Proof of Theorem 3.1. $\mathscr{F}_{nn} \rightsquigarrow \mathscr{R}$ implies $\mathsf{P}_n \Longrightarrow \mathsf{P}_\infty$ in R^∞ as in Corollary 2.5. By Lemma 2.3, $\theta_1, \theta_2, \ldots$ are an infinite sequence of exchangeable random variables under P_∞. By De Finetti's Theorem, as in [6], p. 365, they are therefore conditionally independent. By Lemma 2.3 $\mathsf{S}_k(\theta)$ converges almost P_∞-surely and its limit is distributed like $\mathscr{R}$ by Theorem 2.4. L o è v e shows that its distribution function serves as a conditioning random variable, and for this purpose we may obviously replace the distribution function by the measure itself.

Corollary 3.2. $\mathscr{F}_{nn}$ *converges in probability to a fixed element* F *in* $\mathscr{X}R$ *iff* P_n *converges weakly to an infinite power-product measure* P_∞ *in* $\mathscr{X}R^\infty$; *in other words, iff the singular values* $\theta_1, \theta_2, \ldots$ *are asymptotically independent*.

Proof of Corollary 3.2. Suppose $\mathsf{P}_n \Longrightarrow \mathsf{P}_\infty$ and P_∞ makes the coordinates unconditionally independent and identically distributed. For any bounded, continuous g the strong law of large numbers forces $\int g d\mathscr{F}_{\infty k} = \frac{1}{k}\sum_{i=1}^{k} g(\theta_i)$ to converge to a constant $\int g(\theta_1) d\mathsf{P}_\infty(\theta)$ almost P_∞-surely, so $\mathscr{F}_{\infty\infty}$ is almost P_∞-surely a fixed measure $\mathsf{F} = \mathsf{E}(\mathscr{F}_{\infty\infty}) = \int \mathsf{S}_1(\theta) d\mathsf{P}_\infty(\theta)$. By Theorem 2.4, $\mathscr{F}_{nn}$ converges in distribution to this same fixed element of $\mathscr{X}R$, and convergence in distribution to a degenerate distribution is equivalent to convergence in probability. The converse of the Corollary follows from Theorem 3.1.

Q.E.D.

A word of caution is in order. If $\theta_1, \theta_2, \ldots$ are unconditionally independent under P_∞ and if they are also bounded, then the random distribution functions $\mathscr{F}_{\infty k}(x)$ of $\mathscr{F}_{\infty k}$ have the property that the expression $\sqrt{k}[\mathscr{F}_{\infty k}(x) - \mathscr{F}_{\infty\infty}(x)]$ converges in distribution to a Brownian bridge or pinned Wiener process by the usual arguments for empirical cumulative distribution functions. This convergence should not be confused with any assertion about the raspectrum "error process" $\sqrt{n}[\mathscr{F}_{nn} - \mathsf{E}\mathscr{F}_{nn}]$ which will behave very differently. C.L. Mallows of Bell Telephone Laboratories [personal communication] has derived an asymptotic covariance kernel for this latter process, an expression of considerable complexity which is the first result of its kind known to the author.

In practical applications of Theorem 3.1 it is advantageous to prove beforehand that a sequence of singular value distributions P_n is tight and so has a weakly convergent subsequence. Then one can invoke Theorem 3.1 to describe the behaviour of such subsequence limits, and, knowing this behaviour, one can often return to the original sequence to prove actual convergence. The author has carried through such a program in some generality and is revising the work for publication.

The overall approach of this paper is not restricted to properties of singular value sets depending only on exchangeability. Exploiting stronger continuity properties and boundedness, it is possible to prove, as in Wachter [11] that if $\mathsf{P}_n \Rightarrow \mathsf{P}_\infty$ and P_n are singular value distributions for $p(n) \times m(n)$ marginals of a distribution Q on $R^{\infty\times\infty}$ which makes rows independent and columns exchangeable, then P_∞ is an infinite product measure. Such a result yields immediately a degenerate raspectrum by Corollary 3.2. These sufficient conditions for degenerancy of the raspectrum lead to derivations along the lines of Wachter [10] of limiting raspectra for a large class of random matrix distributions.

REFERENCES

[1] L. Arnold, On the Asymptotic Distribution of the Eigenvalues of Random Matrices, *Journal of Mathematical Analysis and Applications,* 20 (1967), 262-268.

[2] L. Arnold, The possible asymptotic distributions of Eigenvalues of Random Symmetric Matrices, (typescript), 1967.

[3] P. Billingsley, *Convergence of Probability Measures,* John Wiley, New York, 1968.

[4] A.P. Dempster, *Elements of Continuous Multivariate Analysis,* Addison Wesley, Reading, Massachusetts, 1969.

[5] U. Grenander, *Probabilities on Algebraic Structures,* John Wiley, New York, and Almqvist & Wicksell, Stockholm, 1963.

[6] M. Loève, *Probability Theory,* Van Nostrand Reinhold, Cincinnati, Third Edition, 1963.

[7] C. Mallows – K.W. Wachter, The Asymptotic Configuration of Wishart Eigenvalues, (abstract), *Ann. Math. Statist.,* 41 (1970), 1384.

[8] M.L. Mehta, *Random Matrices and the Statistical Theory of Energy Levels,* Academic Press, London, 1967.

[9] W.H. Olson – V.R. Uppuluri, Asymptotic Density of Eigenvalues for a Gaussian Ensemble of Matrices, (abstract), *Ann. Math. Statist.,* 40 (1969), 1512.

[10] K.W. Wachter, Raspectra: A Notion of Spectrum for Random Matrices and its Determination in Important Statistical Cases, (typescript), 1971.

[11] K.W. Wachter, The Singular Values of Infinite Random Matrices and Convergence of Raspectra, (typescript), 1972.

[12] E.P. Wigner, Random Matrices in Physics, *SIAM Review,* 9 (1967), 1-23.

COLLOQUIA MATHEMATICA SOCIETATIS JÁNOS BOLYAI
9. EUROPEAN MEETING OF STATISTICIANS, BUDAPEST (HUNGARY), 1972.

THE DIFFUSION APPROXIMATION TO A BRANCHING PROCESS

T. WILLIAMS

If X is distributed as an exponential with mean a:

(1) $$X \simeq a\,\mathrm{Exp},\quad f(x) = \frac{1}{a} e^{-\frac{x}{a}} \qquad (x > 0);$$

and N is distributed as a Poisson with mean b:

(2) $$N \simeq \text{Poisson}\,\{b\},\quad \mathsf{P}(N = n) = e^{-b}\frac{b^n}{n!} \qquad (n > 0)\,;$$

then we say that the random sum

(3) $$Y \equiv X_1 + X_2 + \ldots + X_N$$

is distributed as a Bessel variate with parameters a and b:

$$Y \simeq \text{Bessel}\,\{a, b\}\,.$$

Clearly, Y is a null sum and so vanishes if and only if $N = 0$; hence we have

(4) $$\mathsf{P}(Y = 0) = \mathsf{P}(N = 0) = e^{-b}\,.$$

Also, when $N = n$ is fixed, Y is distributed as a $\Gamma(n)$ with probability density

$$\frac{y^{n-1}}{a^n(n-1)!} e^{-\frac{y}{a}} .$$

Weighting these against the Poisson distribution (2) and including the atom of probability (4), we obtain

$$
\begin{aligned}
f(y) &= e^{-b}\,\delta(y) + \sum_{n=1}^{\infty} e^{-b}\frac{b^n}{n!}\frac{y^{n-1}}{a^n(n-1)!} e^{-\frac{y}{a}} = \\
&= e^{-b}\left\{\delta(y) + \sqrt{\frac{b}{ay}}\, e^{-\frac{y}{a}} I_1\left(2\sqrt{\frac{by}{a}}\right)\right\},
\end{aligned}
\tag{5}
$$

upon invoking the familiar power series for the Bessel function of imaginary argument.

On the other hand, the well-known result for random sums, applied to (3), shows that

$$M_Y(\theta) = P_N(M_X(\theta)) .$$

But (1) has moment-generator

$$M_X(\theta) = \frac{1}{1 - a\theta} \qquad \left(\theta < \frac{1}{a}\right) ,$$

whilst (2) has probability-generator

$$P_N(\theta) = \exp\,[b(\theta - 1)] ;$$

it therefore follows that for the Bessel variate

$$M_Y(\theta) = \exp\frac{ab\theta}{1 - a\theta} \qquad \left(\operatorname{Re}\theta < \frac{1}{a}\right) . \tag{6}$$

(It is, of course, also possible to retrieve (5) directly from (6) by using the Fourier inversion formula and the contour-integral representation of the Bessel function.) The cumulants are thus given by

$$\kappa_n = n!a^n b ; \tag{7}$$

in particular, the mean and variance are equal to

(8) $\quad \mathsf{E}Y = ab \quad$ and $\quad \operatorname{var} Y = 2a^2 b$.

Now, customary arguments show that the Galton – Watson process may be represented in the limit by a diffusion process with infinitesimal mean and variance

(9) $\quad m(x, t) = \alpha x \qquad V(x, t) = \beta x$,

respectively, and that its moment-generator at time t, $M(\theta, t)$, satisfies the partial differential equation

(10) $$\frac{\partial M}{\partial t} = \theta\left(\alpha + \frac{\beta}{2}\theta\right)\frac{\partial M}{\partial \theta},$$

provided we include the atom of probability, $p_0(t) = \mathsf{P}\{X(t) = 0\}$. The solution of (10) which satisfies the initial condition $X(0) = x_0$, fixed is readily seen to be

(11) $$M(\theta, t) = \exp \frac{\alpha x_0 e^{\alpha t}\theta}{\alpha - \frac{\beta}{2}(e^{\alpha t} - 1)\theta},$$

and comparison with (6) shows that the number $X(t)$ at time t is distributed as

(12) $$X(t) \simeq \text{Bessel}\left\{\frac{\beta}{2\alpha}(e^{\alpha t} - 1), \frac{2\alpha x_0}{\beta(1 - e^{-\alpha t})}\right\},$$

the desired result. Application of (8) and (4) now yields the familiar results

(13) $$\mathsf{E}X(t) = x_0 e^{\alpha t}$$

(14) $$\operatorname{var} X(t) = \frac{\beta}{\alpha} x_0 e^{\alpha t}(e^{\alpha t} - 1),$$

and

(15) $$p_0(t) = \exp\left(-\frac{2\alpha x_0}{\beta(1 - e^{-\alpha t})}\right).$$

All this is straightforward and elementary; what is presented here as being of some interest is the interpretation (12), in terms of (1), (2) and (3); as we have seen, it conveys at once all the information about the solution of (12). It should be remarked, however, that the representation (12) is curious, in that it becomes in the limit,

$$X(t) \simeq \text{Bessel}\left\{\frac{\beta}{2}t,\ \frac{2x_0}{\beta t}\right\} \qquad (t \to 0)\ ;$$

i.e., it displays the fixed initial condition as the Poisson-sum of a large number of small exponentials. Often a problem formulation like (12) will suggest a direct and illuminating probabilistic derivation, but the ulterior roles played by the Poisson and the exponential here is not apparent.